DYNAMICS OF
MACHINERY

McGRAW-HILL SERIES IN MECHANICAL ENGINEERING

KARL H. VESPER, *Stanford University,*
Consulting Editor

DYNAMICS OF MACHINERY

Richard M. Phelan

Professor of Mechanical Engineering
Cornell University

COMMITTEE FOR TECHNICAL EDUCATION

McGraw-Hill Book Company

New York   St. Louis   San Francisco
Toronto   London   Sydney

578   ***Dynamics of Machinery***

Library of Congress Catalog Card Number 67-10878

ISBN 07-049770-2

2 3 4 5 6 7 8 9 – M A M M – 7 6 5 4 3 2 1

To Olive

PREFACE

Dictionaries, encyclopedias, and textbooks are in general agreement that dynamics is that branch of mechanics which treats of the motion of a system of particles under the influence of forces. Although this definition is satisfactory for many engineering purposes, there are also numerous important situations that are truly dynamic if not traditionally a part of dynamics. Thus, to keep the discussion as general as possible, we shall consider dynamics as being concerned with all dynamic situations, where dynamic means "Pertaining to change or process regarded as manifestation of energy or agency, involving or producing alteration; . . ."[1] Thus, even though the body of the text will emphasize applications to machinery and mechanical engineering, the reader should keep in mind that the principles and methods apply equally well to such diverse cases as the dynamics of a heat-treating furnace, a manufacturing plant, and a communications system. In every case the key words are *change* and *energy*.

On the basis of energy considerations, problems in dynamics can be placed into one of the following three categories: (1) systems in which energy can exist in only one form and for which the outside source of energy is unaffected by time variations within the system, (2) systems in which energy can exist in two distinct forms and for which the outside source of energy is unaffected by time variations within the system, and (3) systems in which time variations have a direct effect on the outside source of energy and thereby control the supply of energy to and removal of energy from the system.

Although, in reality, there are no systems for which the outside source of energy is unaffected by variations within the system and categories 1 and 2 do not exist in an absolute sense, there are many problems of major engineering significance for which the errors introduced by assuming they lie in categories 1 or 2 are negligible for all practical purposes.

In relation to machines, category 1 consists primarily of systems made

[1] Webster's New International Dictionary, 2d ed., G. & C. Merriam Company, Springfield, Mass., 1957.

up of rigid bodies in which kinetic energy alone is significant. Typical examples are linkages and rotors.

Category 2 consists primarily of systems in which there can be a continual interchange of energy between the forms of kinetic and potential energy. In relation to machines, the potential energy usually is the result of elastic deformation (as in a spring) or a change in elevation (as for a pendulum). The most important class of engineering problems in this category is that called vibrations.

Category 3 is characterized by "feedback" from within the system to an outside source of energy such that time variations within the system effectively control the delivery of energy to the system. Although in some cases, such as so-called self-excited vibrations, the control is unintentional and usually undesired, in most situations control of the source of energy is the desired end result, and feedback control systems is the term commonly applied to this category.

The dynamics of rigid bodies, or particles, has long been recognized as being important in all fields of engineering, and considerable time has been and is being devoted to it in courses in physics and mechanics. Mechanical engineers are almost universally involved with forces, motion, and mass in machines and have recognized the importance of dynamics, although it has been only in the past several decades that additional work, outside the area of fluid mechanics, has been required beyond the simple determination of accelerations without regard to force considerations in kinematics.

The author believes that *the undergraduate course in mechanical engineering cannot be considered complete unless all students are introduced to all three categories of dynamics of machinery.* The author also believes that this should be done within the framework of a single course in which the entire field of dynamics of machinery is treated as a logical, continuous extension of the introductory dynamics studied by the student in physics and mechanics.

The benefits are manifold. Although making more efficient use of the students' time by minimizing repetition and eliminating a number of inconsistencies found in the current literature is important, *the major advantage of this approach is that one sees all of dynamics of machinery as simply dynamics,* and not as several separate subjects that too often appear to be almost unrelated. This not only simplifies the learning process but adds to an enhanced depth of understanding a sense of perspective relative to the importance and usefulness of many of the almost overwhelmingly large number of special methods and techniques that will be encountered by the students when the need arises for reference to the traditional sources of information.

Even though the major reason for writing this book has been the author's concern for undergraduate education in mechanical engineering, the continuing education of recent and not-so-recent graduates cannot be ignored. One of the major stumbling blocks to present efforts in continuing education is that textbooks currently available are designed primarily to educate specialists in a particular category of dynamics of machinery and many times start off or quickly jump to a level beyond that at which the practicing engineer can comfortably work. This is particularly true for feedback control systems and to a lesser extent for vibrations. The integrated treatment in this text should alleviate this problem by permitting the reader to start at the point appropriate to his background and continue as far as he wishes or needs—through this book and then on to more specialized and advanced books.

The topics and depth of treatment have been selected with two goals in mind. First, the text must introduce all categories in dynamics of machinery at sufficient depth so that the reader will have a fundamental knowledge of what kinds of problems are involved, what the similarities and differences are between the categories, how solutions are reached, and where one can go for more detailed information as the need arises in specific situations. It is to be hoped that elective courses will be available in which the student with the desire for more knowledge can continue his studies in the category or categories in which he is most interested.

Second, through lack of time, opportunity, or interest, many students will not follow up the required course with one or more elective courses and yet will find themselves confronted in professional practice with engineering problems in dynamics that can be solved with only a little more knowledge than might be adequate for introductory purposes only. In general, an attempt has been made to include material sufficient for practically all needs of the mechanical engineer who is *not* a specialist in dynamics of machinery. He should be able to arrive at good engineering answers to many real and significant problems, and he should be able to discuss intelligently with an expert more complex problems if the need arises. Consequently, considerably more material is included than the author has found practical to cover in a single course with four semester-hours credit.

The first category in dynamics of machinery involves the solution of second-order differential equations. However, the zero-order term is missing (no energy storage, as in elastic deformation) and the first-order term can almost always be neglected (friction or damping is so small that energy dissipation is negligible). As a result, only the second-order and constant terms remain and the equation can be integrated twice in closed form or numerically in every case as a problem in simple calculus. In general, the traditional topics related to dynamics of rigid bodies have been retained, although there has been a reduction in the number of special methods that contribute little to understanding and have limited usefulness in narrow areas of decreasing interest. The major addition to the traditional list is a section on the determination of the response of a linkage when acted on by an external force.

Except for a few nonlinear systems in category three, categories 2 and 3 are always concerned with situations in which energy storage and/or dissipation (or their equivalents) are involved and we must work at all times with differential equations.

Chapters 3, 4, 5, and 6 are concerned with the solution of second-order and families of second-order differential equations. The majority of the material, particularly in Chapters 3, 4, and 5, is equally important in both categories 2 and 3. In spite of the high degree of generality that actually exists, most references in these chapters are to the class of problems in category 2 called vibrations—for the simple reason that the great majority of engineering students are better off if the development of a new body of knowledge can be taken in logical steps of increasing complexity and generality. Vibrations is well recognized as making up a class of problems that in themselves are important to mechanical engineers and it is the logical next step from dynamics of rigid bodies. In addition, the physical concepts of inertia, damping or energy dissipation, and spring rate or energy storage—associated with reality in vibrations—become important analogous quantities in studying systems involving feedback in many diverse fields, nonengineering as well as engineering.

The coverage of vibrations far exceeds that in current texts on dynamics of machinery and is comparable to, in some cases even exceeds, that in most books written for a first course in mechanical vibrations. The major differences are (1) almost exclusive use of complex numbers in relation to sinusoidal response, (2) emphasis on physical understanding, (3) emphasis on multi-degree-of-freedom systems including the important practical case of the isolation of a system with six degrees of freedom, and (4) emphasis on relative merits of the many available methods and techniques for arriving at solutions to engineering problems.

In the last four chapters the background developed in the preceding chapters is extended to the more general and complex dynamics of feedback control systems. The coverage concentrates on fundamental concepts and methods for studying system behavior. For example, the reader is introduced to transfer functions, block diagrams, the Laplace transformation, stability analysis of linear systems—by use of Routh's criterion and the root-locus, conformal-mapping, and attenuation-phase methods—elementary compensation methods for improving system performance, and several of the more common nonlinear systems.

Since dynamics of machinery has meaning only when something possessing inertia is accelerated, the author believes that the student should study it after he knows something about accelerations and why components and machines have inertia—in other words, after he has been introduced to mechanical design with respect to kinematics and design of machine elements. At this point in his development, the student has been led through the design process from the specifications of motion requirements, the selection of suitable mechanisms, the determination of velocities and accelerations, and the determination of dimensions that will insure satisfactory performance *under assumed conditions of loading.* In all cases the loads must include those due to the performing of a useful function and, in an ever increasing number of cases, those due to acceleration. Although an allowance should have been made for anticipated dynamic forces, precise calculations cannot be made until the distribution of the mass of the component is known. Consequently, the designer is faced with an iterative process in which the dynamic effect can really be considered only after the component has first been designed.

A particular attempt has been made to minimize confusion between what are the fundamental concepts and what are the tools that can be useful in obtaining answers. As a result, computers and programming are not discussed directly in the body of the text, although the attention of the reader is called many times to situations in which numerical methods and a digital computer or the analog computer would be particularly useful. The important point is that computers enable the engineer to extend his understanding of the physical and mathematical concepts of relatively simple systems, which can be treated thoroughly without the use of computers, to solving problems related to systems of much greater complexity, where "much greater complexity" usually refers principally to the number of computations required.

Most students will have already been exposed to the digital computer, but relatively few will have worked with the electronic differential analyzer or, as it is more commonly called, the analog computer. This lack of balance is regrettable, because the analog computer is particularly useful in studying the time-varying behavior of lumped-parameter systems and, consequently, solving the great majority of problems related to vibrations and control systems. It is to be hoped that all students studying dynamics of machinery will have the opportunity for using the analog computer in both of its major modes of operation, which are to obtain solutions by (1) working directly with the differential equations and (2) simulating the system by letting blocks on the computer correspond directly to blocks in the real system. The first mode is equally useful in vibrations and in control systems. The second mode is particularly useful in studying control systems, because the effects of varying parameters in a given block can be determined quickly and directly and because actual blocks can be combined with computer blocks to achieve a maximum degree of reality without having to work with the entire real system. A brief introduction to the use of the analog computer in both modes is presented in Appendix A.

The 458 problems in the back of the book range in complexity from relatively simple exercises in mathematics with single specific answers to open-end design problems for which there can be many answers. A number, at all degrees of complexity, extend the coverage of the book by introducing topics and concepts that are interesting and worth considering when time is available, although not of sufficient importance to warrant inclusion in the body of the text. Many are well suited as bases for major design studies or projects that can provide opportunity for integrating material from preceding courses and other areas of technology.

Answers to selected problems are provided primarily to enable the reader who is studying by himself to check his own work. An attempt has been made to include a representative example from each of the most important types of problems.

The author would like to acknowledge the many helpful comments and suggestions of his colleagues at Cornell University and elsewhere. Special recognition should be given to Professor B. E. Quinn of Purdue University and Professors R. L. Wehe and J. F. Booker of Cornell University. In particular, the author is most indebted to Professor Booker for his thorough and detailed study of the manuscript and for the many hours of his time so generously spent in discussing points of agreement and disagreement with the author.

Richard M. Phelan

CONTENTS

DYNAMICS OF
MACHINERY

chapter 1

RIGID BODIES

Historically, *dynamics* has been defined as that branch of mechanics concerned with the motion of bodies and the relationships between forces and motions. It has been subdivided into *kinematics*, in which relative motion is considered without regard to the effects of mass and force upon the motion, and *kinetics*, which is concerned with the changes in motion produced by forces. At the present time, however, most engineers find it convenient to use kinetics when referring to the effect of forces upon the motion of particles and dynamics when referring to the effect of forces upon the motion of solid bodies or a gross collection of particles as in a liquid. This text will follow the latter viewpoint and, although a good background in kinematics will be essential, major emphasis will be given to the relationships between force and motion.

The basis for solving all problems in dynamics is deceptively simple—all that is required is a thorough understanding of Newton's laws of motion and the ability to perform the necessary calculations. Previous studies in physics, mathematics, engineering mechanics, and kinematics will have provided the basic tools, and the purpose of this text will be to extend their usefulness by considering in some detail several categories of problems that deserve special attention because of their great importance to the design and performance of machines.

This and the following chapter will be concerned with the application of Newton's laws of motion to systems for which good engineering answers can be obtained by assuming that energy can exist in only one form within the system. For practical purposes this means that we shall neglect changes in potential energy that accompany the elastic deformation of real materials under load and the change of elevation of a body in a gravitational field. In other words, we shall be concerned solely with systems

1

in which the parts can be considered to be rigid, i.e., made of materials with infinite moduli of elasticity, and in which energy can exist only in the form of kinetic energy.

1-1. Newtonian Mechanics With few exceptions, notably those involving space travel, the approximation resulting from using Newtonian mechanics with the surface of the earth as the frame of reference will be sufficient for all problems in dynamics of machinery. Consequently, in this text, whenever terms such as velocity and acceleration are used without qualification, they will be considered as absolute vector quantities, i.e., having magnitude and direction relative to the earth's surface.

The reader will recall that Newton's laws of motion can be expressed as follows:

1 A particle will remain at rest or will continue in a straight line at a constant speed unless acted upon by a force.
2 The time rate of change of momentum of a particle is equal to the magnitude of the applied force and acts in the direction of the force.
3 When two particles exert forces on one another, these forces are equal in magnitude and opposite in direction and act along the line joining the particles.

The second law is expressed mathematically as

$$\frac{d\mathbf{M}}{dt} = \frac{d(m\mathbf{V})}{dt} = m\frac{d\mathbf{V}}{dt} = \mathbf{F} \qquad (1\text{-}1)$$

where $\mathbf{M}$ = momentum of the particle[1]
  m = mass of the particle
  $\mathbf{V}$ = velocity
  t = time
  $\mathbf{F}$ = force
and the first law is seen to be the special case of the second when $\mathbf{F} = 0$.

The third law is often expressed as *action and reaction are equal and opposite,* and it is the basis for statics. The second law combined with the third is the basis for dynamics.

1-2. Rigid Bodies: Plane Motion By definition, a particle has no dimensions and the only motion possible is that of translation[2] in a three-dimensional space. This remains true for each of the infinite number of particles in a solid body; but as a matter of practicality it becomes desirable to develop further Newton's laws to the point where analogous statements can also be made about the gross motion of solid bodies without having to consider each individual particle every time.

---

[1] Throughout this text boldface letters will indicate vector quantities.
[2] Many terms commonly encountered in courses such as mechanism, kinematics, and mechanical design will not be redefined here.

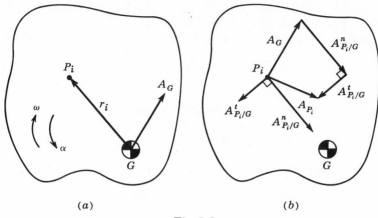

(a)   (b)

Fig. 1-1

The major consideration is that a solid body has three dimensions and it may rotate as well as translate in each.   However, in the great majority of machines, the motion, at least within subassemblies, will be such that all points are constrained to move in paths parallel to a single plane.   If the mass distribution of every part is symmetrical about a single plane, a complete analysis can be made by considering only plane motion.   If the parts are offset from each other, as in many four-link mechanisms, additional calculations may be necessary to determine some of the reactions and moments necessary for design.   The major emphasis here will be given to the dynamics of rigid bodies having plane motion.

Figure 1-1a shows a rigid body that is moving in the plane of the paper with a combination of translation and rotation.   Point G is the center of gravity of the body, and P_i is any particle.   The mass of the particle is m_i, and the radius vector from G to P_i is $\mathbf{r}_i$.   The acceleration of the center of gravity and the angular velocity ω and acceleration α of the body are known.   From kinematics,

$$\mathbf{A}_{P_i} = \mathbf{A}_G + \mathbf{A}_{P_i/G} \tag{1-2}$$

where $\mathbf{A}_{P_i/G}$ means the acceleration of P_i relative to G.   The acceleration of any point in a rigid body relative to any other point in the same body can in turn be broken down into two components, a normal and a tangential component.   In terms of these, Eq. (1-2) becomes

$$\mathbf{A}_{P_i} = \mathbf{A}_G + \mathbf{A}_{P_i/G}^n + \mathbf{A}_{P_i/G}^t \tag{1-3}$$

$A_{P_i/G}^n = r_i\omega^2 = (V_{P_i/G})^2/r_i$, and it is directed from P_i toward G, the center of curvature of the relative path. $A_{P_i/G}^t = r_i\alpha$, and it is directed perpendicular to r_i with the sense of the angular acceleration.   The vector addition in Eq. (1-3) has been performed in Fig. 1-1b.

Applying Newton's second law, Eq. (1-1), to the particle P_i gives

$$\frac{d\mathbf{M}_{P_i}}{dt} = m_i \frac{d\mathbf{V}_{P_i}}{dt} = m_i \mathbf{A}_{P_i} = \mathbf{F}_i \qquad (1\text{-}4)$$

and, in terms of the components given in Eq. (1-3),

$$m_i \mathbf{A}_{P_i} = m_i \mathbf{A}_G + m_i \overrightarrow{r_i \omega^2} + m_i \overrightarrow{r_i \alpha} = \mathbf{F}_i \qquad (1\text{-}5)$$

For the entire body,

$$\mathbf{F} = \Sigma \mathbf{F}_i = \Sigma m_i \mathbf{A}_G + \Sigma m_i \overrightarrow{r_i \omega^2} + \Sigma m_i \overrightarrow{r_i \alpha} \qquad (1\text{-}6)$$

Considering each term on the right-hand side of Eq. (1-6), we find the following:

1   For $\Sigma m_i \mathbf{A}_G$.   These forces are all parallel and the sum can be written as

$$\Sigma m_i \mathbf{A}_G = \mathbf{A}_G \Sigma m_i = m\mathbf{A}_G \qquad (1\text{-}7)$$

where m is the total mass of the body.   If $\mathbf{g}$, the acceleration of gravity, is substituted for $\mathbf{A}_G$, one arrives at the definition of the center of gravity.   Therefore, it can be concluded that the acceleration force $m\mathbf{A}_G$ acts through the center of gravity in the direction of the acceleration.

2   For $\Sigma m_i \overrightarrow{r_i \omega^2}$.   These forces are all directed toward G, and the sum may be rewritten as

$$\Sigma m_i \overrightarrow{r_i \omega^2} = -\omega^2 \Sigma m_i \mathbf{r}_i \qquad (1\text{-}8)$$

As a direct consequence of the properties of the center of gravity,

$$\Sigma m_i \mathbf{r}_i = 0 \qquad (1\text{-}9)$$

and, therefore, the sum of these forces will be zero.   Since the forces all pass through the center of gravity, there can be no resulting torque.

3   For $\Sigma m_i \overrightarrow{r_i \alpha}$.   These forces are all directed perpendicular to their respective radii in the sense of the direction of the angular acceleration.   Since α is a property of the whole body, the magnitude or absolute value of the sum can be written as

$$|\Sigma m_i \overrightarrow{r_i \alpha}| = |\alpha \Sigma m_i r_i| \qquad (1\text{-}10)$$

Again, because of the properties of the center of gravity, this summation must equal zero. However, the forces are now so directed that there is a resulting torque, which is

$$T = \Sigma m_i \overrightarrow{r_i \alpha r_i} = \alpha \Sigma m_i r_i^2 \tag{1-11}$$

The right-hand summation is by definition the moment of inertia of the body. Thus, Eq. (1-11) can be rewritten in more useful form as

$$T = I_G \alpha \tag{1-12}$$

Therefore, for the body in Fig. 1-1 to have the accelerations indicated, it must be acted upon by a *force $F = mA_G$* through the center of gravity in the direction of the acceleration and a *couple* or *torque $T = I_G \alpha$* in the direction of the angular acceleration, or their equivalents. For convenience, the equations will be written simply as

$$F = mA \tag{1-13}$$

and

$$T = I\alpha \tag{1-14}$$

with the understanding that A and I refer to the center-of-gravity values unless otherwise noted.

1-3. D'Alembert's Principle—Inertia Forces Thus far the discussion has been concerned entirely with the forces and torques required to provide specified values of lineal and angular acceleration. Although this approach is fundamental and is of primary interest in many situations, the engineer is most often interested in finding the forces and torques that are present because a rigid body is constrained to move with definite accelerations. The latter may most conveniently be handled by using D'Alembert's principle, which can be stated as *the reversed effective forces and torques and the real forces and torques together give statical equilibrium.*

We arrive at D'Alembert's principle by rewriting Eqs. (1-13) and (1-14) as

$$F + (-mA) = 0 \tag{1-15}$$

and

$$T + (-I\alpha) = 0 \tag{1-16}$$

and then noting that these in turn fulfill the conditions of static equilibrium. That is, Eq. (1-15) can be written as

$$\Sigma F = 0 \tag{1-17}$$

and Eq. (1-16) as

$$\Sigma T = 0 \tag{1-18}$$

The force $-mA$ is known as the *inertia force*, and the torque $-I\alpha$ is known as the *inertia torque* or *inertia couple*.

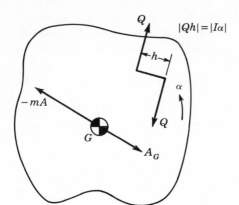

Fig. 1-2   *Inertia force and torque.*

1-4. Inertia Forces and Torques—The Equivalent Offset Force

The lineal and angular accelerations and the corresponding inertia force and inertia couple for a body in plane motion are shown in Fig. 1-2.   The magnitudes of Q and h may have any values as long as their product equals $I\alpha$, and the vectors for Q may be drawn at any point, in any direction, provided they are parallel and the direction of the couple is the same as $-\alpha$.

The most convenient value for Q is mA, and the most convenient location for the couple is that shown in Fig. 1-3a, where one vector $\mathbf{Q}$ just exactly cancels the inertia force $-m\mathbf{A}$ through the center of gravity.   The net result is the equivalent offset force shown in Fig. 1-3b, where the force is $-m\mathbf{A}$ and the offset is

$$h = \frac{I\alpha}{mA} \tag{1-19}$$

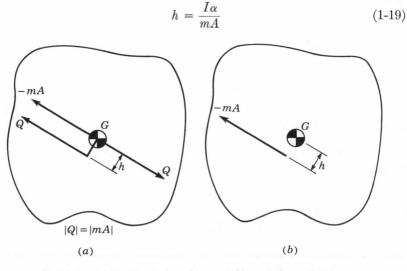

(a)   (b)

Fig. 1-3   (a) *Inertia force and torque;* (b) *equivalent offset force.*

1-5. Linkages   The four-bar linkage is the linkage used most frequently to provide constrained motion in machines.   The particular four-bar linkage with one member infinitely long is known as the slider-crank mechanism and is of major importance because of its use at high speeds in internal-combustion engines, pumps, compressors, punches, etc., where dynamic effects cannot be ignored.   Since the slider is constrained to have only rectilinear translation, it has been practical and convenient to develop special methods for handling this particular type of mechanism. In following sections considerable attention will be given to the slider-crank mechanism; but for the moment let us consider in Example 1-1 the application of the concept of the offset inertia force to the dynamics of a more general four-bar linkage.

Example 1-1   The crank of a four-bar linkage is to rotate at a uniform speed of 300 rpm while the follower oscillates.   The dimensions of the linkage have been determined to provide (1) the desired motion characteristics and (2) adequate strength to carry the working loads on the members.   The working loads were based upon a static-force analysis and include an allowance for dynamic loads.   However, since the linkage is to operate at a relatively high speed, the accelerations, and thus the dynamic effects, may be significant.   To provide information required before a definite statement can be made about the adequacy of the present design dimensions, we are asked to make a dynamic analysis of the linkage.   Specifically, we are asked to determine the following:

(a) The magnitude and direction of the resultant inertia force on the frame.   The force acts to "shake" the foundation and is known as *shaking force*.

(b) The magnitude and direction of the bearing reactions at each connection between links.

(c) The torque that must act on the crank (or crankshaft) to ensure a uniform speed of rotation.

In a real situation all of the above would have to be determined for several phases of operation so that we could be certain that the worst possible combination of magnitudes and directions had been discovered.   If the designer is faced with having to make many routine calculations of this type, he should consider the use of an analytical vector method, such as that based upon complex numbers,[1] and a digital computer. However, since our main objective is to gain understanding of what is happening, we shall use a graphical vector solution for one phase of operation.

The dimensions, weights, and moments of inertia of the links are given, along with the acceleration polygon for the linkage, in Fig. 1-4.   The crank 2 will be counterbalanced so that its center of gravity will be at its center of rotation, point O_2.   Using the information in Fig. 1-4, the inertia forces and offset distances for the moving links are calculated as follows:

For link 2

$$F_{i2} = m_2 A_{G_2} = \frac{W_2}{g} A_{G_2} = \frac{87.2}{386} \times 0 = 0$$

[1] H. H. Mabie and F. W. Ocvirk, "Mechanisms and Dynamics of Machinery," 2d ed., pp. 385–392, John Wiley & Sons, Inc., New York, 1963.

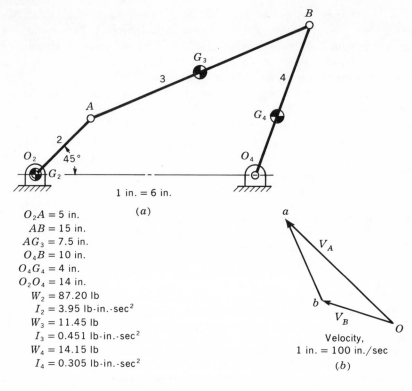

$O_2A = 5$ in. *(a)*
$AB = 15$ in.
$AG_3 = 7.5$ in.
$O_4B = 10$ in.
$O_4G_4 = 4$ in.
$O_2O_4 = 14$ in.
$W_2 = 87.20$ lb
$I_2 = 3.95$ lb-in.-sec^2
$W_3 = 11.45$ lb
$I_3 = 0.451$ lb-in.-sec^2
$W_4 = 14.15$ lb
$I_4 = 0.305$ lb-in.-sec^2

1 in. = 6 in.

Velocity,
1 in. = 100 in./sec
(b)

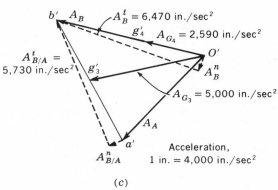

$A_B^t = 6{,}470$ in./sec^2
$A_{G_4} = 2{,}590$ in./sec^2
$A_{B/A}^t = 5{,}730$ in./sec^2
A_B^n
$A_{G_3} = 5{,}000$ in./sec^2
A_A
$A_{B/A}^n$ Acceleration,
1 in. = 4,000 in./sec^2

(c)

Fig. 1-4

where g = acceleration of gravity = 386 in./sec^2

$$\alpha_2 = 0 \qquad \text{given}$$

$$h_2 = \frac{I_2\alpha_2}{m_2 A_{G_2}} = \frac{0}{0} \qquad \text{indeterminate (no significance)}$$

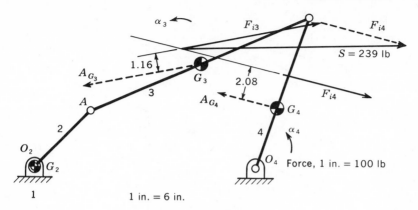

Fig. 1-5

For link 3

$$F_{i3} = m_3 A_{G_3} = \frac{W_3}{g} A_{G_3} = \frac{11.45}{386} \times 5,000 = 148.3 \text{ lb}$$

$$\alpha_3 = \frac{A^t_{B/A}}{BA} = \frac{5,730}{15} = 382 \text{ rad/sec}^2, \text{ ccw}$$

$$h_3 = \frac{I_3\alpha_3}{m_3 A_{G_3}} = \frac{I_3\alpha_3}{F_{i3}} = \frac{0.451 \times 382}{148.3} = 1.16 \text{ in.}$$

For link 4

$$F_{i4} = m_4 A_{G_4} = \frac{W_4}{g} A_{G_4} = \frac{14.15}{386} \times 2,590 = 94.9 \text{ lb}$$

$$\alpha_4 = \frac{A^t_B}{O_4 B} = \frac{6,460}{10} = 646 \text{ rad/sec}^2, \text{ ccw}$$

$$h_4 = \frac{I_4\alpha_4}{m_4 A_{G_4}} = \frac{I_4\alpha_4}{F_{i4}} = \frac{0.305 \times 646}{94.9} = 2.08 \text{ in.}$$

Shaking Force. Since there are only two nonzero inertia forces, the resultant shaking force will act through the point of intersection of the lines of action of F_{i3} and F_{i4}. This is illustrated in Fig. 1-5. The magnitude of the shaking force **S** is seen to be 239 lb.

Bearing Reactions. Since determining the external couple on link 2 is one of the goals of this analysis, we must consider link 2 last. Both link 3 and link 4 are three-force members, the forces being the offset inertia force and the bearing reactions at each end. Equilibrium requires only that $\Sigma F = 0$, which, in turn for three forces, requires that the lines of action of the three forces intersect at a point.

Considering link 3 first, we find that for equilibrium

$$\Sigma \mathbf{F}_3 = 0$$
$$\mathbf{F}_{23} + \mathbf{F}_{43} + \mathbf{F}_{i3} = 0 \tag{1-20}$$

where $\mathbf{F}_{23}$ = force exerted by link 2 on link 3
$\mathbf{F}_{43}$ = force exerted by link 4 on link 3
$\mathbf{F}_{i3}$ = offset inertia force on link 3

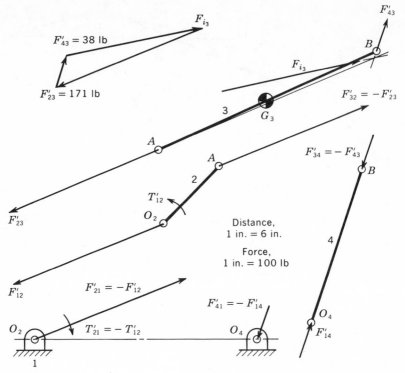

Fig. 1-6

The solution of Eq. (1-20) requires knowing at least four of the six values (three magnitudes and three directions) associated with the three vectors; and only two, the magnitude and direction of the inertia force $\mathbf{F}_{i3}$, are already known. To find the additional information, we must take a closer look at how the bearing reactions are developed. There are several ways of considering this problem, but the most straightforward method, which is also the one most readily adaptable for solution by use of a digital computer, is to determine the reactions due to each inertia force alone and then use the principle of superposition to find the resultant reactions at each bearing. For example, if we consider $\mathbf{F}_{i3}$ acting alone on the linkage, link 4 can carry a force only in tension or compression. In this case, the line of action of $\mathbf{F}_{43}$ must lie along O_4B and we can solve Eq. (1-20). The procedure is illustrated in Fig. 1-6, where $\mathbf{F}'_{43}$ and $\mathbf{F}'_{23}$ are the reactions of links 4 and 2, respectively, required on link 3 to balance the inertia force $\mathbf{F}_{i3}$.

By use of Newton's third law we know that $\mathbf{F}'_{32} = -\mathbf{F}'_{23}$ and $\mathbf{F}'_{34} = -\mathbf{F}'_{43}$, and the forces on links 2 and 4 are next determined. As shown, a couple, $\mathbf{T}'_{12}$ in the ccw direction, must be provided for equilibrium of link 2. The magnitude of the couple is not important at this time.

Finally, the reactions on the frame, $\mathbf{F}'_{21}$, $\mathbf{T}'_{21}$, and $\mathbf{F}'_{41}$, are determined by use, again, of Newton's third law.

The solution for the reactions, $\mathbf{F}''_{34}$, etc., due to $\mathbf{F}_{i4}$ acting by itself are shown in Fig. 1-7. The bearing reactions are determined in Fig. 1-8 by adding the components from Figs. 1-6 and 1-7 and by use of Newton's third law.   For example,

$$\mathbf{F}_{43} = \mathbf{F}'_{43} + \mathbf{F}''_{43} \tag{1-21}$$

and

$$\mathbf{F}_{34} = -\mathbf{F}_{43} \tag{1-22}$$

Crankshaft Torque.   As shown in Fig. 1-8, the forces $\mathbf{F}_{32}$ and $\mathbf{F}_{12}$ are equal and opposite but not collinear.   Thus, they result in a cw couple that must be balanced by a ccw couple or crankshaft torque if the angular velocity of link 2 is to remain constant.   The magnitude of the couple is the product of one of the forces and the perpendicular distance between them.   Therefore,

$$T_{12} = T_{21} = 248 \times 1.93 = 479 \text{ lb-in.}$$

The couple T_{21} acts to rotate or "rock" the frame and is called the *rocking couple*.

Closure.   In reality it is impossible to maintain a constant speed of rotation of the crank.   The use of flywheels to decrease the speed variation will be discussed in some detail in Sec. 1-7, but at this time we can note that, to ensure a constant speed of rotation of link 2, either we must have a flywheel with infinite inertia or the crankshaft must be driven by a device or system with the nonexistent characteristic of a torque output that is independent of speed.   For practical purposes, the speed variation can be kept within any reasonable limits by choosing the right combination of flywheel and motor.

If we were interested in the behavior of the present linkage in the absence of all external forces, we would find that the cw couple of 479 lb-in. on link 2 would result

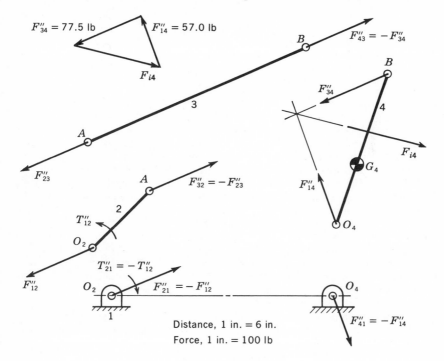

Fig. 1-7

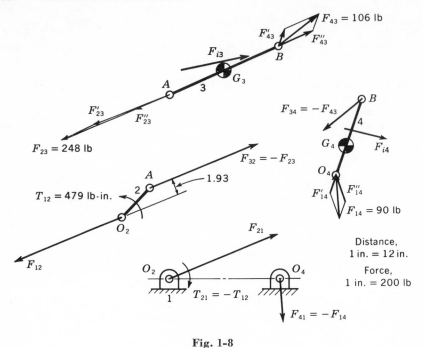

Fig. 1-8

in, from Eq. (1-14), an angular acceleration of

$$\alpha_2 = \frac{T}{I_2} = \frac{479}{3.95} = 121.3 \text{ rad/sec, cw}$$

Since $\alpha_2 \neq 0$, the acceleration polygon in Fig. 1-4c is no longer correct. To determine the true acceleration of link 2, as well as the bearing reactions throughout the linkage, we must use an iterative process. The general procedure would be to use the value of α_2 (or something slightly lower in magnitude) to draw a new acceleration polygon to be used in a new inertia-force analysis, which would result in still another value for α_2. The process would need to be repeated until the value of α_2 determined from the force analysis and the value used in finding the accelerations agree closely enough for purposes of the analysis.

If the moving parts, links 2, 3, and 4, are taken as a free body, we find the linkage is acted upon by an external torque ($\mathbf{T}_{12}$) and four external forces (the two offset inertia forces and the bearing reactions $\mathbf{F}_{12}$ and $\mathbf{F}_{42}$). In Fig. 1-9 the inertia forces have been combined into the shaking force S (as in Fig. 1-5) and the reactions from the frame have been combined into a force S'. As to be expected, for equilibrium the forces S and S' are equal and opposite and the couple resulting from the offset of 2.00 in. is equal and opposite to $\mathbf{T}_{12}$.

The analysis completed above is for only one phase of the linkage, and the operational forces have hardly been mentioned. In a real situation, the dynamic analysis would have to be repeated for many more positions so that the variations of the forces and torques with phase would be available for an entire cycle of operation. Similarly, a number of static-force analyses, using the operational loads, would have to be made

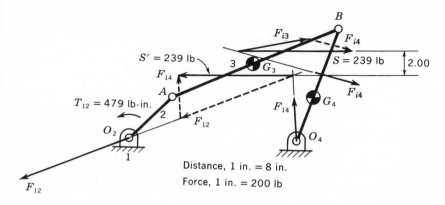

Fig. 1-9

and the results combined (superimposed) with the dynamic analyses so that the resultant of the dynamic and operational forces would be known for all phases during a complete cycle.   These resultants, particularly the worst possible combinations, would be used in checking the strength calculations and in selecting bearings for each of the joints.   The resultant forces and couples acting on the frame become important because they can induce vibrations and noise that may be disagreeable if not destructive.   This will be given more consideration in Chaps. 4 through 6.

1-6. Slider-crank Mechanism   The dynamic analysis of the slider-crank mechanism may be performed in the same manner as illustrated in Example 1-1 for the general case of a four-bar linkage.   But, since the slider is constrained to reciprocate in a straight line and since the most important operating condition involves the crank rotating at an essentially uniform speed, the procedure can be simplified to a considerable extent by using an analytical approach involving two approximations: one in relation to the expression for the acceleration of the slider and the other in arriving at an approximately equivalent connecting rod.

Slider acceleration   The displacement x of the slider from the "head end" dead-center phase of the slider-crank mechanism in Fig. 1-10 can be

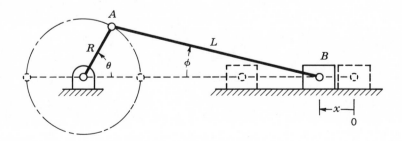

Fig. 1-10   *Slider-crank mechanism.*

expressed as

$$\begin{aligned}
x &= R + L - R \cos\theta - L \cos\phi \\
&= R(1 - \cos\theta) + L(1 - \cos\phi) \\
&= R(1 - \cos\theta) + L(1 - \sqrt{1 - \sin^2\phi}) \\
&= R(1 - \cos\theta) + L\left[1 - \sqrt{1 - \left(\frac{R}{L}\right)^2 \sin^2\theta}\right]
\end{aligned} \tag{1-23}$$

where positive displacement is to the left. This is the exact solution, but it is more convenient for our purposes to expand the quantity under the radical in accordance with the binomial theorem and then to keep only those terms that are significant. After expansion, Eq. (1-23) becomes

$$x = R\left[(1 - \cos\theta) + \frac{1}{2}\frac{R}{L}\sin^2\theta + \frac{1}{2 \times 4}\left(\frac{R}{L}\right)^3 \sin^4\theta \right. \\
\left. + \frac{1 \times 3}{2 \times 4 \times 6}\left(\frac{R}{L}\right)^5 \sin^6\theta + \cdots\right] \tag{1-24}$$

Differentiating Eq. (1-24) with respect to time gives for the velocity of the slider

$$\frac{dx}{dt} = R \sin\theta \frac{d\theta}{dt} + R\frac{R}{L}\sin\theta\cos\theta\frac{d\theta}{dt} + \frac{R}{2}\left(\frac{R}{L}\right)^3 \sin^3\theta\cos\theta\frac{d\theta}{dt} \\
+ \frac{3R}{8}\left(\frac{R}{L}\right)^5 \sin^5\theta\cos\theta\frac{d\theta}{dt} + \cdots \tag{1-25}$$

Substituting ω for $d\theta/dt$ and considering the case when ω is constant, differentiation of Eq. (1-25) with respect to time leads to, term by term,

$$R\omega\frac{d\sin\theta}{dt} = R\omega\cos\theta\frac{d\theta}{dt} = R\omega^2\cos\theta \tag{1-26}$$

$$R\omega\frac{R}{L}\frac{d\sin\theta\cos\theta}{dt} = R\omega\frac{R}{L}\left(-\sin^2\theta\frac{d\theta}{dt} + \cos^2\theta\frac{d\theta}{dt}\right) \\
= R\omega^2\frac{R}{L}(\cos^2\theta - \sin^2\theta) \tag{1-27}$$

$$\tfrac{1}{2}R\omega\left(\frac{R}{L}\right)^3\frac{d\sin^3\theta\cos\theta}{dt} = \tfrac{1}{2}R\omega\left(\frac{R}{L}\right)^3\left(-\sin^4\theta\frac{d\theta}{dt} + 3\sin^2\theta\cos^2\theta\frac{d\theta}{dt}\right) \\
= \tfrac{1}{2}R\omega\left(\frac{R}{L}\right)^3(3\sin^2\theta\cos^2\theta - \sin^4\theta) \tag{1-28}$$

$$\tfrac{3}{8}R\omega\left(\frac{R}{L}\right)^5\frac{d\sin^5\theta\cos\theta}{dt} = \tfrac{3}{8}R\omega\left(\frac{R}{L}\right)^5\left(-\sin^6\theta\frac{d\theta}{dt} + 5\sin^4\theta\cos^2\theta\frac{d\theta}{dt}\right) \\
= \tfrac{3}{8}R\omega^2\left(\frac{R}{L}\right)^5(5\sin^4\theta\cos^2\theta - \sin^6\theta) \tag{1-29}$$

Utilizing trigonometric identities to simplify the right-hand side of Eqs. (1-27), (1-28), and (1-29), we find, respectively,

$$R\omega^2 \frac{R}{L} (\cos^2 \theta - \sin^2 \theta) = R\omega^2 \frac{R}{L} \cos 2\theta \tag{1-30}$$

$$\frac{1}{2}R\omega^2 \left(\frac{R}{L}\right)^3 (3 \sin^2 \theta \cos^2 \theta - \sin^4 \theta)$$
$$= R\omega^2 \left(\frac{R}{L}\right)^3 (\frac{1}{4} \cos 2\theta - \frac{1}{4} \cos 4\theta) \tag{1-31}$$

and $\frac{3}{8}R\omega^2 \left(\frac{R}{L}\right)^5 (5 \sin^4 \theta \cos^2 \theta - \sin^6 \theta)$
$$= R\omega^2 \left(\frac{R}{L}\right)^5 (\frac{15}{128} \cos 2\theta - \frac{3}{16} \cos 4\theta + \frac{9}{128} \cos 2\theta) \tag{1-32}$$

Now, arranging the multiple-angle terms back into a series and simplifying, we find

$$\frac{d^2x}{dt^2} = R\omega^2 \left\{ \cos \theta + \left[\frac{R}{L} + \frac{1}{4} \left(\frac{R}{L}\right)^3 + \frac{15}{128} \left(\frac{R}{L}\right)^5 + \cdots \right] \cos 2\theta \right.$$
$$- \left[\frac{1}{4} \left(\frac{R}{L}\right)^3 + \frac{3}{16} \left(\frac{R}{L}\right)^5 + \cdots \right] \cos 4\theta$$
$$\left. + \left[\frac{9}{128} \left(\frac{R}{L}\right)^5 + \cdots \right] \cos 6\theta - \cdots \right\} \tag{1-33}$$

The $\cos \theta$ term would be the only one if the slider had simple harmonic motion. The remaining terms are due to the use of a finite-length connecting rod and the $\cos 2\theta$ term is the second harmonic, the $\cos 4\theta$ term is the fourth, etc. It should be noted that after the first (or fundamental) only even harmonics appear.

From a practical viewpoint, one seldom encounters values of L/R much less than 3.5. Using this value in Eq. (1-33) gives

$$\frac{d^2x}{dt^2} = R\omega^2 [\cos \theta + (0.2845 + 0.0058 + 0.0002 + \cdots) \cos 2\theta$$
$$- (0.0058 + 0.0004 + \cdots) \cos 4\theta$$
$$+ (0.0001 + \cdots) \cos 6\theta - \cdots] \tag{1-34}$$

As can be seen, even for this relatively small value of L/R, the terms beyond $\cos 2\theta$ can be neglected without appreciable loss. In fact, only minor error is introduced if everything beyond the L/R term in the coefficient for $\cos 2\theta$ is neglected. This is commonly done, and Eq. (1-33) becomes

$$\frac{d^2x}{dt^2} = R\omega^2 \left(\cos \theta + \frac{R}{L} \cos 2\theta \right) \tag{1-35}$$

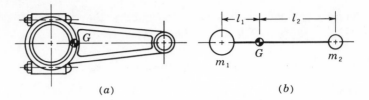

Fig. 1-11   (a) *Connecting rod;* (b) *equivalent link.*

It should be noted that Eq. (1-35) can be reached directly by dropping all terms beyond $\sin^2 \theta$ in Eq. (1-24) and then differentiating twice with respect to time.

Connecting rod—equivalent links   For two bodies to be dynamically equivalent, the forces and torques associated with accelerated motion must be identical for both bodies.   Thus, from Eqs. (1-13) and (1-14), we find that the following conditions must be met:

1   Both bodies have the same mass.
2   The center of gravity is located at the same place in both bodies.
3   Both bodies have the same moment of inertia.

Usually we are interested in finding an equivalent link having point (lumped) masses so that dynamic analyses can be simplified, since only forces acting through the point masses would need to be considered. Figure 1-11a shows a typical connecting rod that might be used in an internal-combustion engine, a compressor, or a pump. . The point-mass connecting rod in Fig. 1-11b will be its equivalent, provided the three conditions given above are met.   These conditions require, respectively,

$$1 \qquad\qquad m_1 + m_2 = m \qquad \text{or} \qquad W_1 + W_2 = W \qquad\qquad (1\text{-}36)$$
$$2 \qquad\qquad\qquad m_1 l_1 = m_2 l_2 \qquad\qquad\qquad (1\text{-}37)$$
$$3 \qquad\qquad\qquad m_1 l_1^2 + m_2 l_2^2 = I \qquad\qquad\qquad (1\text{-}38)$$

Thus, we have three equations with four unknowns (l_1, l_2, m_1, and m_2), and one can be chosen arbitrarily.   It is usually most convenient to choose one of the lengths to locate one of the masses at a point with a readily determined motion.   For example, for the connecting rod in Fig. 1-10, one of the masses is usually located at point B, where it reciprocates with the slider.   Assuming that one of the lengths will be chosen, the other length may be found by combining Eqs. (1-37) and (1-38) as follows: Rewriting Eq. (1-38) gives

$$m_1 l_1 l_1 + m_2 l_2 l_2 = I \qquad\qquad (1\text{-}39)$$

Then, substituting from Eq. (1-37) $m_2 l_2$ for $m_1 l_1$ and vice versa into Eq. (1-39) gives

$$m_2 l_2 l_1 + m_1 l_1 l_2 = I \qquad\qquad (1\text{-}40)$$

which can then be rearranged to give

$$(m_1 + m_2)l_1l_2 = ml_1l_2 = I = m\rho^2 \qquad (1\text{-}41)$$

where ρ = radius of gyration.

From Eq. (1-41) we can write

$$l_1l_2 = \frac{I}{m} = \rho^2 \qquad (1\text{-}42)$$

from which, given one length, the other may be found. Solving Eqs. (1-36) and (1-37) for m_1 gives

$$m_1 = \frac{l_2}{l_1 + l_2}\, m \qquad (1\text{-}43)$$

and then, from Eq. (1-36),

$$m_2 = m - m_1 \qquad (1\text{-}44)$$

Example 1-2 The dimensions, weight, and moment of inertia for a connecting rod in an automobile engine are given in Fig. 1-12a. We are asked to find the kinetically equivalent connecting rod when one of the point masses is to be located at B.

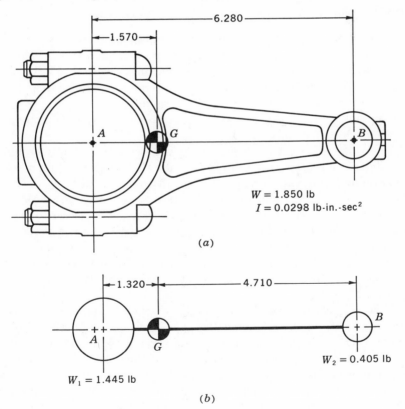

$$W = 1.850 \text{ lb}$$
$$I = 0.0298 \text{ lb-in.-sec}^2$$

(a)

$$W_1 = 1.445 \text{ lb}$$
$$W_2 = 0.405 \text{ lb}$$

(b)

Fig. 1-12   *(a) Connecting rod; (b) kinetically equivalent link.*

From given information,

$$l_2 = 6.280 - 1.570 = 4.710 \text{ in.}$$

and

$$m = \frac{W}{g} = \frac{1.850}{386} = 0.00479 \text{ lb-sec}^2/\text{in.}$$

From Eq. (1-42),

$$l_1 = \frac{I}{l_2 m} = \frac{0.0298}{4.710 \times 0.00479} = 1.320 \text{ in.}$$

From Eq. (1-43),

$$m_1 = \frac{l_2}{l_1 + l_2} m = \frac{4.710}{1.320 + 4.710} \, 0.00479 = 0.00374 \text{ lb-sec}^2/\text{in.}$$

$$W_1 = m_1 g = 0.00374 \times 386 = 1.445 \text{ lb}$$

From Eq. (1-44),

$$m_2 = m - m_1 = 0.00479 - 0.00374 = 0.00105 \text{ lb-sec}^2/\text{in.}$$

$$W_2 = W - W_1 = 1.850 - 1.445 = 0.405 \text{ lb}$$

The kinetically equivalent connecting rod is shown in Fig. 1-12b.

The ideal equivalent connecting rod would be with the masses located at points such as A and B in Fig. 1-4, where the mass at A would then rotate in a circle and the mass at B could be lumped with the inertia of the follower 4. Unfortunately, this is not generally possible with well-proportioned connecting rods and there is little to be gained by applying the concept of a kinetically equivalent link to the analysis of a four-bar mechanism such as in Fig. 1-4.

The kinetically equivalent connecting rod is also not directly useful for the slider-crank mechanism; but in this case the practical requirements in relation to bearings and strength and rigidity of crankshafts usually result in a connecting rod, as in Fig. 1-12, with one end much larger than the other and thus with the center of gravity located much closer to one end than to the other. In this case, if one mass is specified to be at the small end, the other mass will be located relatively close to the center of the joint at the large end—in fact, so close that placing it there would not be too bad an approximation and would greatly simplify the rest of the analysis.

However, calculating the magnitude of the second mass by use of Eqs. (1-39) through (1-41) and then placing it at the "big end" of the connecting rod results in an equivalent rod with neither the correct moment of inertia nor the correct location of the center of gravity. The usual approach is to divide the mass so that the centers of gravity of the approximately equivalent and the actual connecting rods are in the same place. This then means that only condition 3, that both bodies have the same moment of inertia, is violated. The net result is that the inertia

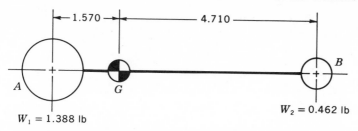

Fig. 1-13   *Statically equivalent link for connecting rod in Fig. 1-12a.*

force and its direction are the same for both but the line of action will be slightly different.   The conditions for the *approximately equivalent* link become those for static equilibrium, that is,

$$m_1 + m_2 = m \tag{1-45}$$
and
$$m_1 l_1 = m_2 l_2 \tag{1-46}$$

Applying Eqs. (1-45) and (1-46) to the connecting rod in Fig. 1-12a leads to the approximately equivalent connecting rod in Fig. 1-13.   This should be compared with the kinetically equivalent rod in Fig. 1-12b. The "approximately equivalent" connecting rod is more appropriately known as the "statically equivalent" rod because it meets the conditions for static equilibrium, as expressed in Eqs. (1-45) and (1-46).

Shaking forces   The crank of the mechanism in Fig. 1-14 is rotating at a uniform speed.   Its inertia force acts radially outward and has a magnitude

$$F_2 = m_2 O_2 G_2 \omega_2{}^2 \tag{1-47}$$

Since there is no angular acceleration of the crank, its moment of inertia does not enter the problem and a statically equivalent link, i.e., meeting the conditions specified by Eqs. (1-45) and (1-46), will give exactly the same inertia force as the real link.   For convenience, it is customary to use the equivalent link with one mass at the crank pin A and the other at the crank center O_2.   Since only the mass at A will contribute to the

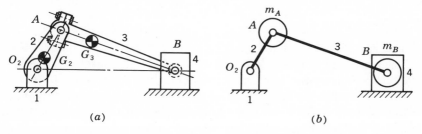

Fig. 1-14   (a) *Slider-crank mechanism;* (b) *equivalent-link representation.*

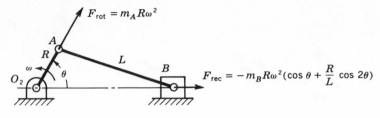

Fig. 1-15   *Slider-crank mechanism inertia forces.*

inertia force, only it must be calculated.   Thus, the lumped mass at A due to the crank 2 is

$$m_{A_2} = \frac{m_2 O_2 G_2}{R} \tag{1-48}$$

and
$$F_2 = m_{A_2} R \omega_2^2 \tag{1-49}$$

Applying Eqs. (1-45) and (1-46) to the connecting rod results in a mass m_{A_3} at A and a mass m_{B_3} at B.   In terms of the lumped masses, the mechanism in Fig. 1-14a can be simplified to that in Fig. 1-14b, where

$$m_A = m_{A_2} + m_{A_3} \tag{1-50}$$

and
$$m_B = m_{B_3} + m_{B_4} \tag{1-51}$$

The mass at A is rotating at a constant speed, and the inertia force will be a rotating vector with a magnitude of

$$F_{\text{rot}} = m_A R \omega^2 \tag{1-52}$$

and directed as shown in Fig. 1-15.

The mass at B is reciprocating, and the inertia force will be

$$\mathbf{F}_{\text{rec}} = -m_B \mathbf{A}_B \tag{1-53}$$

which, from Eq. (1-35), becomes

$$\mathbf{F}_{\text{rec}} = -m_B R \omega^2 \left(\cos \theta + \frac{R}{L} \cos 2\theta \right) \tag{1-54}$$

and acts as shown in Fig. 1-15.

The discussion in Example 1-1 can be applied to the mechanism in Fig. 1-15 to determine the bearing reactions, shaking force, and torques or couples.   It will be found that the solution for the slider-crank mechanism is considerably simpler, because the use of the statically equivalent link eliminates all forces on the connecting rod except the tension or compression force required to provide $\mathbf{F}_{34}$ on the slider.

The shaking force acts on the frame (link 1) and acts as a forcing function that can result in vibrations and noise that may damage the machine or create discomfort for persons nearby. Forced vibrations will be considered in detail in Chaps. 4 and 6, but at this time we should note that the first step in minimizing the problems associated with a forced vibration is to minimize the magnitude of the disturbing force. In general, we shall try to add mass or remove mass in appropriate positions to eliminate or minimize the magnitude of the shaking force. This is called *balancing of machinery* and will be the subject for Chap. 2. In the case of the slider-crank mechanism, the discussion of the balancing of inertia forces can be simplified greatly if each of the inertia force terms is considered separately.

In particular, it becomes desirable to consider the reciprocating inertia force $\mathbf{F}_{rec}$, Eq. (1-54), in terms of the first two harmonics[1] as the sum of the primary and secondary inertia or, more commonly, shaking forces. Thus,

$$\mathbf{F}_{rec} = \mathbf{F}_{pri} + \mathbf{F}_{sec} \tag{1-55}$$

where

$$\mathbf{F}_{pri} = -m_B R \omega^2 \cos \theta \tag{1-56}$$

and

$$\mathbf{F}_{sec} = -m_B \frac{R^2}{L} \omega^2 \cos 2\theta \tag{1-57}$$

Physically, the primary inertia or shaking force may be considered to be the projection on the line of centers of a vector $m_B R \omega^2$ that rotates at the speed of the crank. Rewriting the secondary shaking force as

$$\mathbf{F}_{sec} = -m_B \frac{R^2}{4L} (2\omega)^2 \cos 2\theta \tag{1-58}$$

leads to the physical concept of the projection on the line of centers of a vector $m_B(R^2/4L)(2\omega)^2$ rotating at twice the speed of the crank.

Therefore, each of the inertia forces in Fig. 1-15 can be represented by a rotating vector or the projection of a rotating vector as shown in Fig. 1-16. The three components can be combined into a single diagram, as shown in Fig. 1-17, by adding the radii to give concentric circles. This permits the rapid determination of the magnitude and direction of the resultant shaking force for all crank angles.

1-7. Flywheels The most important and most frequently encountered motion in machinery is rotation. In many cases both the power input and the work output are continuous and the torque and/or load variations are negligible for relatively long periods of time. Examples of this type of operation would be a turbine driving a generator, an electric

---

[1] The reader should recall that $R\omega^2(R/L) \cos 2\theta$ is only an approximation for the second harmonic.

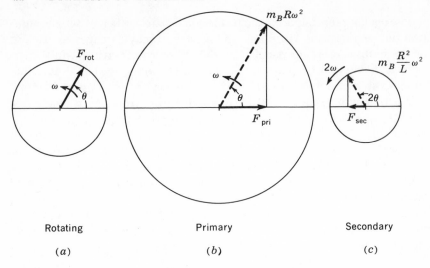

Rotating   Primary   Secondary

(a)   (b)   (c)

Fig. 1-16   *Rotating-vector representation of slider-crank mechanism inertia forces.*

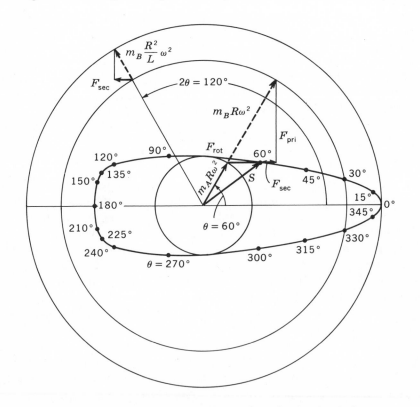

Fig. 1-17   *Polar diagram of slider-crank mechanism inertia forces.*

motor driving a centrifugal pump, and an electric motor driving a lathe or a milling machine.

In numerous other cases the power input or output or both may vary greatly and frequently, possibly during each cycle of operation. This type of operation generally involves converting reciprocation into rotation, as in the internal-combustion engine, or rotation into reciprocation or oscillation, as in reciprocating pumps and compressors, punch and other types of presses, shapers, and toggle-type rock crushers. In these operations some method must be provided for storing and delivering energy at high rates. For example, a single-cylinder 4-cycle gasoline engine receives its energy from the burning fuel-air mixture during about one-quarter of a cycle and, at the other extreme, a large punch press may be required to deliver many thousands of foot-pounds of energy in a fraction of a second. Both of these machines require appreciable "flywheel effect" or rotational inertia in order to work.

If there were no rotational inertia from the crankshaft, flywheel, etc., the single-cylinder engine could not pass through dead center and the engine would come to a sudden stop after the first power stroke. In a practical case, we not only wish to get by the dead-center positions but also to be able to have a useful output of power throughout the cycle. The rotational inertia acts somewhat as a storage battery, but the energy exists as kinetic energy rather than chemical.

From previous studies the reader will recall that the kinetic energy of a rotating body can be expressed as

$$E_K = \tfrac{1}{2}I\omega^2 \tag{1-59}$$

Thus, the change in kinetic energy resulting from a change in speed is

$$\Delta E_K = \tfrac{1}{2}I(\omega_2{}^2 - \omega_1{}^2) \tag{1-60}$$

Equation (1-60) can be used directly in many cases to determine the moment of inertia required when a definite amount of energy is to be delivered or absorbed without the speed of rotation decreasing or increasing beyond a specified value. For example, many punch presses consist of a toggle linkage connected to a flywheel by means of a clutch. The flywheel is usually driven by an electric motor and the flywheel rotates continuously. The clutch is engaged during the punching operation, and the greater part of the energy comes from the flywheel. The motor is usually a high-slip motor,[1] and it needs only to bring the flywheel back up to speed during the relatively long time interval between press operations.

In many other situations, the torque acting on the flywheel may vary in magnitude and/or direction a number of times each revolution. These

[1] C. C. Libby, "Motor Selection and Application," chap. 5, McGraw-Hill Book Company, New York, 1960.

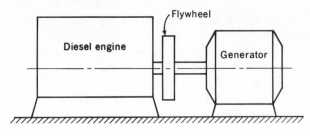

Fig. 1-18

variations may result from external or internal variations in load or driving effort and from varying inertia forces resulting from oscillating and reciprocating parts, as in Example 1-1 and in Fig. 1-18, where a diesel engine is driving a 60-cycle alternating-current generator. In the case of the diesel engine the speed, and thus the frequency, must be held constant to a high degree of accuracy. An automatic control system, or governor, eliminates the effect of relatively long-term variations in load, and the inertia of the rotating parts (crankshaft, flywheel, generator rotor, etc.) minimizes the short-term variations due to the reciprocating and oscillating parts of the engine and sudden changes in load.

For convenience in flywheel calculations, the permissible speed variation is expressed as a *coefficient of fluctuation K*, which is defined as the difference between the maximum and minimum speeds divided by the average or nominal speed. Thus,

$$K = \frac{\omega_{max} - \omega_{min}}{\omega} \tag{1-61}$$

Although not generally true, it is convenient to assume that the variation in speed is equally above and below the nominal speed. Using this assumption, we find

$$\omega = \frac{\omega_{max} + \omega_{min}}{2} \tag{1-62}$$

Multiplying Eq. (1-61) by Eq. (1-62) and rearranging gives

$$K\omega^2 = \frac{1}{2}(\omega_{max}^2 - \omega_{min}^2) \tag{1-63}$$

Multiplying both sides of Eq. (1-63) by I gives

$$KI\omega^2 = \frac{1}{2}I(\omega_{max}^2 - \omega_{min}^2) \tag{1-64}$$

which, from Eq. (1-60), is seen to be

$$KI\omega^2 = \Delta E_{K,max} \tag{1-65}$$

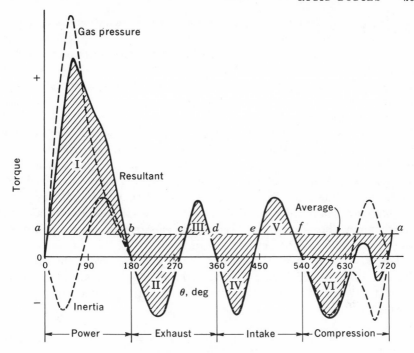

Fig. 1-19 *Typical torque-crank-angle diagram for a single-cylinder 4-cycle diesel engine.*

Rearranging Eq. (1-65) to solve for the moment of inertia required, we find

$$I = \frac{\Delta E_{K,\max}}{K\omega^2} \tag{1-66}$$

and the remaining problem is to determine during which part of the cycle the maximum variation in speed takes place.

Figure 1-19 shows the variation of torque on the crankshaft of a single-cylinder 4-cycle diesel engine. As shown, the resultant is the sum of torques due to the gas pressure and inertia forces.   The area under the torque-crank-angle curve may be expressed as

$$\int T \, d\theta = \text{work or energy} \tag{1-67}$$

and the integral for the entire cycle becomes the output work per cycle. In this case the output is positive and has an average value as shown. By definition, power is the rate of doing work, and the horsepower output of this engine may be calculated from

$$\text{hp} = \frac{T_{\text{av}}\omega}{12 \times 550} = \frac{T_{\text{av}}n}{63,000} \tag{1-68}$$

where T_{av} is in pound-inches, ω is in radians per second, and n is in rpm.

When the torque curve is above or below the T_{av} line, the engine delivers more or less torque, respectively, than the load requires and the excess or deficiency results in a change in speed. Thus, the areas labeled I, III, and V represent energy that must be absorbed by the rotating inertia, and the areas labeled II, IV, and VI represent energy that must be supplied by the rotating inertia.

In Fig. 1-19 it can be seen that there is a large excess of energy available between a and b and that the flywheel speed would increase during this interval. The flywheel will slow down between b and c and will speed up again between c and d. However, comparison of the areas involved leads to the conclusion that the speed at d will be less than at b and the speed at f will be less than at d. Finally, the minimum speed will occur at a and the maximum speed at b. Therefore, the energy represented by area I is the value of $\Delta E_{K,max}$ to be used in Eq. (1-66).

1-8. Gyroscopes Although not stated in these terms, the preceding section was concerned with the change of angular momentum given to a rotating disk or flywheel by a torque acting in the plane of rotation, and Newton's second law could be expressed as

$$\mathbf{T} = \frac{d\mathbf{H}}{dt} = \frac{d(I\omega)}{dt} = I\frac{d\omega}{dt} = I\alpha \qquad (1\text{-}69)$$

where $\mathbf{H}$ = angular momentum. However, in the more general case with a couple acting in a plane other than perpendicular to the axis of rotation, the situation becomes a problem in three dimensions and may often be handled best by use of analytical vector methods.[1] One important special case in three dimensions that can be handled rather simply is when a disk that is rotating at a high speed about its axis of symmetry is subjected to a couple in a plane containing the axis of rotation.

Figure 1-20a shows a disk or wheel rotating about its axis of symmetry, the xx axis. The momentum vector $\mathbf{H}$ has been drawn in accordance with the right-hand rule. The x, y, and z axes are mutually perpendicular. Now, when a couple Fl is applied in the xz plane, there will be a change in momentum which, by use of the right-hand rule, will result in a vector along the y axis in the direction shown in Fig. 1-20b. The change in momentum is related to the impulse by

$$\Delta H = Fl\,\Delta t = T\,\Delta t \qquad (1\text{-}70)$$

[1] I. H. Shames, "Engineering Mechanics-Dynamics," chap. 18, Prentice-Hall, Inc., Englewood Cliffs, N.J., 1960.

G. W. Housner and D. E. Hudson, "Applied Mechanics-Dynamics," 2d ed., pp. 202–252, D. Van Nostrand Company, Inc., Princeton, N.J., 1959.

D. F. Gunder and D. A. Stuart, "Engineering Mechanics," chap. 10, John Wiley & Sons, Inc., New York, 1959.

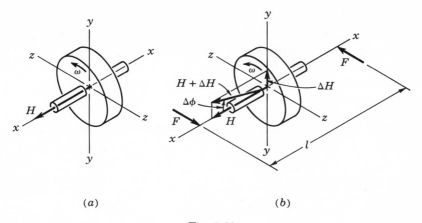

(a)   (b)

Fig. 1-20

The resulting vector diagram is also shown in Fig. 1-20b. Since, for the case under consideration, the angular velocity of the disk is high and the momentum **H** is much greater than the change in momentum **ΔH**, we can write

$$\Delta H = H\,\Delta\phi = I\omega\,\Delta\phi \tag{1-71}$$

Dividing both sides of Eqs. (1-70) and (1-71) by Δt and then combining them leads to

$$T = I\omega\,\frac{\Delta\phi}{\Delta t} \tag{1-72}$$

which, in the limit, as $\Delta t \to 0$, becomes

$$T = I\omega\,\frac{d\phi}{dt} \tag{1-73}$$

Also, as $\Delta t \to 0$, we find that $H + \Delta H \to H$. Thus, the only effect of the couple has been to give the angular-momentum vector, *and therefore the spin axis of the disk*, an angular velocity $d\phi/dt$ in the xy plane. This motion is called *precession*, and we have what is known as the *gyroscopic effect*. For convenience, we shall replace $d\phi/dt$ by Ω, the velocity of precession, and write Eq. (1-73) as

$$T = I\omega\Omega \tag{1-74}$$

It should be noted that T, ω, and Ω act about mutually perpendicular axes. Although the directional relationships may always be found as above, it is often convenient to have a quick method for determining the directions. One simple way is to remember that precession tends to rotate the spin vector **ω** in the direction to line up with the torque axis. Another scheme is illustrated in Fig. 1-21, where the arrows on the arcs

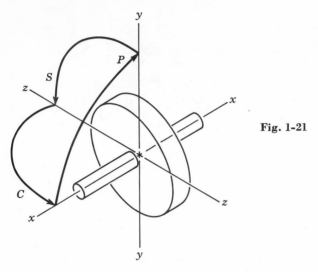

Fig. 1-21

C, P, and S indicate the directions of the couple, precession, and spin, respectively.   The arcs form a closed path describing a 90° segment of a sphere, and the feature that makes the method so simple is that the arcs are drawn in repeating alphabetical order, for example, CPS, SCP, or PSC.

The gyroscopic effect is applied in several different ways.   For example, when mounted in gimbals, as in Fig. 1-22, the spin axis tends to remain fixed in space and we have the familiar toy gyroscope, as well as the heart of the gyroscopic compass used on board ships.   One of the important present and future uses of the gyroscopic effect is in inertial guidance control systems for missiles and space travel, where gyroscopes are used to sense the angular motion of the body.[1]   In all situations where extreme accuracy is required for a long period of time, the major problems arise from unwanted couples introduced by friction in the gimbal bearings.

1-9. Linkage Response   Thus far the discussion of the dynamics of linkages has been limited almost entirely to the determination of forces resulting from or required to ensure a constant speed of rotation of one member.   This is the most frequently encountered situation in machines, but there are still many important cases where the motion results from a rapid input or release of energy; for example, the opening and closing of a shutter of a camera and the opening of a circuit breaker.   In these cases we are interested in determining the response of the linkage to a given input.

[1] F. H. Raven, "Automatic Control Engineering," chap. 14, McGraw-Hill Book Company, New York, 1961.

J. E. Gibson and F. B. Tuteur, "Control System Components," pp. 343–361, McGraw-Hill Book Company, New York, 1958.

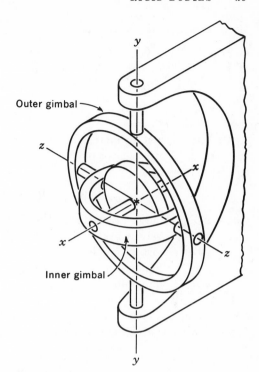

Fig. 1-22 *Gyroscope mounted in gimbals.*

In most cases the energy comes from a spring or a compressed gas and the transfer of energy to the mechanism is a function of the position of the input link.   Since the relationship between time and energy is not known, it is normally simpler to use energy considerations rather than Newton's laws, and only this approach will be discussed here.

In Fig. 1-23 the withdrawal of the pin releases the energy stored in the compressed spring.   At $t = 0+$ (the instant the pin is withdrawn) all parts are at rest and the energy balance simply states that all of the energy exists as potential energy in the spring.   A short time after the pin is withdrawn the linkage will have moved to a new phase and the energy given up by the spring has gone into doing useful work, being dis-

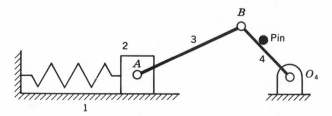

Fig. 1-23

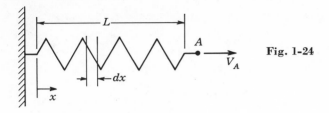

Fig. 1-24

sipated by friction, or increasing the potential energy of other parts, or it appears as kinetic energy of the moving parts. In equation form, this can be expressed as

$$\Delta E_s = \Delta W + \Delta W_\mu + \Delta E_P + \Delta E_K \qquad (1\text{-}75)$$

where ΔE_s = energy given up by the spring
ΔW = useful work
ΔW_μ = work of friction
ΔE_P = change in potential energy
ΔE_K = change in kinetic energy

In most cases no useful work is performed and both the loss due to friction and the change in potential energy are negligibly small in comparison with the change in kinetic energy. Thus, for most practical purposes, Eq. (1-75) becomes

$$\Delta E_s = \Delta E_K \qquad (1\text{-}76)$$

In many problems the mass of the spring is so small that its contribution to the total kinetic energy can be neglected. But in many other cases, particularly where large accelerations and thus large forces are required, serious error will be introduced by ignoring the mass of the spring. Generally, one end of the spring is fixed and the other end moves in a straight line, as in Fig. 1-23, and the kinetic energy of an elemental length of the spring is

$$dE_K = \tfrac{1}{2} V^2 \, dm \qquad (1\text{-}77)$$

If it is assumed that the velocity varies linearly from zero at the fixed end to V_A at the moving end[1] and that the mass is distributed uniformly over the length L, for the spring in Fig. 1-24,

$$V = \frac{x}{L} V_A \qquad (1\text{-}78)$$

and $$dm = \frac{m_s}{L} \, dx \qquad (1\text{-}79)$$

where m_s = mass of entire spring.

[1] This assumption is valid only as long as the velocity V_A is much less than the velocity of wave propagation, considering the spring as a continuous member. See Sec. 6-11 for discussion of longitudinal vibrations of uniform rods.

Substituting from Eqs. (1-78) and (1-79) into Eq. (1-77) and integrating gives

$$E_K = \int dE_K = \int_0^L \frac{1}{2}\left(\frac{x}{L}V_A\right)^2 \frac{m_s}{L}\,dx = \tfrac{1}{6}m_s V_A{}^2 \qquad (1\text{-}80)$$

The kinetic energy of an equivalent mass at A is

$$E_K = \tfrac{1}{2}m_{\text{equiv}}V_A{}^2 \qquad (1\text{-}81)$$

Thus, for the kinetic energies to be the same,

$$m_{\text{equiv}} = \frac{m_s}{3} \qquad (1\text{-}82)$$

This result is also of use in some vibration problems, although to a lesser degree because the mass of the spring is usually a much smaller fraction of the total than it is in the type of problem being considered here.

The kinetic energy of a body moving in plane motion can be expressed as

$$E_K = \tfrac{1}{2}mV_G{}^2 + \tfrac{1}{2}I\omega^2 \qquad (1\text{-}83)$$

where m = mass of the link

V_G = velocity of the center of gravity

I = mass moment of inertia of the link about the center of gravity

ω = angular velocity of the link

In these terms, the kinetic energy of the linkage in Fig. 1-23 is

$$E_K = \frac{1}{2}\left(\frac{m_s}{3} + m_2\right)V_A{}^2 + \tfrac{1}{2}m_3 V_{G_3}{}^2 + \tfrac{1}{2}I_3\omega_3{}^2 + \tfrac{1}{2}m_4 V_{G_4}{}^2 + \tfrac{1}{2}I_4\omega_4{}^2 \qquad (1\text{-}84)$$

Equation (1-84) may be simplified somewhat by noting that since link 4 rotates about a fixed axis, its kinetic energy can be expressed as

$$E_{K,4} = \tfrac{1}{2}I_{O_4}\omega_4{}^2 \qquad (1\text{-}85)$$

where, by use of the parallel-axis theorem,

$$I_{O_4} = I_4 + m_4(O_4 G_4)^2 \qquad (1\text{-}86)$$

Substituting from Eq. (1-85) into Eq. (1-84) gives

$$E_K = \frac{1}{2}\left(\frac{m_s}{3} + m_2\right)V_A{}^2 + \tfrac{1}{2}m_3 V_{G_3}{}^2 + \tfrac{1}{2}I_3\omega_3{}^2 + \tfrac{1}{2}I_{O_4}\omega_4{}^2 \qquad (1\text{-}87)$$

From previous courses the reader should recall that the velocities in a linkage can be studied and a complete velocity polygon[1] or vector dia-

[1] R. M. Phelan, "Fundamentals of Mechanical Design," 2d ed., pp. 38–41, McGraw-Hill Book Company, New York, 1962.

gram drawn for any phase without knowing the actual magnitude of the velocity of any point. The important fact here is that if the velocity of one point is doubled, the velocities of all points and links are similarly doubled, and, as long as neither the masses nor moments of inertia of the links change with velocity, the kinetic energy of the linkage would be quadrupled. Consequently, if we let E_{K1} be the kinetic energy calculated by assuming the velocity of any particular point or link to be 1 in./sec or 1 rad/sec (as appropriate), we find that

$$E_K = K^2 E_{K1} \qquad (1\text{-}88)$$

or
$$K = \left(\frac{E_K}{E_{K1}}\right)^{1/2} \qquad (1\text{-}89)$$

where K is the scale factor by which the magnitudes from the velocity polygon must be multiplied for the linkage to have the correct amount of kinetic energy.

Using the relationships in Eqs. (1-76), (1-87), and (1-89), we can determine the actual velocities for all phases. In general, one is interested mainly in the motion of one particular member, such as link 4 in Fig. 1-23. In this case it is most convenient to assume $\omega_4 = 1$ rad/sec for enough phases to permit drawing a smooth curve of the actual values of ω_4 versus θ_4, where θ_4 is the angular position of link 4.

The solution is not yet complete, because one usually wishes to know the motion characteristics as functions of time, not position. Time may be introduced by considering the definition of angular velocity, which is

$$\omega = \lim_{\Delta t \to 0} \frac{\Delta \theta}{\Delta t} = \frac{d\theta}{dt} \qquad (1\text{-}90)$$

Rearranging Eq. (1-90) gives

$$dt = \frac{1}{\omega} d\theta \qquad (1\text{-}91)$$

Integrating both sides of Eq. (1-91) gives

$$t_2 - t_1 = \Delta t = \int_{\theta_1}^{\theta_2} \frac{1}{\omega} d\theta \qquad (1\text{-}92)$$

Therefore, Δt is the area under the $1/\omega$ versus θ curve, and it may be found by using a planimeter or a numerical method for integration, such as the trapezoidal rule or Simpson's rule.

In this type of problem the most important information is usually the output position as a function of time. This, as well as the velocity as a function of time, is available upon performing the integration in Eq. (1-92). If, however, there is a need for acceleration information, we must

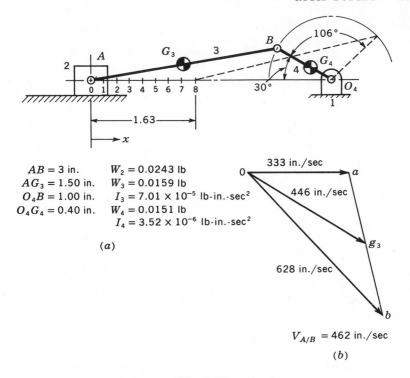

AB = 3 in. $W_2 = 0.0243$ lb
AG_3 = 1.50 in. $W_3 = 0.0159$ lb
O_4B = 1.00 in. $I_3 = 7.01 \times 10^{-5}$ lb-in.-sec^2
O_4G_4 = 0.40 in. $W_4 = 0.0151$ lb
                    $I_4 = 3.52 \times 10^{-6}$ lb-in.-sec^2

(a)

$V_{A/B}$ = 462 in./sec

(b)

Fig. 1-25

now resort to graphical differentiation[1]—the simplest approach being to differentiate the velocity-time curve.

Example 1-3 A mechanism similar to that in Fig. 1-23 is being designed. The specifications are that when link 4 has rotated through an angle of 106° it should have an angular velocity of 100 rps and the spring compression should be zero. Preliminary calculations have indicated that the dimensions, weights, and moments of inertia will be as given in Fig. 1-25a. We are asked to determine (a) the spring rate required by the compression spring and (b) the time required after the pin is pulled for link 4 to rotate through the 106° angle.

Solution. For convenience, the slider displacement has been divided into eight equal parts and the phases are numbered 0 through 8. If we neglect the minor losses due to friction and the change in gravitational potential energy of links 3 and 4, we can say that all of the potential energy stored elastically in the spring has been transformed into kinetic energy by the time link 4 has rotated 106°. Therefore, we can determine the spring requirements by calculating the kinetic energy of the linkage and spring when the linkage is in phase 8 and link 4 is rotating at 100 rps. Under these conditions,

$$V_B = R\omega = 1 \times 100 \times 2\pi = 628 \text{ in./sec}$$

and the velocity polygon for this phase is shown in Fig. 1-25b.

[1] *Ibid.*, pp. 50–52.

The pertinent equation is Eq. (1-87),

$$E_K = \frac{1}{2}\left(\frac{m_s}{3} + m_2\right) V_A{}^2 + \frac{1}{2}m_3 V_{G_3}{}^2 + \frac{1}{2}I_3\omega_3{}^2 + \frac{1}{2}I_{O_4}\omega_4{}^2$$

With the exception of m_s, which cannot be specified until the spring is designed, the remaining information is available in Fig. 1-25.

Neglecting the spring,

$$
\begin{aligned}
E_K &= \frac{1}{2}\frac{W_2}{g} V_A{}^2 + \frac{1}{2}\frac{W_3}{g} V_{G_3}{}^2 + \frac{1}{2}I_3\omega_3{}^2 + \frac{1}{2}[I_4 + m_4(O_4G_4)^2]\omega_4{}^2 \\
&= \frac{0.0243 \times 333^2}{2 \times 386} + \frac{0.0159 \times 446^2}{2 \times 386} + \frac{7.01 \times 10^{-5} \times 154^2}{2} \\
&\qquad\qquad + \frac{1}{2}\left(3.52 \times 10^{-6} + \frac{0.0151 \times 0.4^2}{386}\right)628^2 \\
&= 3.49 + 4.10 + 0.83 + 1.93 = 10.35 \text{ in.-lb}
\end{aligned}
$$

The potential energy in a compressed linear spring is

$$E_P = \frac{1}{2}P\delta \qquad\qquad (1\text{-}93)$$

where P = force

δ = deflection under force P

From Fig. 1-25a, $\delta = 1.63$ in. Therefore, since $E_P = E_K$,

$$E_P = 10.35 = \frac{1}{2}P \times 1.63$$

and

$$P = 12.70 \text{ lb}$$

The spring rate is

$$\frac{P}{\delta} = \frac{12.70}{1.63} = 7.79 \text{ lb/in.}$$

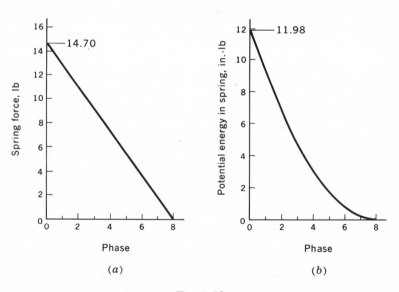

(a)

(b)

Fig. 1-26

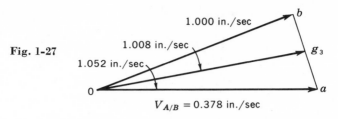

Fig. 1-27

$V_{A/B} = 0.378$ in./sec

A spring that will carry a maximum force of 12.70 lb with a spring rate of 7.79 lb/in. will weigh about 0.03 lb. Realizing that a recalculation, including one-third the weight of the spring, will result in a somewhat heavier spring because of the increased force and an increased wire diameter, let us estimate that the final spring will weigh about 0.034 lb. Using this estimate of spring weight, the recalculation leads to

$$E_K = E_P = 11.98 \text{ in.-lb}$$
$$P_{\max} = 14.70 \text{ lb}$$

and

$$\frac{P}{\delta} = 9.02 \text{ lb/in.}$$

Although the weight of the spring is important, there is little point in making another trial solution at this time because the weights and inertias of the other links must also be considered only estimates until a complete force analysis has been made after the accelerations have been determined.

The spring force and the potential energy in the spring are shown as functions of the phase of the mechanism in Fig. 1-26a and b, respectively. The next step is to find the angular velocity of link 4 as a function of the phase by use of Eqs. (1-87) and (1-88). This requires determining the linear velocity of the centers of gravity of links 2 and 3 and the angular velocity of link 3 for $\omega_4 = 1$ rad/sec for several phases, the final accuracy being closely related to the number of phases used. The following is based upon calculations at phases 0.1, 0.3, and 1 through 8, but sample numerical results will be presented for only phase 3.

The velocity polygon for phase 3 with $\omega_4 = 1$ rad/sec is given in Fig. 1-27. From the velocity polygon,

$$V_A = 1.052 \text{ in./sec}$$
$$V_{G_3} = 1.008 \text{ in./sec}$$
$$\omega_3 = \frac{V_{A/B}}{AB} = \frac{0.378}{3} = 0.1260 \text{ rad/sec}$$

Using Eq. (1-87), the kinetic energy E_K in phase 3 for $\omega_4 = 1$ rad/sec is found to be 8.20×10^{-5} in.-lb. Neglecting friction, etc., the actual kinetic energy must equal the potential energy given up by the spring, which, from Fig. 1-26b, is found to be

$$E_K = \Delta E_{P_3} = E_{P_0} - E_{P_3} = 11.98 - 4.67 = 7.31 \text{ in.-lb}$$

Using Eq. (1-89) to find the velocity scale factor gives

$$K = \left(\frac{E_K}{E_{K1}}\right)^{\frac{1}{2}} = \left(\frac{7.31}{8.20 \times 10^{-5}}\right)^{\frac{1}{2}} = 299$$

Thus, all velocities in phase 3 are 299 times those in Fig. 1-27. For example, $V_A = 299 \times 1.052 = 314$ in./sec and $\omega_4 = 299 \times 1 = 299$ rad/sec. In a similar manner the velocities are found for all phases.

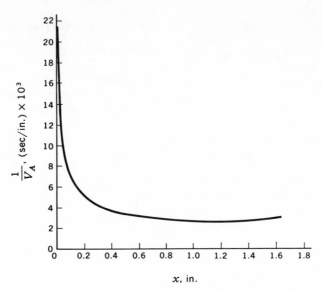

Fig. 1-28

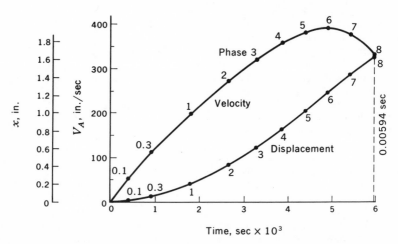

Fig. 1-29

Time can be introduced by applying Eq. (1-92) to the data for link 4. But, since the phases were designated by dividing the displacement of A into equal divisions, it will be more convenient to adapt Eq. (1-92) for rectilinear motion. Thus,

$$\Delta t = \int_0^x \frac{1}{V_A} dx \tag{1-94}$$

Using the results of the previous step, values of $1/V_A$ have been calculated and are plotted as a function of the displacement in Fig. 1-28. Since the mechanism starts from rest, $1/V_A = \infty$ at $x = 0$ and a small error may result because of the necessity for approximating the area between $x = 0$ and $x = 0+$. Increased accuracy could be gained by using smaller increments in change of phase or by applying Newton's laws to a trial-and-error solution for the acceleration at $t = 0$, when both x and V_A are zero. In a practical situation the extra work is not justified because more important effects have already been ignored.

The velocity-time and displacement-time graphs for the slider are presented in Fig. 1-29. As noted, the time required to reach phase 8 is about 0.00594 sec.

Closure. The answers to the original questions are (*a*) the spring rate should be 9.02 lb/in. and (*b*) 0.00594 sec will be required for link 4 to rotate through 106°.

In a real situation the design cannot be considered complete until it has been shown that the proposed dimensions provide adequate strength and rigidity and that the proposed bearings will be satisfactory under the worst possible combination of forces on each member. This will require a series of inertia-force analyses, as discussed in Sec. 1-5, starting with the graphical differentiation of the velocity-time curve in Fig. 1-29 to arrive at acceleration-time data for the slider. This information can then be used in combination with the available velocity-time data to calculate the accelerations of other points and links.

It should also be noted that a somewhat stiffer spring will have to be used to provide the energy that is dissipated by friction. Since the friction loss in properly designed joints is quite small and since the calculations would be unduly lengthy and inherently inaccurate, the best approach would be to determine experimentally the response of the mechanism using several different springs. For example, one might try spring rates of 9.10, 9.20, 9.30, etc., lb/in. until the actual performance is as close as desired to that specified.

chapter **2**

BALANCING OF MACHINERY

Inertia forces exist wherever parts having finite mass are accelerated. The forces are important internally because the parts themselves must be designed to perform satisfactorily under all combinations of inertia and service loads and externally because the resultant external or shaking force becomes a disturbing force on the frame and associated parts. In both cases varying forces acting on elastic bodies can give rise to serious, even destructive, vibrations of the parts or the complete machine and adjacent structures and equipment. The presence of vibration and the accompanying noise can be serious problems with respect to the physical and mental well-being of the machine operators and other persons nearby.

In most respects it is convenient to make a distinction between those inertia forces resulting from the motion of rigid bodies and those resulting from vibrations of elastic bodies. Although, in reality, all bodies are elastic, the combination of mass, rigidity, and speed will often be such that vibrations of the parts will be insignificant and the methods outlined in the preceding chapter can be used with good results to determine the magnitudes and directions of the inertia forces at any and all times. This chapter will be concerned solely with the balancing of machinery in which the bodies can be considered to be rigid, and the balancing of inertia forces in elastic members will be left for later sections.[1]

2-1. Rigid Bodies In general, little can be done to decrease the magnitude of internal inertia forces resulting from the motion of rigid members except to select materials giving the highest possible ratio of strength to weight, such as aluminum and magnesium alloys and heat-

---

[1] The vibration problems most closely related to balancing of inertia forces are the phenomenon of whirling of shafts (Sec. 4-20) and the theory of vibration absorbers (Secs. 6-2 through 6-4).

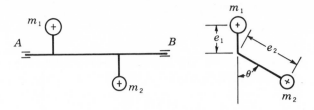

Fig. 2-1 *Unbalanced rigid body.*

treated alloy steels, and to use shapes that also give the maximum ratio of strength to weight, such as I sections and hollow cylinders. Vibrations will be considered in some detail in following chapters, and it will be seen that the external effects will be directly related to the magnitude of the shaking force. Thus, efforts directed toward decreasing the shaking forces are always worthwhile and in many cases even essential.

The general approach to the minimization of the magnitude of the shaking forces is to balance the effect by introducing another shaking force that, in so far as possible, is equal in magnitude and opposite in direction to the original shaking force. This process is called balancing of machinery, and the following sections will consider the principles used in balancing (1) rotating and (2) reciprocating machinery.

2-2. Rotating Machinery The shaking forces in rotating machinery are due almost entirely to the inertia forces and couples associated with unbalanced rotating members.

Figure 2-1 shows, schematically, a rotating member with two unbalanced masses m_1 and m_2 whose centers of gravity are located at distances e_1 and e_2, respectively, from the axis of rotation. The inertia (centrifugal) forces will act radially through the centers of gravity and will be equal to $m_1e_1\omega^2$ and $m_2e_2\omega^2$, as shown in Fig. 2-2. For the orientation shown, $m_1e_1\omega^2$ acts in the vertical plane and $m_2e_2\omega^2$ can be resolved into components $m_2e_2\omega^2 \cos \theta$ in the vertical plane and $m_2e_2\omega^2 \sin \theta$ in the horizontal plane. Considering each plane separately leads to the free-body diagrams in Fig. 2-3, and the problem becomes that of determining what should be done to eliminate the unbalanced forces.

Obviously, the simplest approach would be to locate and remove m_1 and

Fig. 2-2 *Inertia forces on body in Fig. 2-1 when rotating at constant speed.*

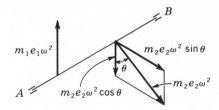

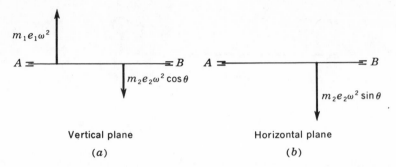

Vertical plane

(a)

Horizontal plane

(b)

Fig. 2-3   *Inertia forces in Fig. 2-2 resolved into components in the vertical and horizontal planes.*

m_2. However, in general this is not possible in real problems and it becomes necessary to add or remove mass at other points.

The conditions for static equilibrium ($\Sigma \mathbf{M} = 0$ and $\Sigma \mathbf{F} = 0$) can be used to find the reactions at bearings A and B. But the goal of balancing is to provide zero forces on the bearings. Since the forces in the vertical plane are not collinear and, consequently, cannot be resolved into a single force, there will also be a couple and it is necessary to introduce corrections in two places. Practical considerations for removing or adding mass will often determine where corrections can actually be made. Here we shall assume that the balancing planes labeled L (left) and R (right) in Fig. 2-4 will be satisfactory.

Considering the vertical plane first, we can determine the inertia-force correction required in the L balancing plane by summing moments about the R plane:

$$\Sigma \mathbf{M}_R = 0$$

$$m_{LV}e_{LV}\omega^2 l - m_1e_1\omega^2 a_1 + m_2e_2\omega^2(\cos\theta)a_2 = 0 \qquad (2\text{-}1)$$

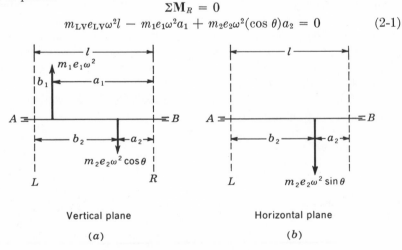

Vertical plane

(a)

Horizontal plane

(b)

Fig. 2-4   *Location of balancing planes for body in Fig. 2-1.*

Since ω is the same for all terms, we can conclude that the speed of rotation is not a factor in determining the degree to which a *rigid* rotating body is balanced. Thus, the unbalances, m_1e_1, etc., can be considered as equivalent vector forces and Eq. (2-1) can be rewritten as

$$m_{\mathrm{LV}}e_{\mathrm{LV}}l - m_1e_1a_1 + m_2e_2(\cos\theta)a_2 = 0 \qquad (2\text{-}2)$$

from which

$$m_{\mathrm{LV}}e_{\mathrm{LV}} = \frac{m_1e_1a_1 - m_2e_2(\cos\theta)a_2}{l} \qquad (2\text{-}3)$$

The correction required in the R balancing plane can be found by summing moments about the L plane in the same manner as above, but now only one force $m_{\mathrm{RV}}e_{\mathrm{RV}}$ remains to be found and it is usually simpler to sum the forces. Thus,

$$\Sigma F_V = 0$$
$$-m_{\mathrm{LV}}e_{\mathrm{LV}} + m_1e_1 - m_2e_2\cos\theta + m_{\mathrm{RV}}e_{\mathrm{RV}} = 0 \qquad (2\text{-}4)$$

from which

$$m_{\mathrm{RV}}e_{\mathrm{RV}} = m_{\mathrm{LV}}e_{\mathrm{LV}} - m_1e_1 + m_2e_2\cos\theta \qquad (2\text{-}5)$$

In the same manner as above, the balance corrections required in the horizontal plane can be shown to be as follows:

$$m_{\mathrm{LH}}e_{\mathrm{LH}} = -\frac{m_2e_2(\sin\theta)a_2}{l} \qquad (2\text{-}6)$$

and

$$m_{\mathrm{RH}}e_{\mathrm{RH}} = m_2e_2\sin\theta - m_{\mathrm{LH}}e_{\mathrm{LH}} \qquad (2\text{-}7)$$

When the balance corrections are included, the free-body diagram of the rotor will appear as shown in Fig. 2-5, where the horizontal and vertical components have been combined into their resultants m_{Le_L} and m_{Re_R} in the two balancing planes. It is interesting and important to note that *no matter how many unbalanced masses there may be, complete balance of a rigid rotor can always be obtained by making a correction in each of two arbitrarily chosen balancing planes.*

Fig. 2-5 *Free-body diagram showing forces in balancing planes.*

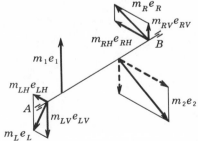

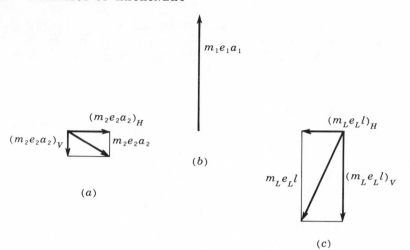

Fig. 2-6 *Graphical representation of moments of balance-correction and inertia forces.*

The rather involved process of resolving the unbalances into their components in two planes, such as the horizontal and vertical, solving for the corrections in each of these planes, and then combining the component corrections to find the resultant corrections in the two balancing planes can be avoided by using a graphical vector solution. For example, $m_2e_2a_2$ is the moment of the unbalance in plane 2 about the R correction plane, and it can be resolved into its components in the horizontal and vertical planes, as in Fig. 2-6a. In a similar manner the moments $m_1e_1a_1$ and m_Le_Ll can be resolved into the components shown in Fig. 2-6b and c, respectively. The vector sum of the vertical components is the same as solving Eq. (2-2), and the vector sum of the horizontal components is the

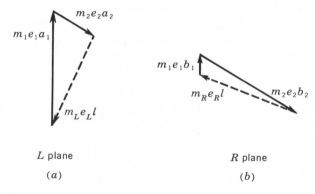

Fig. 2-7 *Graphical solutions for equilibrium of moments of balance-correction and inertia forces.*

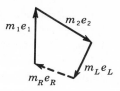

Fig. 2-8 *Graphical solution for equilibrium of balance-correction and inertia forces.*

same as solving the moment equation for the horizontal plane. Since equilibrium requires the vector sum of the components to equal zero, it follows that the sum of the original vectors must also equal zero. Therefore, we can say

$$\Sigma m_i \mathbf{e}_i a_i = 0 \qquad\qquad (2\text{-}8)$$

Figure 2-7 illustrates how Eq. (2-8) would be used to determine the corrections required for the rotor in Fig. 2-5. As before, after one balancing correction has been obtained by taking moments about the other balancing plane, only one inertia force remains to be found; and, following the reasoning given above, we can say

$$\Sigma m_i \mathbf{e}_i = 0 \qquad\qquad (2\text{-}9)$$

Application of Eq. (2-9) to determining the correction in the R balancing plane is illustrated in Fig. 2-8.

2-3. Static and Dynamic Unbalance   As the terms imply, static unbalance refers to an object at rest and dynamic unbalance refers to a rotating body. *The condition for static balance is simply that the axis of rotation pass through the center of gravity of the body.* Thus, static balance requires only that

$$\Sigma m_i \mathbf{e}_i = 0 \qquad\qquad (2\text{-}10)$$

Figure 2-9 shows how Eq. (2-10) can be used to determine the static-balance correction $m_s e_s$ for the rotor in Fig. 2-1. It should be noted that *static balance can always be achieved by making only one correction* and the amount of correction is independent of the plane in which it is to be made.

Dynamic balance requires not only that the axis of rotation pass through the center of gravity but also that it be a principal axis of inertia. This is the case discussed in the preceding section, and *corrections in two planes will be required.*

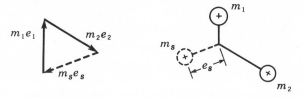

Fig. 2-9 *Graphical solution for balance correction for statically unbalanced rigid body.*

2-4. Practical Considerations   In the discussion in Sec. 2-2 the corrections were always indicated as the addition of inertia forces. Actually, a correction can be made either by adding mass at the proper radius or eccentricity or by removing material diametrically opposite from where the addition was indicated.   In numerous cases it may be more convenient to add mass in one correction plane and remove it in the other.

Many rotors, such as centrifugal pumps, narrow gears, flywheels, and automobile wheels, are so thin in relation to their diameter that the unbalance exists principally in one plane.   In such cases satisfactory operation can often be obtained by considering only static balancing. However, even here the balancing[1] may be done under dynamic conditions because of the increased accuracy that is possible because of the multiplying effect of the ω^2 term.

Many other rotors require dynamic balancing even though the unbalance is largely static.   In these cases, it is often convenient to balance the rotor statically as the first step.   When this is done, only a small couple remains and the rotor is then dynamically balanced by making two more corrections.   Thus, performing static balancing first requires a total of three corrections rather than just the two that would be sufficient if the first step involved taking moments about one of the correction planes.

Long rotors, such as found in axial-flow compressors, steam and gas turbines, and electric generators, must often be considered as a number of flexibly connected unbalanced disks.   The unbalanced inertia forces can deflect the shaft with respect to its desired axis of rotation and thus change the unbalance as the speed changes.   To ensure smooth running over a wide range of speeds it is necessary to make corrections in many planes.   The methods involved are beyond the scope of this text, and the reader is referred to the literature[2] for further information.

2-5. Reciprocating Machinery   Reciprocating machinery consists of those machines based on the slider-crank mechanism and its inversions. Of the many possibilities, the slider crank itself is by far the most important because of its use in high-speed engines, pumps, and compressors, where the inertia forces are too great to be neglected.   Only the slider crank will be considered in detail here, although the principles and methods will be generally applicable to all machines with reciprocating parts.

2-6. Single-slider Machines   As discussed in Sec. 1-6, for most purposes the slider-crank mechanism can be represented by the equivalent

---

[1] See Sec. 4-15 for discussions of the principles of operation of balancing machines and field balancing methods.

[2] J. P. Den Hartog, "Mechanical Vibrations," 4th ed., pp. 243–246, McGraw-Hill Book Company, New York, 1956.

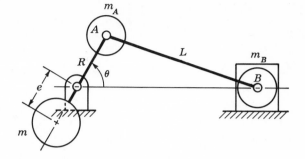

Fig. 2-10 *Slider-crank mechanism with counterbalance for rotating inertia force.*

linkage with concentrated masses in Fig. 1-14b, for which the resultant inertia force is

$$\mathbf{F}_i = \mathbf{F}_{\text{rot}} + \mathbf{F}_{\text{pri}} + \mathbf{F}_{\text{sec}} \tag{2-11}$$

where $F_{\text{rot}} = m_A R \omega^2$

$F_{\text{pri}} = m_B R \omega^2 \cos \theta$

$F_{\text{sec}} = m_B \dfrac{R^2}{L} \omega^2 \cos 2\theta$

and m_A = total rotating unbalance (crank + equivalent rod)

m_B = total reciprocating unbalance (slider or piston + equivalent rod)

As shown in Fig. 1-17, the resultant inertia force in this case varies from a maximum horizontal force to the right, through a minimum vertical force, to a relatively large horizontal force to the left.

The vertical force, and part of the horizontal force, is due solely to the rotating unbalance. This can be eliminated, or counterbalanced, by simply adding a counterbalance $me = m_A R$ diametrically opposite to A, as shown in Fig. 2-10. If the rotating unbalance is eliminated, only the horizontal forces due to the reciprocating unbalance remain. Unfortunately, there is no direct method for counterbalancing the primary and secondary forces, and the best one can do is to cancel part of the horizontal force by overbalancing the crank, i.e., making $me > m_A R$, thereby providing a rotating force that acts opposite to $m_A R \omega^2$.

If $me = (m_A + m_B)R$, the net rotating force is a vector $-m_B R \omega^2$. Thus, as shown in Fig. 2-11a, only the secondary inertia force acts in the horizontal direction and the primary shaking force has been eliminated, but at the expense of the introduction of a vertical shaking force of the same magnitude.

It should be noted that the secondary force cannot be affected except by introducing a counterbalance that rotates at twice the speed of the crankshaft.

In some cases better results are obtained with an intermediate degree of overbalancing. For example, if $me = (m_A + \tfrac{2}{3}m_B)R$, the rotating

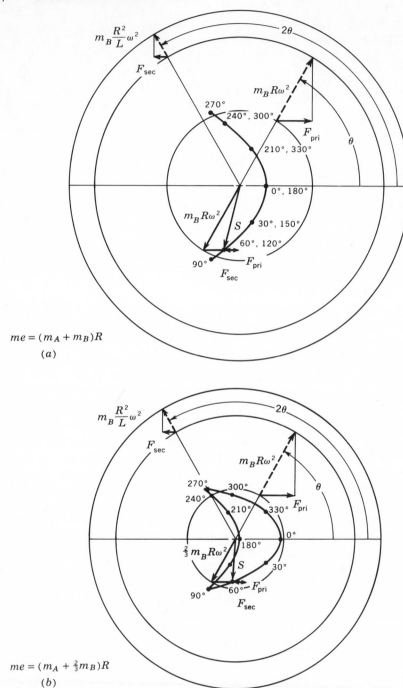

$$me = (m_A + m_B)R$$
(a)

$$me = (m_A + \tfrac{2}{3}m_B)R$$
(b)

Fig. 2-11 *Polar diagrams of slider-crank mechanism inertia forces.   (a) 100 percent overbalanced crankshaft; (b) two-thirds overbalanced crankshaft.*

counterbalance will be $-\frac{2}{3}m_B\mathbf{R}$ and the result will be that shown in Fig. 2-11*b*, where the horizontal and vertical inertia forces are relatively small and almost equal.

2-7. Multislider Machines   A multislider machine is one consisting of two or more slider-crank mechanisms whose cranks are fixed with respect to each other. This is most often achieved by offsetting the sliders along the axis of a crankshaft having several throws.   The internal-combustion engine is the most common multislider machine, and several widely used configurations are shown schematically in Fig. 2-12.

As for the case of rotating machinery (Sec. 2-2), balance requires that the resultant forces and couples be zero.   However, the situation now is

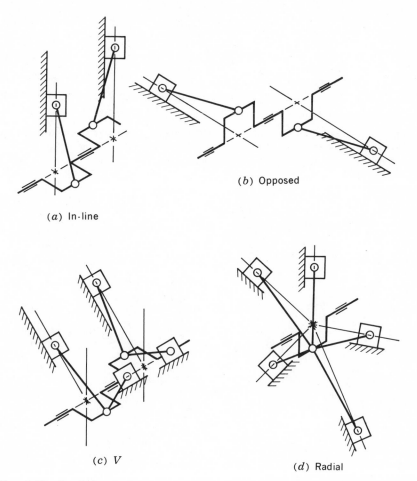

(*a*) In-line

(*b*) Opposed

(*c*) V

(*d*) Radial

Fig. 2-12 *Typical cylinder and crankshaft configurations for internal-combustion engines.*

considerably more complex, because the forces are no longer constant-magnitude rotating vectors but are the varying-magnitude harmonics of the reciprocating inertia forces (assuming that the crankshaft itself is already balanced). Also, in general, it is not possible to achieve balance of any arbitrary arrangement by the simple addition or subtraction of rotating weights, and the designer is usually limited to analyzing configurations to determine whether or not they are inherently balanced and, if not, whether or not balance can be achieved in a relatively simple manner.

Assuming that the crankshaft is balanced and that only the primary and secondary inertia forces are significant, the conditions for balance are[1]

1	$\Sigma F_{\text{pri}} = 0$	(2-12)
2	$\Sigma F_{\text{sec}} = 0$	(2-13)
3	$\Sigma M_{F\text{pri}} = 0$	(2-14)
4	$\Sigma M_{F\text{sec}} = 0$	(2-15)

The summations in Eqs. (2-12) through (2-15) can be performed graphically, but since only a few configurations are likely to be encountered, it is often simpler and more convenient to use numerical methods.

---

[1] For a discussion of engine balancing when higher harmonics are important, see C. M. Harris and C. E. Crede (eds.), "Shock and Vibration Handbook," vol. III, pp. 47-18–47-30, McGraw-Hill Book Company, New York, 1961.

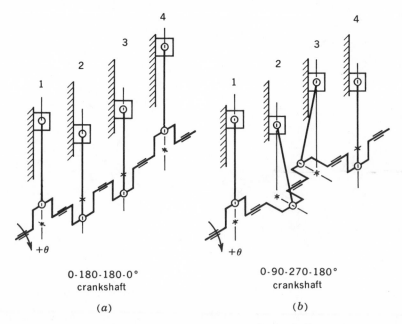

0-180-180-0°
crankshaft

(*a*)

0-90-270-180°
crankshaft

(*b*)

Fig. 2-13   *Crankshaft configurations for 4-cylinder in-line engines.*

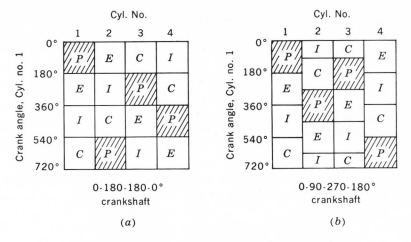

P = power; E = exhaust; I = intake; and C = compression

Fig. 2-14 *Firing-order diagrams for the engines in Fig. 2-13.*

In general, it is desirable first to combine the forces acting in the plane of rotation at each of the crank throws into resultant primary and secondary inertia forces before performing the summations in Eqs. (2-12) through (2-15). These princip'es will be applied to several types of internal-combustion engines in the following sections.

2-8. The 4-cylinder In-line 4-cycle Engine   Figure 2-13 shows two possible configurations for 4-cylinder in-line 4-cycle engines, and Fig. 2-14 shows the corresponding firing-order diagrams.

The smoothness of the power flow, i.e., the evenness in spacing of the power pulses, is as important as the degree of balance in selecting the best crankshaft configuration for a given type of engine.   As can be seen, the power strokes are uniformly spaced for only the engine with the crankshaft with the crank throws lying in a plane (Fig. 2-13*a*) and, all other things being equal, this will be the smoother-running engine.

Balance of engine with 0-180-180-0° crankshaft   The summation of primary inertia forces becomes[1]

$$\Sigma \mathbf{F}_{\text{pri}} = m_1 R_1 \omega^2 \cos \theta_1 + m_2 R_2 \omega^2 \cos \theta_2 + m_3 R_3 \omega^2 \cos \theta_3$$
$$+ m_4 R_4 \omega^2 \cos \theta_4 \quad (2\text{-}16)$$

where m_1, m_2, m_3, and m_4 = total reciprocating mass at sliders 1 through 4, respectively, and R_1, R_2, R_3, and R_4 = crank radii.

[1] The directions of rotations in Fig. 2-13 have been shown as clockwise, and the cylinders are numbered from the front to the rear to correspond with common automotive practice.   *In this and remaining sections in this chapter positive angles will be measured in the clockwise direction.*

Generally, $m_1R_1 = m_2R_2 = m_3R_3 = m_4R_4 = mR$, and Eq. (2-16) can then be rewritten as

$$\Sigma\mathbf{F}_{\text{pri}} = mR\omega^2(\cos\theta_1 + \cos\theta_2 + \cos\theta_3 + \cos\theta_4) \quad (2\text{-}17)$$

But,
$$\cos\theta_2 = \cos(\theta_1 - 180°) = -\cos\theta_1$$
$$\cos\theta_3 = \cos(\theta_1 - 180°) = -\cos\theta_1 \quad (2\text{-}18)$$
and
$$\cos\theta_4 = \cos\theta_1$$

Substituting from Eqs. (2-18) into Eq. (2-17) gives

$$\Sigma\mathbf{F}_{\text{pri}} = mR\omega^2(\cos\theta_1 - \cos\theta_1 - \cos\theta_1 + \cos\theta_1) = 0 \quad (2\text{-}19)$$

and, therefore, the primary inertia forces are balanced.

The summation of secondary inertia forces becomes

$$\Sigma\mathbf{F}_{\text{sec}} = m_1\frac{R_1{}^2}{L_1}\omega^2\cos 2\theta_1 + m_2\frac{R_2{}^2}{L_2}\omega^2\cos 2\theta_2 + m_3\frac{R_3{}^2}{L_3}\omega^2\cos 2\theta_3$$
$$+ m_4\frac{R_4{}^2}{L_4}\omega^2\cos 2\theta_4 \quad (2\text{-}20)$$

where L_1, L_2, L_3, and L_4 = lengths of the connecting rods.

Assuming $L_1 = L_2 = L_3 = L_4 = L$ as well as

$$m_1R_1 = m_2R_2 = m_3R_3 = m_4R_4 = mR$$

Eq. (2-20) can be written as

$$\Sigma\mathbf{F}_{\text{sec}} = m\frac{R^2}{L}\omega^2(\cos 2\theta_1 + \cos 2\theta_2 + \cos 2\theta_3 + \cos 2\theta_4) \quad (2\text{-}21)$$

But,
$$\cos 2\theta_2 = \cos 2(\theta_1 - 180°) = \cos 2\theta_1$$
$$\cos 2\theta_3 = \cos 2(\theta_1 - 180°) = \cos 2\theta_1 \quad (2\text{-}22)$$
and
$$\cos 2\theta_4 = \cos 2\theta_1$$

Substituting from Eqs. (2-22) into Eq. (2-21) gives

$$\Sigma\mathbf{F}_{\text{sec}} = m\frac{R^2}{L}\omega^2(\cos 2\theta_1 + \cos 2\theta_1 + \cos 2\theta_1 + \cos 2\theta_1)$$
$$= 4m\frac{R^2}{L}\omega^2\cos 2\theta_1 \quad (2\text{-}23)$$

which, in general, does not equal zero. Therefore, the secondary inertia forces are not balanced.

For calculating moments it is convenient to use cylinder 1 as reference and let a_2, a_3, and a_4 be the distances from the center line of cylinder 1 to the center lines of cylinders 2, 3, and 4, respectively. The summation of moments of the primary inertia forces becomes

$$\Sigma\mathbf{M}_{\text{pri}} = m_2R_2\omega^2(\cos\theta_2)a_2 + m_3R_3\omega^2(\cos\theta_3)a_3 + m_4R_4\omega^2(\cos\theta_4)a_4 \quad (2\text{-}24)$$

Assuming, as before, that $m_2R_2 = m_3R_3 = m_4R_4 = mR$ and substituting for the angles from Eqs. (2-18) into Eq. (2-24), we find

$$\Sigma\mathbf{M}_{\text{pri}} = mR\omega^2(\cos\theta_1)(-a_2 - a_3 + a_4) \tag{2-25}$$

from which we can see that the moment of the primary forces will be balanced if

$$a_2 + a_3 = a_4 \tag{2-26}$$

Equation (2-26) will be satisfied if the cylinder spacing is symmetrical about the center of the engine.   One of the simplest ways to ensure this is to separate the cylinders by identical distances a so that

$$a_2 = a \qquad a_3 = 2a \qquad a_4 = 3a \tag{2-27}$$

and thus
$$a_2 + a_3 = a + 2a = 3a = a_4 \tag{2-28}$$

The summation of moments of the secondary forces becomes

$$\Sigma\mathbf{M}_{\text{sec}} = m_2\frac{R_2{}^2}{L_2}\omega^2(\cos 2\theta_2)a_2 + m_3\frac{R_3{}^2}{L_3}\omega^2(\cos 2\theta_3)a_3$$
$$+ m_4\frac{R_4{}^2}{L_4}\omega^2(\cos 2\theta_4)a_4 \tag{2-29}$$

Assuming $m_2R_2/L_2 = m_3R_3/L_3 = m_4R_4/L_4 = mR/L$, and substituting from Eqs. (2-22) for the angles, Eq. (2-29) can be written as

$$\Sigma\mathbf{M}_{\text{sec}} = m\frac{R^2}{L}\omega^2(\cos 2\theta_1)(a_2 + a_3 + a_4) \tag{2-30}$$

which cannot equal zero at all times, except for the trivial case of

$$a_2 = a_3 = a_4 = 0$$

Therefore, it appears that the secondary inertia force and the moment of the secondary inertia force are not balanced.   However, since there is a resultant secondary inertia force as well as a resultant moment of the secondary inertia forces, we can conclude that the moment is not a couple, but, rather, is the result of the resultant inertia force acting at a distance z from cylinder 1.   By definition,

$$M_{\text{res}} = F_{\text{res}}z \tag{2-31}$$

where $M_{\text{res}} = |\Sigma\mathbf{M}_{\text{sec}}|$ and $F_{\text{res}} = |\Sigma\mathbf{F}_{\text{sec}}|$.   Combining Eqs. (2-23), (2-30), and (2-31), we find

$$z = \frac{M_{\text{res}}}{F_{\text{res}}} = \frac{a_2 + a_3 + a_4}{4} \tag{2-32}$$

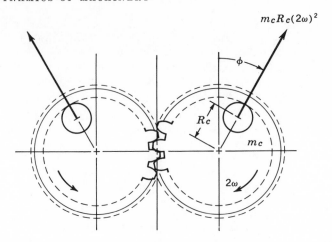

Fig. 2-15   *Generation of a sinusoidally varying inertia force by use of contrarotating masses.*

which is constant.   If $a_2 + a_3 = a_4$, the conditions for balance of the moment of the primary inertia forces, Eq. (2-32) becomes

$$z = \frac{a_4}{2} \tag{2-33}$$

Thus, in this case, the only unbalance is the secondary shaking force acting through the center of the engine.

One device that has been used to balance the secondary shaking force is illustrated in Fig. 2-15.   The gears are driven at twice the speed of rotation of the crankshaft and the resultant inertia force is

$$F_c = 2m_c R_c (2\omega)^2 \cos \phi \tag{2-34}$$

But since the gears rotate at twice the crankshaft speed,

$$\phi = 2\theta \tag{2-35}$$

where θ is the corresponding angle of rotation of the crankshaft.   Equation (2-34) can be written as

$$F_c = 8m_c R_c \omega^2 \cos 2\theta \tag{2-36}$$

Comparing Eq. (2-36) with Eq. (2-23) shows that if $8m_c R_c = 4mR^2/L$ and $\theta = -\theta_1 = \theta_1 + 180°$,

$$\mathbf{F}_c = -\Sigma \mathbf{F}_{sec} \tag{2-37}$$

and the engine will be completely balanced when the line of action of $\mathbf{F}_c$ coincides with that of $\Sigma \mathbf{F}_{sec}$.

Because of the complexity and additional expense, this device has not been used in many engines.   However, this is a relatively simple way to generate a sinusoidally varying force, and the principle has been used in vibration exciters to shake bridges and other structures in vibration studies and in unloaders to shake coal and similar materials out of railroad cars and other containers.

Balance of engine with 0-90-270-180° crankshaft   Assuming, as before, that $m_1R_1 = m_2R_2 = m_3R_3 = m_4R_4 = mR$, the summation of the primary inertia forces will again be given by Eq. (2-17). However, now

$$\cos \theta_2 = \cos (\theta_1 - 270°) = \sin \theta_1$$
$$\cos \theta_3 = \cos (\theta_1 - 90°) = - \sin \theta_1 \qquad (2\text{-}38)$$
$$\cos \theta_4 = \cos (\theta_1 - 180°) = - \cos \theta_1$$

Substituting from Eqs. (2-38) into Eq. (2-17) gives

$$\Sigma \mathbf{F}_{\text{pri}} = mR\omega^2(\cos \theta_1 + \sin \theta_1 - \sin \theta_1 - \cos \theta_1) = 0 \qquad (2\text{-}39)$$

and we find that the primary inertia forces are balanced.

If we assume that the connecting rods have the same lengths, the summation of the secondary forces will be given by Eq. (2-21). Now,

$$\cos 2\theta_2 = \cos 2(\theta_1 - 270°) = - \cos 2\theta_1$$
$$\cos 2\theta_3 = \cos 2(\theta_1 - 90°) = - \cos 2\theta_1 \qquad (2\text{-}40)$$
$$\cos 2\theta_4 = \cos 2(\theta_1 - 180°) = \cos 2\theta_1$$

Substituting from Eqs. (2-40) into Eq. (2-21) gives

$$\Sigma \mathbf{M}_{\text{pri}} = mR\omega^2(\sin \theta_1)(+a_2 - a_3 - a_4) \qquad (2\text{-}42)$$

and the secondary inertia forces are also balanced.

The summation of the moments of the primary inertia forces is given by Eq. (2-24). Assuming that $m_1R_1 = m_2R_2 = m_3R_3 = m_4R_4 = mR$ and substituting for the angles from Eqs. (2-38), Eq. (2-24) becomes

$$\Sigma \mathbf{M}_{\text{pri}} = mR\omega^2(\sin \theta_1)(-a_2 + a_3 - a_4) \qquad (2\text{-}42)$$

which cannot be zero at all times. Since the primary shaking force is zero, the moment must be a couple in the plane of the cylinders. If the cylinders are equally spaced, the couple, from Eq. (2-42), is

$$C_{\text{pri}} = 2amR\omega^2 \sin \theta_1 \qquad (2\text{-}43)$$

where a = distance between cylinders.

The summation of the moments of the secondary inertia forces is given by Eq. (2-29). Assuming that

$$\frac{m_1R_1}{L_1} = \frac{m_2R_2}{L_2} = \frac{m_3R_3}{L_3} = \frac{m_4R_4}{L_4} = \frac{mR}{L}$$

and substituting from Eqs. (2-40) for the angles, Eq. (2-29) becomes

$$\Sigma \mathbf{M}_{sec} = m \frac{R^2}{L} \omega^2 (\cos 2\theta_1)(-a_2 - a_3 + a_4) \qquad (2\text{-}44)$$

Therefore, the moment of the secondary shaking forces will be balanced whenever

$$a_2 + a_3 = a_4 \qquad (2\text{-}45)$$

As before, this can be achieved by locating the cylinders to be symmetrical about the center of the engine.

Thus, the only unbalance is the primary couple. This could be balanced by introducing an equal and opposite couple by using two sets of contrarotating weights in a manner similar to that discussed above. In this case the weights would have to rotate at the crankshaft speed. This has not been a matter of great practical concern, since this crankshaft arrangement has found practically no use in 4-cylinder engines because of the uneven spacing of the power pulses.

2-9. V-type Engines As shown in Fig. 2-12c, V-type engines are characterized by having two banks of in-line cylinders and a single crankshaft. The angle between the banks of cylinders is not arbitrary but, rather, its choice is based upon considerations of balance and spacing of power pulses.

When each bank of cylinders, considered by itself, is completely balanced, it is obvious that the engine will be balanced for all bank angles and the bank angle will be chosen to give the smoothest delivery of power. With proper arrangement of throws on the crankshaft and spacing of the cylinders, both the 6- and 8-cylinder in-line engines are balanced. Thus, a V-12 engine should have a 60° bank angle to provide 12 equally spaced power impulses per cycle and a V-16 engine should have a 45° bank angle to provide 16 equally spaced power impulses per cycle.

For automobiles, the short length and low height given by the V-8 engine has made this the most popular type of engine in the higher-powered models. With a 90° bank angle, either of the crankshaft arrangements in Fig. 2-13 will provide eight equally spaced power impulses and considerations of balance become the determining factors in selection of the arrangement to use.

As discussed in the preceding section, the 0-180-180-0° crankshaft results in an unbalanced secondary shaking force and the 0-90-270-180° crankshaft results in an unbalanced moment of the primary inertia forces. Thus, it will not be necessary to carry out a complete analysis, but only to see what happens to the unbalanced force and moment when two 4-cylinder in-line engines are combined to make a V-8 engine.

0-180-180-0° crankshaft As discussed in the preceding section, the secondary shaking force of a 4-cylinder in-line engine with a 0-180-180-0°

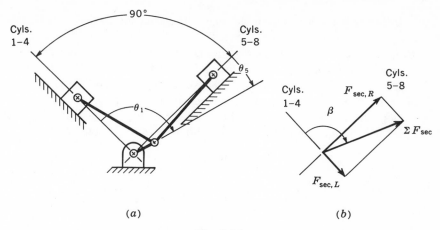

Fig. 2-16

crankshaft is a vector

$$\Sigma F_{sec} = 4m \frac{R^2}{L} \omega^2 \cos 2\theta_1 \qquad (2\text{-}46)$$

that acts through the center of the engine. Each bank of cylinders in the V-8 engine will have such a force, and if cylinders 1, 2, 3, and 4 are in the left-hand bank and 5, 6, 7, and 8 are in the right-hand bank (looking from the front of the engine) the forces and angles will be as shown in Fig. 2-16. The magnitude of the resultant secondary force is

$$\Sigma F_{sec} = \sqrt{(F_{sec,L})^2 + (F_{sec,R})^2}$$
$$= \sqrt{\left(4m \frac{R^2}{L} \omega^2 \cos 2\theta_1\right)^2 + \left(4m \frac{R^2}{L} \omega^2 \cos 2\theta_5\right)^2} \qquad (2\text{-}47)$$

and it is directed, relative to cylinder 1, at an angle

$$\beta = \tan^{-1} \frac{F_{sec,R}}{F_{sec,L}} = \tan^{-1} \frac{4m(R^2/L)\omega^2 \cos 2\theta_5}{4m(R^2/L)\omega^2 \cos 2\theta_1} \qquad (2\text{-}48)$$

But $\theta_5 = \theta_1 - 90°$ and thus, by trigonometric identities,

$$\cos 2\theta_5 = \cos 2(\theta_1 - 90°) = -\cos 2\theta_1 \qquad (2\text{-}49)$$

Substituting from Eq. (2-49) into Eqs. (2-47) and (2-48) and simplifying gives, respectively,

$$\Sigma F_{sec} = \sqrt{2} \times 4m \frac{R^2}{L} \omega^2 \cos 2\theta_1 \qquad (2\text{-}50)$$

and

$$\beta = \tan^{-1}(-1) = 135° \text{ or } -45° \qquad (2\text{-}51)$$

Therefore, the V-8 engine with a 0-180-180-0° crankshaft will have a resultant secondary force that acts through the center of the engine along a line that is perpendicular to the plane bisecting the angle between the banks of cylinders, as shown in Fig. 2-17.

0-90-270-180° crankshaft Although the unbalanced summation of moments of the primary inertia forces for the two banks could be combined in a manner similar to that used above for the secondary forces, it will be more useful to consider in turn each pair of cylinders that uses a common throw on the crankshaft. For example, the primary inertia forces for cylinders 1 and 5 can be combined into a resultant force

$$\Sigma \mathbf{F}_{pri,1+5} = \mathbf{F}_{pri,1} + \mathbf{F}_{pri,5} \qquad (2\text{-}52)$$

which has a magnitude of

$$F_{pri,1+5} = \sqrt{(mR\omega^2 \cos\theta_1)^2 + (mR\omega^2 \cos\theta_5)^2} \qquad (2\text{-}53)$$

and, relative to cylinder 1, acts at an angle of

$$\beta = \tan^{-1} \frac{mR\omega^2 \cos\theta_5}{mR\omega^2 \cos\theta_1} \qquad (2\text{-}54)$$

But, $\theta_5 = \theta_1 - 90°$, and thus

$$\cos\theta_5 = \cos(\theta_1 - 90°) = \sin\theta_1 \qquad (2\text{-}55)$$

Substituting from Eq. (2-55) into Eqs. (2-53) and (2-54) and simplifying gives, respectively,

$$F_{pri,1+5} = mR\omega^2 \qquad (2\text{-}56)$$

and $$\beta = \tan^{-1} \frac{\sin\theta_1}{\cos\theta_1} = \tan^{-1} \tan\theta_1 = \theta_1 \qquad (2\text{-}57)$$

Therefore, the resultant of the primary inertia forces of cylinders 1 and 5 is a vector that rotates with the crank and has a magnitude equal to the maximum value of the individual inertia forces. Consequently, the resultant can be balanced by adding a counterbalance with a magnitude of mR directly opposite the throw. In a similar manner each pair of

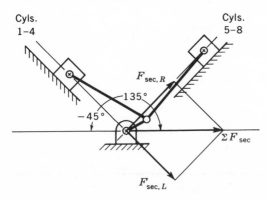

Cyls. 1–4

Cyls. 5–8

$F_{sec, R}$

$-135°$

$-45°$

ΣF_{sec}

$F_{sec, L}$

Fig. 2-17 *Resultant of secondary shaking forces for a V-type engine with a 90° bank angle and a 0-180-180-0° crankshaft.*

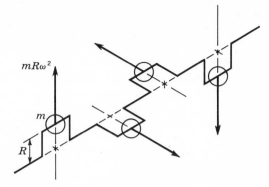

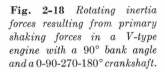

Fig. 2-18 *Rotating inertia forces resulting from primary shaking forces in a V-type engine with a 90° bank angle and a 0-90-270-180° crankshaft.*

cylinders (2 and 6, etc.) will be found to have a resultant unbalance $mR\omega^2$ rotating with the corresponding crank throws.

Thus, the resultants of the inertia forces from the four pairs of cylinders act as four rotating unbalanced weights, as shown in Fig. 2-18. Complete balance can be obtained by counterbalancing each throw or by adding two weights to provide dynamic balance, as discussed in Sec. 2-2. From a practical viewpoint the former method is preferred because, except in the immediate vicinity of the crank throws, there are no moments acting on the crankshaft from the primary inertia forces.

As a matter of interest, it should be noted that the counterbalances required are a function of the reciprocating mass of the piston and connecting rod. Consequently, the crankshaft of a V-8 engine may require rebalancing if the connecting rods and/or pistons are replaced.

chapter 3

VIBRATIONS

Vibrations is the name given to the important class of problems in dynamics in which the energy in a system is continuously changing back and forth between the forms of potential and kinetic energy. The great majority of cases involve potential energy in the elastic deformation of solid bodies or in the compressibility of fluids; but potential energy as a function of position is also encountered, as for a pendulum.

Vibration problems may be divided into a number of groups and subgroups in several different ways. From a mathematician's viewpoint the major classifications might be (1) linear and (2) nonlinear. But from the designer's viewpoint the major classifications will be (1) lumped-mass and (2) continuously distributed-mass systems. The most important combination of these is that of lumped-mass, linear systems, and it will receive the most attention in the following sections.

3-1. System Elements The basic vibration system consisting of a mass, a spring, and a damper, is shown schematically in Fig. 3-1. Both the physical and mathematical characteristics of the elements are deeply involved in the understanding and solving of problems in vibrations. Consequently, we shall first consider the elements and their relationship

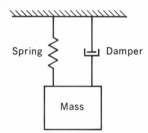

Fig. 3-1 *Basic vibration system.*

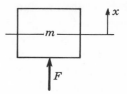

Fig. 3-2 *Schematic representation of an inertial element (mass).*

to the equations of motion for the systems in some generality before taking a closer look at specific types of problems in the following chapters.

3-2. Mass   As in the preceding chapters, the important attributes of mass will be inertia and its relationship to kinetic energy.   In general, the mass can be considered to be constant (the speeds involved are much less than that of light) and the significant relationship is again expressed by Newton's second law.   Thus, in general terms, for rectilinear motion

$$m\mathbf{A} = \mathbf{F} \qquad (3\text{-}1)$$

Although acceleration and force are vector quantities, as indicated in Eq. (3-1), vibrations are usually concerned with rectilinear or rotational motion where the directions can be adequately described as positive or negative.   Thus, Eq. (3-1) is usually written simply as

$$mA = F \qquad (3\text{-}2)$$

and the directions are indicated algebraically.   This scheme will be followed throughout the remainder of this book, and the reader is cautioned to keep in mind that we shall actually be dealing with vector quantities even though not so indicated.   For the mass shown in Fig. 3-2, Eq. (3-2) becomes

$$m\frac{d^2x}{dt^2} = F \qquad (3\text{-}3)$$

As a matter of convenience a number of simplified notations have been used to eliminate the necessity for writing out completely dx/dt, d^2x/dt^2, etc.   Several of the more common schemes are

$$\dot{x} = \frac{dx}{dt} \qquad \ddot{x} = \frac{d^2x}{dt^2} \qquad \text{etc.} \qquad (3\text{-}4)$$

$$Dx = \frac{dx}{dt} \qquad D^2x = \frac{d^2x}{dt^2} \qquad \text{etc.} \qquad (3\text{-}5)$$

and $$px = \frac{dx}{dt} \qquad p^2x = \frac{d^2x}{dt^2} \qquad \text{etc.} \qquad (3\text{-}6)$$

In Eqs. (3-5) and (3-6), D and p are called *operators* because they signify the operation of differentiation—in this case with respect to time. Historically, the dot notation has been commonly used in the study of vibrations and the operator p has been widely used in the study of feedback control systems. None of the "shorthand" notations has any particular advantage over the others, and the reader will encounter all of them in his future reading outside this text. Preference will be given here to the operator D notation because most readers are already familiar with it from previous mathematics courses and because it has not been so closely associated with any particular area in dynamics. Thus, for the time being we shall express the relationship between the force and the acceleration of the mass in Fig. 3-2 as

$$m D^2 x = F \qquad (3\text{-}7)$$

It should be noted that the force and acceleration act in the same direction.

3-3. Springs The important characteristic of the spring is its ability to store energy as potential energy in elastic deformation. The force required to deflect the spring can be expressed as

$$F = f(\delta) \qquad (3\text{-}8)$$

where δ is the deflection or deformation of the spring. For tension-compression springs the variation of force with deformation can often be expressed as

$$F = C x^n \qquad (3\text{-}9)$$

where, as shown in Fig. 3-3a, x is the compression or extension of the spring. Force-deformation curves for representative values of the

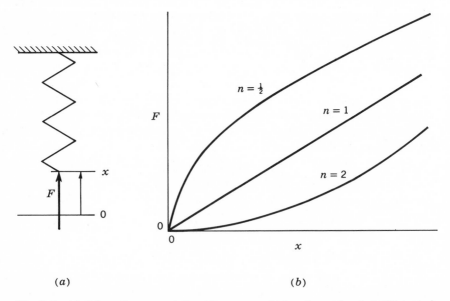

(*a*)   (*b*)

Fig. 3-3 (*a*) *Schematic representation of an energy-storing element (spring); (b) typical force-deflection characteristics of springs.*

exponent n are illustrated in Fig. 3-3b.

When $n = 1$,

$$F = Cx \tag{3-10}$$

The curve is a straight line and we have what is called a *linear* spring. Although the term "linear" implies a straight-line relationship, it will be used in later sections in somewhat different, but consistent, ways. For springs, the simplest approach is to consider that a linear spring is one whose *spring rate* dF/dx is the same at all deflected positions, that is,

$$\frac{dF}{dx} = \text{const} = k \tag{3-11}$$

and, thus, Eq. (3-9) becomes

$$F = kx \tag{3-12}$$

Of the commonly used metallic springs,[1] the helical and leaf springs are usually linear and the conical and volute springs, air springs, and those using materials such as rubber, cork, and felt are usually nonlinear. In many real situations, the mass itself may also be the elastic element and we have what is called a *continuous* system. Considerable useful information can often be gained by approximating a continuous system with lumped masses and springs, but this subject will be left for Sec. 4-19.

A nonlinear spring, such as for $n = 2$ in Fig. 3-3b, is often used with small displacements about an initial deformation, such as might be given by the dead weight of an object. In such a situation it is often convenient to approximate the nonlinear spring by assuming a linear spring with the same spring rate as the actual spring at the initially deformed position, as in Example 5-4.

Thus far in this section, only magnitudes have been discussed, while in actuality we are interested in vector quantities. In fact, we shall be most interested in the forces exerted by the spring on whatever is connected to its ends. For example, assuming that the spring in Fig. 3-3a is linear, the force exerted on the ground is

$$F = kx \tag{3-13}$$

and the force exerted on the moving element (whatever is compressing the spring) is

$$F = -kx \tag{3-14}$$

---

[1] A. M. Wahl, "Mechanical Springs," 2d ed., McGraw-Hill Book Company, New York, 1963.

N. P. Chironis (ed.), "Spring Design and Application," McGraw-Hill Book Company, New York, 1961.

R. M. Phelan, "Fundamentals of Mechanical Design," 2d ed., chap. 9, McGraw-Hill Book Company, New York, 1962.

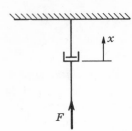

Fig. 3-4 *Schematic representation of an energy-dissipating element (dashpot).*

3-4. Dampers The distinguishing feature of damping is that of dissipation of energy by friction. The common classifications of damping are as follows: (1) viscous, (2) Coulomb or dry friction, and (3) solid or internal. All types act to oppose motion and dissipate energy, but they differ greatly in other important respects.

For a true viscous damper, the force developed is proportional to the velocity of one part relative to the other. Thus, for the damper in Fig. 3-4,

$$F = c\,Dx \qquad (3\text{-}15)$$

where c = damping coefficient.

However, as for the spring, we are interested primarily in the force exerted *by* the damper on whatever is connected to its ends. Therefore, for the force acting on the fixed member in Fig. 3-4, we have

$$F = c\,Dx \qquad (3\text{-}16)$$

and for the force acting on the moving member we have

$$F = -c\,Dx \qquad (3\text{-}17)$$

For a true Coulomb damper, the force has a constant magnitude at all times.[1] Thus, for the damper in Fig. 3-5,

$$|F| = \mu N \qquad (3\text{-}18)$$

[1] In reality, true Coulomb friction does not exist. In many cases the assumption of a constant-magnitude friction force will be satisfactory, but in many other important cases more realistic information will be required. For discussions of the theories related to dry friction see, for example, D. D. Fuller, "Theory and Practice of Lubrication for Engineers," chap. 10, John Wiley & Sons, Inc., New York, 1956.

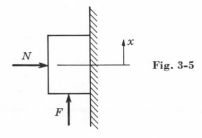

Fig. 3-5

where μ is the coefficient of friction and N is the normal force between the sliding surfaces. The force again reverses direction when the velocity changes direction, but Eq. (3-18) does not include velocity and we must resort to an artifice if we are to include direction in the equation for the force. Using one scheme, the force acting on the fixed member in Fig. 3-5 is

$$F = \mu N \frac{Dx}{|Dx|} \tag{3-19}$$

where $|Dx|$ = absolute value, or magnitude, of Dx, and the force exerted on the moving member is

$$F = -\mu N \frac{Dx}{|Dx|} \tag{3-20}$$

Solid or internal damping is due to the friction within the spring material itself, and it gives rise to the effect known as hysteresis. The mechanism of solid damping is complicated and not completely defined. The magnitude of the damping force has been found to be a function of many variables, such as the chemical constitution and structure of the material; the type, state, and magnitude of stress; the frequency of stress variations; and temperature.[1]   As a first approximation we can assume that the damping force is proportional to stress.   Since, in most members, stress is proportional to displacement, we can write

$$|F| = \gamma x \tag{3-21}$$

where γ = solid damping coefficient.

Now, Eq. (3-21) appears to be quite similar to Eq. (3-13).   However, there is a big difference in that the spring force opposes displacement whereas the damping force opposes velocity, and the direction must be kept straight by using some scheme as used above for Coulomb damping. Therefore, for the spring in Fig. 3-3a, the damping force exerted on the fixed member can be expressed as

$$F = \gamma |x| \frac{Dx}{|Dx|} \tag{3-22}$$

and the damping force on the moving member becomes

$$F = -\gamma |x| \frac{Dx}{|Dx|} \tag{3-23}$$

It is not possible to represent the complicated relationship in Eq. (3-23) in any simple diagram, but Eqs. (3-17) and (3-20) can be shown, as in Fig. 3-6.   The most important classes of damping are viscous and Coulomb.

The major advantage of viscous damping is that it is the type most

[1] C. M. Harris and C. E. Crede (eds.), "Shock and Vibration Handbook," vol. II, chap. 36, McGraw-Hill Book Company, New York, 1961.

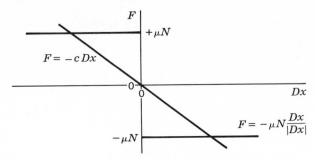

Fig. 3-6 *Force-velocity characteristics for viscous and Coulomb dampers.*

easily handled analytically because it is the only one that is a continuous, linear function of the velocity.   As a result, most analytical studies in vibrations assume viscous damping even though it is only an approximation to real situations in which it is difficult to provide true viscous damping.

Although Coulomb damping cannot actually be represented by simple straight lines as shown in Fig. 3-6, dry or rubbing friction provides an easy way to introduce damping into systems—in fact, it is almost impossible to keep it out.   When the straight lines are a sufficiently good approximation to the real situation, the damping term becomes *piecewise linear* and, as discussed in Sec. 5-3, analytical studies can be carried out readily without the use of analog or digital computers.

Solid damping is relatively small and generally becomes significant only when it is the only type of damping present and the situation is extremely critical.   Normally it can be neglected without introducing any appreciable error.

3-5. Equations of Motion   The motion of the mass or masses is usually the most important characteristic of a vibrating system.   In general, Newton's laws apply and we can write the equation or equations in terms of the forces acting on the mass or masses.

The system in Fig. 3-7*a*, consisting of a mass, a damper, and a spring, is constrained (without additional friction) to move in a horizontal direction when acted upon by the external force $F(t)$. $F(t)$ can be any arbitrary function of time and is known as the *forcing function.*

As discussed in Secs. 3-3 and 3-4, the direction and magnitude of the spring force are functions of the deformation of the spring and the direction of the damper force is a function of the velocity of one part relative to the other, even though the magnitude may be related to other parameters.

Thus, in somewhat general terms, the free-body diagram in Fig. 3-7*b* shows the forces acting on the mass when it is displaced and moving in the positive direction.   Using Newton's second law, the equation of

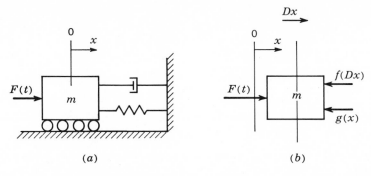

Fig. 3-7   (a) *Single-degree-of-freedom system;* (b) *free-body diagram for system with positive displacement and velocity.*

motion is

$$m\, D^2x = \Sigma F = f(Dx) + g(x) + F(t) \tag{3-24}$$

The system in Fig. 3-8 is slightly more complicated in that there are two masses and subscripts are necessary for keeping track of the positions, velocities, accelerations, and forces.   Assuming that both masses are displaced and moving in the positive direction so that $x_1 > x_2$ and $Dx_1 > Dx_2$, the free-body diagrams for the masses will be as shown in Fig. 3-9.   Now, two equations are required to describe the motion of the system.   They are

$$m_1\, D^2x_1 = \Sigma F_1 \tag{3-25}$$

and

$$m_2\, D^2x_2 = \Sigma F_2 \tag{3-26}$$

From Fig. 3-9, Eqs. (3-25) and (3-26) are respectively

$$m_1\, D^2x_1 = f_1(Dx_1 - Dx_2) + g_1(x_1 - x_2) + F(t) \tag{3-27}$$

and

$$m_2\, D^2x_2 = f_1(Dx_2 - Dx_1) + g_1(x_2 - x_1) + f_2(Dx_2) + g_2(x_2) \tag{3-28}$$

Although only the systems in Fig. 3-7a and Fig. 3-8 have been considered thus far, more complicated systems can be handled in exactly the same manner and we shall find a system of differential equations with an

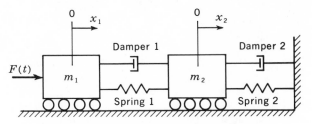

Fig. 3-8   *System with two degrees of freedom.*

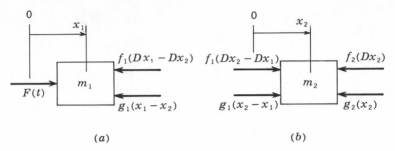

(a)   (b)

Fig. 3-9 *Free-body diagrams for the system in Fig. 3-8 when $x_1 > x_2 > 0$ and $Dx_1 > Dx_2 > 0$.*

equation for each mass. We shall also find that a position coordinate for each mass x_1, x_2, etc., would be necessary and sufficient to describe the position of every mass and, thus, the entire system at all times. Each *necessary and sufficient coordinate represents a degree of freedom,* and systems are often classified by the number of degrees of freedom they contain. For example, in Fig. 3-7a we see a single-degree-of-freedom system and in Fig. 3-8 a two-degrees-of-freedom system.

The coordinates can be angular as well as rectilinear. For example, the system in Fig. 3-10a has six degrees of freedom, using the coordinates x, y, z, θ, ϕ, and ψ. A continuous system, such as the beam in Fig. 3-10b, requires an infinite number of coordinates to describe its deflected shape and is, therefore, a system with an infinite number of degrees of freedom.

As a practical matter, the method of solving the equations of motion is closely related to the degrees of freedom of the system and to whether the system is linear or nonlinear. The next three chapters will consider some of the more important types of problems, starting with the single-degree-of-freedom, linear system in Chap. 4.

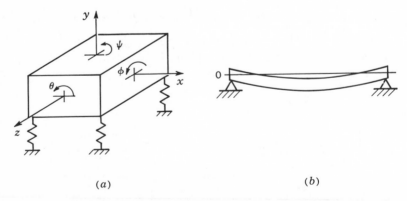

(a)   (b)

Fig. 3-10 (a) *System with six degrees of freedom;* (b) *system with an infinite number of degrees of freedom.*

chapter 4

LINEAR
SINGLE-DEGREE-OF-FREEDOM
SYSTEMS

The single-degree-of-freedom systems in Fig. 4-1[1] are not only the simplest to work with but are also quite satisfactory as models for use in solving the great majority of vibration problems encountered by the design engineer. In general, the physical concepts and ideas related to systems with a single degree of freedom carry over to systems with many degrees of freedom, although the difficulty of obtaining a solution increases rapidly with the number of degrees of freedom of lumped-mass systems.

4-1. The Effect of Gravity The linear systems in Fig. 4-1 are shown in their "at-rest" or free positions. Since the mass in Fig. 4-1*a* is con-

[1] All of the discussion in this chapter can be applied directly to torsional, or rotational, systems by replacing each term with its torsional equivalent. However, since there are few torsional systems of interest that are truly linear single-degree-of-freedom systems, their treatment will be left for Chap. 5, where nonlinear systems will be considered, and to Chap. 6, where multi-degrees-of-freedom systems will be considered.

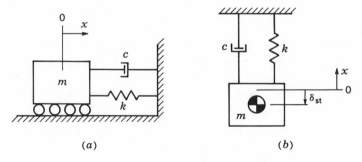

(*a*)   (*b*)

Fig. 4-1 *Linear single-degree-of-freedom systems.*

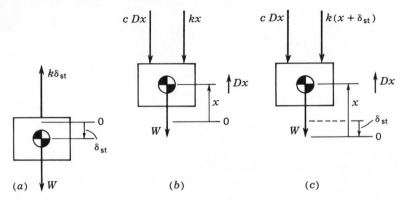

Fig. 4-2 *Free-body diagrams.* (a) *Static equilibrium;* (b) *displacement measured from zero deflection of spring;* (c) *displacement measured from position of static equilibrium.*

strained to move horizontally, it is apparent that the force of gravity does not enter the problem.   The equation of motion is

$$m\,D^2x = -c\,Dx - kx \tag{4-1}$$

which can be rearranged as

$$m\,D^2x + c\,Dx + kx = 0 \tag{4-2}$$

The system in Fig. 4-1b is a schematic representation of a system in which the mass is constrained to move vertically.   The free-body diagram for the mass when hanging motionless is given in Fig. 4-2a.

From Newton's second law,

$$m\,D^2x = \Sigma F \tag{4-3}$$

Since the system is motionless, that is, $Dx = D^2x = 0$,

$$m\,D^2x = 0 \tag{4-4}$$

and, therefore, $$0 = \Sigma F = -W - k\delta_{st} \tag{4-5}$$

From Eq. (4-5),

$$W = -k\delta_{st} \tag{4-6}$$

where W = weight of the mass and δ_{st} = static deformation or deflection of the spring.   *Note:* W acts in the negative direction and δ_{st} is a negative distance.

Figure 4-2b shows the free-body diagram for the mass when both the displacement and velocity are positive.   The equation of motion is

$$m\,D^2x = -c\,Dx - kx - W \tag{4-7}$$

which, after rearranging, becomes

$$m\,D^2x + c\,Dx + kx = -W \tag{4-8}$$

and it appears that the force of gravity is a factor that must be considered.

However, if we measure the displacement from the static-deflection position, as shown in Fig. 4-2c, the forces are exactly the same as before but the spring force must now be written as

$$F_s = -k(x + \delta_{st}) \tag{4-9}$$

and Eq. (4-8) becomes

$$m\,D^2x + c\,Dx + k(x + \delta_{st}) = -W \tag{4-10}$$

Since $k\delta_{st}$ is a constant for a given system, we can rearrange Eq. (4-10) as

$$m\,D^2x + c\,Dx + kx = -W - k\delta_{st} \tag{4-11}$$

But, from Eq. (4-5),

$$-W - k\delta_{st} = 0 \tag{4-12}$$

Therefore,

$$m\,D^2x + c\,Dx + kx = 0 \tag{4-13}$$

which agrees exactly with Eq. (4-2). Thus, *gravity can be neglected when studying linear systems if the displacement is measured from the static-deflection position.* This procedure will be followed throughout the remainder of this book.

It should be noted that, in general, the effect of gravity cannot be treated so simply when dealing with nonlinear springs or with damping that is a function of position as well as velocity.

4-2. Classical Method—Differential Equations The equation of motion for the mass of the linear single-degree-of-freedom system in Fig. 4-3 is

$$m\,D^2x = -c\,Dx - kx + F(t) \tag{4-14}$$

or

$$m\,D^2x + c\,Dx + kx = F(t) \tag{4-15}$$

Equation (4-15) is a linear, ordinary differential equation with constant coefficients, and its solution is discussed thoroughly in almost all textbooks on calculus or differential equations. Thus, the discussion below will be only a brief review that emphasizes those aspects of the solution that are most important in the study of vibrations.

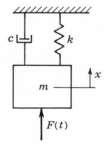

Fig. 4-3 *General linear single-degree-of-freedom system.*

The reader will recall that the general solution of a differential equation, such as Eq. (4-15), consists of the sum of all functions $x = f(t)$ that fulfill the equality. However, we are usually interested in the simplest expression, with the fewest terms, that will contain complete information without redundancy. The usual approach is to consider the problem in two parts by finding a solution, called the complementary function, to that part of the equation involving only the dependent variable, e.g.,

$$m\,D^2x + c\,Dx + kx = 0 \tag{4-16}$$

from which

$$x_1 = f_1(t) \tag{4-17}$$

and a solution, called a particular solution, of the entire equation that is based on a "particular" choice of function

$$x_2 = f_2(t) \tag{4-18}$$

The complete solution is then

$$x = x_1 + x_2 \tag{4-19}$$

or

$$x = f_1(t) + f_2(t) \tag{4-20}$$

From a practical point of view some problems can be handled satisfactorily by considering only the complementary function and others by considering only the particular solution; still others require the complete solution.

When $F(t) = 0$, no external forces other than those generated by the motion itself are present. The motion is called *free vibration*, and the complete solution is based on only the complementary function of the more general equation.

When $F(t)$ is a periodic function, such as $F = F_0 \sin \omega t$, we are usually interested in the behavior of the system over a relatively long period of time. As will be shown, damping—no matter how small it may be—quickly dissipates the energy associated with free vibration. Thus, the free or *transient* vibration, introduced during the starting up of the system, soon dies out and the vibration that remains depends upon the *forcing function* $F(t)$ as its source of energy to replace that dissipated by damping. As a result, the vibration soon becomes a steady-state periodic vibration that is the particular solution of the differential equation. This type of motion is called a *steady-state forced vibration*.

In situations in which $F(t)$ lasts for only a short time, such as in impact or shock loading, the system is usually motionless both before and shortly after the application of $F(t)$. The *entire motion will be transient*, although at the instant $F(t)$ disappears it becomes free vibration. Both the complementary function and a particular solution must be obtained for the time during which $F(t)$ acts on the mass.

4-3. Free Vibration   The motion of the system in Fig. 4-3 will be free vibration in the absence of a forcing function, i.e., when $F(t) = 0$.   For this case, the equation of motion is

$$m\, D^2x + c\, Dx + kx = 0 \qquad (4\text{-}21)$$

Dividing Eq. (4-21) through by m gives

$$D^2x + \frac{c}{m}\, Dx + \frac{k}{m}\, x = 0 \qquad (4\text{-}22)$$

The reader will recall from differential equations that the method of obtaining a solution involves assuming

$$x = e^{st} \qquad (4\text{-}23)$$

and determining the values of s that will satisfy Eq. (4-22).   Differentiating Eq. (4-23) with respect to t gives

$$Dx = se^{st} \qquad (4\text{-}24)$$

and differentiating Eq. (4-24) with respect to t gives

$$D^2x = s^2e^{st} \qquad (4\text{-}25)$$

Substituting from Eqs. (4-23) through (4-25) in Eq. (4-22) and factoring out the common factor e^{st} gives

$$\left(s^2 + \frac{c}{m}\, s + \frac{k}{m}\right) e^{st} = 0 \qquad (4\text{-}26)$$

which is satisfied when

$$s^2 + \frac{c}{m}\, s + \frac{k}{m} = 0 \qquad (4\text{-}27)$$

Equations derived in the manner of Eq. (4-27) are called auxiliary or characteristic equations.   These equations are the most important equations that will be encountered in the remainder of this book.   They contain the system parameters that provide the information that permits us to describe the most significant characteristics of the behavior of dynamic systems.   For this reason we shall use the term *characteristic equation* exclusively.   The importance of the characteristic equation cannot be overemphasized.   The basic information to be obtained is the values of s, which are the roots of the characteristic equation.   The roots of Eq. (4-27) are

$$s_1 = -\frac{c}{2m} + \sqrt{\left(\frac{c}{2m}\right)^2 - \frac{k}{m}} \qquad (4\text{-}28)$$

and

$$s_2 = -\frac{c}{2m} - \sqrt{\left(\frac{c}{2m}\right)^2 - \frac{k}{m}} \qquad (4\text{-}29)$$

When $(c/2m)^2 \neq k/m$, the roots are distinct and the complete solution of Eq. (4-22) is

$$x = C_1 e^{s_1 t} + C_2 e^{s_2 t} \tag{4-30}$$

or

$$x = C_1 \exp\left\{ \left[-\frac{c}{2m} + \sqrt{\left(\frac{c}{2m}\right)^2 - \frac{k}{m}} \right] t \right\}$$

$$+ C_2 \exp\left\{ \left[-\frac{c}{2m} - \sqrt{\left(\frac{c}{2m}\right)^2 - \frac{k}{m}} \right] t \right\} \tag{4-31}$$

When $(c/2m)^2 = k/m$, the roots are identical, that is,

$$s_1 = s_2 = s = -\frac{c}{2m}$$

and the solution of Eq. (4-22) is

$$x = C_1 e^{st} + C_2 t e^{st} \tag{4-32}$$

or

$$x = C_1 e^{-(c/2m)t} + C_2 t e^{-(c/2m)t} \tag{4-33}$$

where C_1 and C_2 are arbitrary constants that depend on the initial conditions, i.e., the displacement and velocity at time $t = 0$.

Depending upon the degree of damping, the motion given by Eqs. (4-31) and (4-33) will correspond to one of the following classifications and subclassifications:

1　Undamped free vibrations
2　Damped free vibrations
　　a. Critically damped
　　b. Underdamped
　　c. Overdamped

The following sections will be given to the solution of Eqs. (4-31) and (4-33) and to discussions of the physical meaning of the terms and related topics for these classifications of free vibrations.

4-4. Zero Damping　When $c = 0$, we have undamped free vibration. The equation of motion, Eq. (4-22), becomes

$$D^2 x + \frac{k}{m} x = 0 \tag{4-34}$$

and Eq. (4-31) becomes

$$x = C_1 e^{\sqrt{-k/m}\,t} + C_2 e^{-\sqrt{-k/m}\,t} \tag{4-35}$$

Since k and m are positive values, the exponents involve an imaginary number, the square root of a negative number. The coefficient of the exponent can be written in several ways, as follows:

$$\sqrt{-\frac{k}{m}} \quad \text{or} \quad \sqrt{-1} \times \sqrt{\frac{k}{m}} \quad \text{or} \quad j\sqrt{\frac{k}{m}} \tag{4-36}$$

where, by definition, $j = \sqrt{-1}$.

Substituting from Eq. (4-36) into Eq. (4-35) gives

$$x = C_1 e^{j\sqrt{(k/m)}t} + C_2 e^{-j\sqrt{(k/m)}t} \qquad (4\text{-}37)$$

which, by use of the Euler identity,[1] can be written as

$$x = C_1(\cos \omega t + j \sin \omega t) + C_2(\cos \omega t - j \sin \omega t) \qquad (4\text{-}45)$$

Rearranging Eq. (4-45) gives

$$x = C_3 \cos \omega t + C_4 \sin \omega t \qquad (4\text{-}46)$$

where $C_3 = C_1 + C_2$ and $C_4 = j(C_1 - C_2)$.

The values of C_3 and C_4 will depend upon the initial conditions for a particular system. The curves in Fig. 4-4 show, for arbitrary values of C_3 and C_4, the motion described by Eq. (4-46). The reader will recognize

---

[1] The Euler identity can be derived in the following manner: Since the series expansion of e^y is

$$e^y = 1 + y + \frac{y^2}{2!} + \frac{y^3}{3!} + \frac{y^4}{4!} + \frac{y^5}{5!} + \cdots \qquad (4\text{-}38)$$

we can write

$$e^{j\sqrt{(k/m)}t} = 1 + j\sqrt{\frac{k}{m}}\,t - \frac{1}{2!}\left(\sqrt{\frac{k}{m}}\,t\right)^2 - \frac{1}{3!}j\left(\sqrt{\frac{k}{m}}\,t\right)^3 + \frac{1}{4!}\left(\sqrt{\frac{k}{m}}\,t\right)^4$$
$$+ \frac{1}{5!}j\left(\sqrt{\frac{k}{m}}\,t\right)^5 - \cdots \qquad (4\text{-}39)$$

and

$$e^{-j\sqrt{(k/m)}t} = 1 - j\sqrt{\frac{k}{m}}\,t - \frac{1}{2!}\left(\sqrt{\frac{k}{m}}\,t\right)^2 + \frac{1}{3!}j\left(\sqrt{\frac{k}{m}}\,t\right)^3 + \frac{1}{4!}\left(\sqrt{\frac{k}{m}}\,t\right)^4$$
$$- \frac{1}{5!}j\left(\sqrt{\frac{k}{m}}\,t\right)^5 - \cdots \qquad (4\text{-}40)$$

The series expansion for $\sin \omega t$ is

$$\sin \omega t = \omega t - \frac{(\omega t)^3}{3!} + \frac{(\omega t)^5}{5!} - \cdots \qquad (4\text{-}41)$$

and the series expansion for $\cos \omega t$ is

$$\cos \omega t = 1 - \frac{(\omega t)^2}{2!} + \frac{(\omega t)^4}{4!} - \frac{(\omega t)^6}{6!} + \cdots \qquad (4\text{-}42)$$

Comparing Eq. (4-39) with Eqs. (4-41) and (4-42) shows that

$$e^{j\sqrt{(k/m)}t} = e^{j\omega t} = \cos \omega t + j \sin \omega t \qquad (4\text{-}43)$$

provided $\omega = \sqrt{k/m}$. Similarly, it can be seen that if $\omega = \sqrt{k/m}$,

$$e^{-j\sqrt{(k/m)}t} = e^{-j\omega t} = \cos \omega t - j \sin \omega t \qquad (4\text{-}44)$$

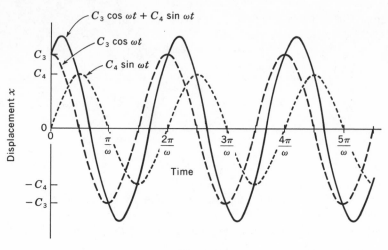

Fig. 4-4

that all three curves are *simple harmonic motion* with the same rotational frequency, ω rad per unit time.

Since the motion described by Eq. (4-46) is a "free vibration" that will continue indefinitely as a natural effect without the application of any external force, the frequency ω is called the *undamped natural frequency* and is given the symbol ω_n. In fact, ω_n *will be reserved exclusively for the undamped natural frequency* even when the system in question has damping and, as will be shown below, a different frequency of free or natural vibration. Thus, Eq. (4-46) can be rewritten as

$$x = C_3 \cos \omega_n t + C_4 \sin \omega_n t \qquad (4\text{-}47)$$

where, from Eq. (4-43),

$$\omega_n = \sqrt{\frac{k}{m}} \qquad \text{rad per unit time} \qquad (4\text{-}48)$$

In many cases it is more convenient to work with frequencies in cycles per unit time. Since one cycle corresponds to 2π rad,

$$f_n = \frac{1}{2\pi} \sqrt{\frac{k}{m}} \qquad \text{cycles per unit time} \qquad (4\text{-}49)$$

The period, or time for one complete cycle, is

$$T = \frac{1}{f_n} = \frac{2\pi}{\sqrt{k/m}} \qquad (4\text{-}50)$$

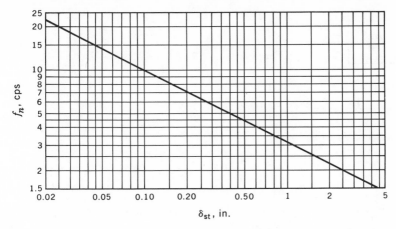

Fig. 4-5   *Natural frequency in cycles per second as a function of the static deflection in inches.*

In terms of the weight and static deflection of the spring, the spring rate can be rewritten as

$$k = \frac{W}{\delta_{st}} \tag{4-51}$$

Substituting from Eq. (4-51) for k and substituting W/g for m in Eq. (4-49) gives

$$f_n = \frac{1}{2\pi} \sqrt{\frac{g}{\delta_{st}}} \tag{4-52}$$

If, as is usual in vibrations, the units are pounds, inches, and seconds, $g = 386$ in./sec² and Eq. (4-52) becomes

$$f_n = 3.13 \sqrt{\frac{1}{\delta_{st}}} \qquad \text{cps} \tag{4-53}$$

where δ_{st} is in inches.   The relationship in Eq. (4-53) is quite useful both in analyzing a given system and in designing or selecting springs to give the desired natural frequency.   Figure 4-5 is a plot of this relationship. It should be noted that if a low natural frequency is desired, a very soft spring (or springs) must be used.

Example 4-1   The system in Fig. 4-6 consists of four identical springs that support a mass that is constrained to move vertically.   The mass has a weight of 50 lb, and the static deflection of the springs has been measured to be 0.309 in.

(*a*) What will be the natural frequency of the system in cycles per minute?   (*b*) What is the spring rate, in pounds per inch, of each spring?   (*c*) What is the equation for the resulting motion if the mass is pushed down in such a manner that, when released, the displacement is −0.250 in. and the velocity is +10.0 in./sec?

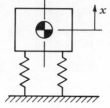

Fig. 4-6

(*a*) Using Eq. (4-53), we find

$$f_n = 3.13 \sqrt{\frac{1}{\delta_{st}}} = 3.13 \sqrt{\frac{1}{0.309}} = 5.63 \text{ cps}$$

or $f_n = 60 \times 5.63 = 338 \text{ cpm}$

(*b*) Using Eq. (4-51), we find for all four springs together

$$k = \frac{W}{\delta_{st}} = \frac{50}{0.309} = 161.8 \text{ lb/in.}$$

Since the springs are in parallel, the spring rates add and, thus,

$$k \text{ per spring} = \frac{k}{4} = \frac{161.8}{4} = 40.45 \text{ lb/in.}$$

(*c*) The general equation, Eq. (4-47), is

$$x = C_3 \cos \omega_n t + C_4 \sin \omega_n t \tag{4-54}$$

which, when differentiated with respect to time, gives

$$Dx = -C_3 \omega_n \sin \omega_n t + C_4 \omega_n \cos \omega_n t \tag{4-55}$$

The initial conditions are $x = -0.250$ in. and $Dx = +10$ in./sec when $t = 0+$ (just released).   Therefore,

$$x \Big|_0 = -0.250 = C_3 \cos (\omega_n \times 0) + C_4 \sin (\omega_n \times 0)$$

from which $C_3 = -0.250$

and $$Dx \Big|_0 = +10 = -\omega_n C_3 \sin (\omega_n \times 0) + \omega_n C_4 \cos (\omega_n \times 0)$$

from which $$C_4 = \frac{10}{\omega_n}$$

Substituting the values of C_3 and C_4 into Eq. (4-54) gives

$$x = -0.250 \cos \omega_n t + \frac{10}{\omega_n} \sin \omega_n t \tag{4-56}$$

But, from (*a*), $f_n = 5.63$ cps.   Therefore,

$$\omega_n^{\cdot} = 2\pi f_n = 2\pi \times 5.63 = 35.4 \text{ rad/sec}$$

and Eq. (4-56) becomes

$$x = -0.250 \cos 35.4t + \frac{10}{35.4} \sin 35.4t$$

$$= -0.250 \cos 35.4t + 0.283 \sin 35.4t$$

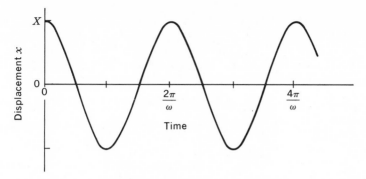

Fig. 4-7 *Simple harmonic motion.*

4-5. Rotating-vector Representation of Simple Harmonic Motion The concept of simple harmonic motion as the projection on the diameter of a point moving at a constant speed on the circumference of a circle should be familiar to all, and the use of the projection on a diameter of a rotating vector to represent a force that varies as a sine or cosine function of time was found to be convenient in earlier chapters. Here we shall extend the use of rotating vectors to include velocity and acceleration as well as displacement and force.

The curve in Fig. 4-7 describes the motion

$$x = X \cos \omega t \tag{4-57}$$

where X is the amplitude of the displacement. Equation (4-57) can also be represented, as shown in Fig. 4-8, by the projection on the horizontal axis of a vector with a magnitude (length) of X rotating at an angular speed of ω rad per unit time. The velocity, found by differentiating Eq. (4-57) with respect to time, is

$$Dx = -\omega X \sin \omega t \tag{4-58}$$

Equation (4-58) can also be represented by a rotating vector having a magnitude ωX; but, in terms of the coordinate system in Fig. 4-8, $\sin \omega t$ is the projection on the vertical axis. To express the velocity as a func-

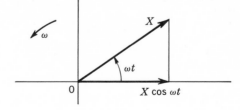

Fig. 4-8 *Rotating-vector representation of simple harmonic motion.*

tion of the cosine, let us rewrite Eq. (4-58) as

$$Dx = -\omega X \sin \omega t = \omega X \cos (\omega t + \theta) \qquad (4\text{-}59)$$

where θ is the angle, called a *phase angle*, of the velocity vector relative to the displacement vector.   For Eq. (4-59) to be valid,

$$-\sin \omega t = \cos (\omega t + \theta) \qquad (4\text{-}60)$$

which is satisfied when $\theta = \pi/2$ rad, or 90°.   Thus, Eq. (4-58) can be written as

$$Dx = \omega X \cos \left(\omega t + \frac{\pi}{2}\right) \qquad (4\text{-}61)$$

In a similar manner, the acceleration can be shown to be given by

$$D^2 x = -\omega^2 X \cos \omega t \qquad (4\text{-}62)$$

or $\qquad\qquad D^2 x = \omega^2 X \cos (\omega t + \pi) \qquad (4\text{-}63)$

The rotating vectors representing Eqs. (4-57), (4-61), and (4-63) are shown in Fig. 4-9.   It should be noted that the three vectors rotate together at the same speed with the acceleration *leading* the displacement by π rad and the velocity *leading* the displacement by $\pi/2$ rad.   From a physical viewpoint the fact that the phase angles are positive, or *leading*, means simply that acceleration must exist before there can be a velocity and velocity must exist before there can be a displacement.

The rotating-vector representation is also useful when the motion is described by an equation containing both cosine and sine terms.   For

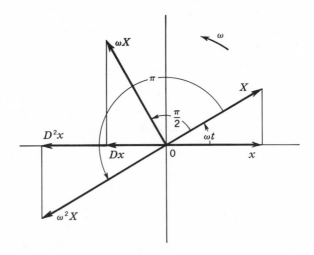

Fig. 4-9 *Rotating-vector representation of displacement, velocity, and acceleration for simple harmonic motion.*

Fig. 4-10 *Rotating-vector representation of*
$X \cos (\omega t + \theta) = C_3 \cos \omega t$

$$+ C_4 \cos \left(\omega t - \frac{\pi}{2} \right)$$

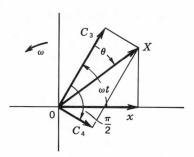

example, Eq. (4-46) can be rewritten as

$$x = C_3 \cos \omega t + C_4 \cos \left(\omega t - \frac{\pi}{2} \right) \tag{4-64}$$

and the rotating vectors will be as shown in Fig. 4-10. The vectors C_3 and C_4 can be added to give a resultant X which is shown to lag θ radians (or degrees) behind ωt.   In terms of the resultant single rotating vector,

$$x = X \cos (\omega t + \theta) \tag{4-65}$$

where

$$X = \sqrt{C_3{}^2 + C_4{}^2} \tag{4-66}$$

and

$$\theta = - \tan^{-1} \frac{C_4}{C_3} \tag{4-67}$$

The above procedure can be applied to any number of harmonic terms *having the same frequency* and the resultant will be a single rotating vector. *Thus, the sum of any number of harmonic terms of the same frequency will also be simple harmonic motion.*   As will be discussed in Sec. 4-11, the result will be quite different if the frequencies are not all the same.

Example 4-2 Determine the magnitude and phase angle for the single rotating vector that will describe the motion of the spring-supported mass in Example 4-1.
Solution A.   From Example 4-1,

$$x = -0.250 \cos 35.4t + 0.283 \sin 35.4t$$

By use of Eq. (4-66),

$$X = \sqrt{C_3{}^2 + C_4{}^2} = \sqrt{(-0.250)^2 + (0.283)^2} = 0.378 \text{ in.}$$

By use of Eq. (4-67),

$$\theta = - \tan^{-1} \frac{0.283}{-0.250} = - \tan^{-1} (-1.13)$$

But,

$$\tan^{-1} (-1.13) = -48.5° \text{ or } +(180 - 48.5) = 131.5°$$

and apparently

$$\theta = +48.5° \text{ or } -131.5°$$

However, the complete equation will be

$$x = X \cos (\omega_n t + \theta)$$

and since x is negative when $t = 0$, the phase angle must be negative.   Therefore,

$$\theta = -131.5°$$

and the equation for the rotating-vector solution is

$$x = 0.378 \cos (35.4t - 131.5°) \tag{4-68}$$

Solution B.   The solution can also be accomplished by writing the general solution directly as

$$x = X \cos (\omega_n t + \theta)$$

and then using the initial conditions to solve for X and θ.
Thus,

$$x \Big|_0 = -0.250 = X \cos (35.4 \times 0 + \theta) = X \cos \theta \tag{4-69}$$

and $\qquad Dx \Big|_0 = 10 = -35.4X \sin (35.4 \times 0 + \theta) = -35.4X \sin \theta \tag{4-70}$

Dividing $Dx \Big|_0$ by $x \Big|_0$ gives

$$-\frac{10}{0.250} = \frac{-35.4X \sin \theta}{X \cos \theta} = -35.4 \tan \theta$$

from which $\qquad\qquad \tan \theta = \dfrac{10}{0.250 \times 35.4} = 1.13$

and, again, $\qquad\qquad\qquad \theta = +48.5° \text{ or } -131.5°$

To solve for X, we rearrange Eq. (4-69) to give

$$X = \frac{x \Big|_0}{\cos \theta} = \frac{-0.250}{\cos \theta}$$

which becomes, for $\theta = +48.5°$,

$$X = \frac{-0.250}{\cos 48.5°} = -0.378$$

and for $\theta = -131.5°$

$$X = \frac{-0.250}{\cos (-131.5°)} = 0.378$$

Thus, the solution can be either

$$x = -0.378 \cos (35.4t + 48.5°) \tag{4-71}$$
or $\qquad\qquad x = 0.378 \cos (35.4t - 131.5°) \tag{4-72}$

To avoid a negative radius, Eq. (4-71) can be rewritten as

$$x = 0.378[- \cos (35.4t + 48.5°)]$$
$$= 0.378 \cos (35.4t + 48.5° \pm 180°)$$
$$= 0.378 \cos (35.4t - 131.5°)$$

and the two solutions are seen to be identical.

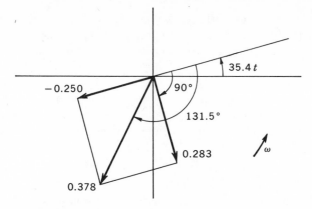

Fig. 4-11   *Rotating-vector representation of*

$$0.378 \cos (35.4t - 131.5°) = -0.250 \cos 35.4t + 0.283 \cos (35.4t - 90°)$$

Solution C.   To solve this problem graphically, the displacement is rewritten, using cosine terms only, as

$$x = -0.250 \cos 35.4t + 0.283 \cos (35.4t - 90°)$$

and the vectors are drawn as shown in Fig. 4-11.

4-6. Complex-number Representation of Simple Harmonic Motion   The concept of a rotating vector and the relationship between a complex number and the sine and cosine of an angle, as expressed in Eqs. (4-43) and (4-44), can be combined to give what is probably the most useful tool for working with periodic functions: the use of a complex number to represent harmonic motion.

Considering only Eq. (4-43), we have

$$e^{j\omega t} = \cos \omega t + j \sin \omega t \qquad (4\text{-}73)$$

which can be given physical significance by considering $e^{j\omega t}$ as a *unit vector* rotating at a frequency ω in a complex plane, as shown in Fig. 4-12.

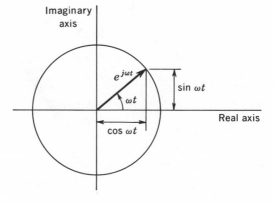

Fig. 4-12 *Complex-number representation of simple harmonic motion.*

The real component is the projection on the horizontal (or real) axis, and the imaginary component is the projection on the vertical (imaginary) axis. Therefore, Eq. (4-57) can be considered as the real part of $Xe^{j\omega t}$ or

$$x = X \cos \omega t = \text{Re } Xe^{j\omega t} \qquad (4\text{-}74)$$

In general, the rotating vector in the complex plane can be used to represent any harmonic motion. Both the real and imaginary components are harmonic, and either component can be used by itself. Thus, we can have

$$x = \text{Re } Xe^{j\omega t} = X \cos \omega t \qquad (4\text{-}75)$$

or $\qquad\qquad x = \text{Im } Xe^{j\omega t} = X \sin \omega t \qquad (4\text{-}76)$

As a practical matter, the displacement is usually written simply as

$$x = Xe^{j\omega t} \qquad (4\text{-}77)$$

and it is understood that one is considering only one of the components.

To find the velocity and acceleration, Eq. (4-77) is differentiated once and twice, respectively, to give

$$Dx = j\omega Xe^{j\omega t} \qquad (4\text{-}78)$$

and $\qquad D^2x = (j)^2\omega^2 Xe^{j\omega t} = -\omega^2 Xe^{j\omega t} \qquad (4\text{-}79)$

Since the rotating vector in the complex plane (Fig. 4-12) must give the same answers as the rotating vector in Fig. 4-8, it follows that if we consider the real components of the rotating vectors, Eqs. (4-78) and (4-79) must be identical with Eqs. (4-61) and (4-63), respectively. Thus,

$$Dx = j\omega Xe^{j\omega t} = \omega X \cos\left(\omega t + \frac{\pi}{2}\right) \qquad (4\text{-}80)$$

and $\qquad D^2x = -\omega^2 Xe^{j\omega t} = \omega^2 X \cos(\omega t + \pi) \qquad (4\text{-}81)$

Rewriting the right-hand terms of Eqs. (4-80) and (4-81) as complex numbers gives, respectively,

$$\omega X \cos\left(\omega t + \frac{\pi}{2}\right) = \omega Xe^{j(\omega t + \pi/2)} \qquad (4\text{-}82)$$

and $\qquad \omega^2 X \cos(\omega t + \pi) = \omega^2 Xe^{j(\omega t + \pi)} \qquad (4\text{-}83)$

From Eqs. (4-80) and (4-82) we can write

$$j\omega Xe^{j\omega t} = \omega Xe^{j(\omega t + \pi/2)} \qquad (4\text{-}84)$$

and, similarly, from Eqs. (4-81) and (4-83),

$$-\omega^2 Xe^{j\omega t} = \omega^2 Xe^{j(\omega t + \pi)} \qquad (4\text{-}85)$$

Thus, multiplication of a complex vector by j has the effect of rotating the vector $\pi/2$ rad ahead in the positive direction, or, in other words, the introduction of a 90° leading phase angle, as shown in Fig. 4-13.

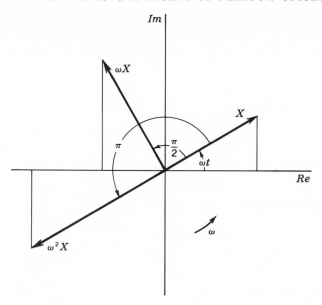

Fig. 4-13 *Complex-number representation of displacement, velocity, and acceleration for simple harmonic motion.*

4-7. Critical Damping   The characteristics of the free vibration of a mass-spring system with damping are closely related to the degree of the damping.   For example, when $(c/2m)^2 < k/m$, the roots of the character-istic equation, Eq. (4-27), are imaginary; but when $(c/2m)^2 > k/m$, the roots are real and negative.   The transition between the imaginary and real roots occurs when

$$\left(\frac{c}{2m}\right)^2 = \frac{k}{m} \qquad (4\text{-}86)$$

The value of damping satisfying Eq. (4-86) is called *critical damping c_c,* which is

$$c_c = 2m \sqrt{\frac{k}{m}} = 2 \sqrt{mk} = 2m\omega_n \qquad (4\text{-}87)$$

As can be seen, the critical value of damping is a function of the mass and the spring rate and has no particular physical significance except in a particular case.   In many engineering situations, general significance can be gained by using dimensionless groups.   Here the dimensionless group of interest is the ratio of the actual damping to the critical value of damp-ing and is called simply *the damping ratio ζ.*   Thus,

$$\zeta = \frac{c}{c_c} \qquad (4\text{-}88)$$

Substituting from Eq. (4-87) into Eq. (4-88) leads to

$$\zeta = \frac{c}{2\sqrt{mk}} = \frac{c}{2m\omega_n} \tag{4-89}$$

Substituting $2\zeta\omega_n$ for c/m, from Eq. (4-89), and $\omega_n{}^2$ for k/m, the equation of motion, Eq. (4-22), becomes

$$D^2x + 2\zeta\omega_n \, Dx + \omega_n{}^2 x = 0 \tag{4-90}$$

The characteristic equation, Eq. (4-27), becomes

$$s^2 + 2\zeta\omega_n s + \omega_n{}^2 = 0 \tag{4-91}$$

and the roots become

$$s_1 = -\zeta\omega_n + \omega_n \sqrt{\zeta^2 - 1} \tag{4-92}$$

and
$$s_2 = -\zeta\omega_n - \omega_n \sqrt{\zeta^2 - 1} \tag{4-93}$$

Vibrations and vibration systems are conveniently classified, with respect to the degree of damping, as critically damped when $\zeta = 1.0$, underdamped when $\zeta < 1.0$, and overdamped when $\zeta > 1.0$. The underdamped case ($\zeta < 1.0$) is the one most frequently encountered and is the one usually meant when a vibration is described simply as damped.

4-8. Underdamped Free Vibrations   When $\zeta < 1$, the roots of the characteristic equation, Eqs. (4-92) and (4-93), have imaginary terms[1] and are written as

$$s_1 = -\zeta\omega_n + j\omega_n \sqrt{1 - \zeta^2} \tag{4-94}$$

and
$$s_2 = -\zeta\omega_n - j\omega_n \sqrt{1 - \zeta^2} \tag{4-95}$$

The complete solution of Eq. (4-90) is

$$x = C_1 \exp\left[(-\zeta\omega_n + j\omega_n \sqrt{1 - \zeta^2})t\right] \\ + C_2 \exp\left[(-\zeta\omega_n - j\omega_n \sqrt{1 - \zeta^2})t\right] \tag{4-96}$$

which can be rewritten as

$$x = e^{-\zeta\omega_n t}[C_1 \exp(j\omega_n \sqrt{1 - \zeta^2}\, t) + C_2 \exp(-j\omega_n \sqrt{1 - \zeta^2}\, t)] \tag{4-97}$$

or
$$x = e^{-\zeta\omega_n t}(C_3 \cos \omega_n \sqrt{1 - \zeta^2}\, t + C_4 \sin \omega_n \sqrt{1 - \zeta^2}\, t) \tag{4-98}$$

or
$$x = Xe^{-\zeta\omega_n t} \cos(\omega_n \sqrt{1 - \zeta^2}\, t + \theta) \tag{4-99}$$

where C_1, C_2, C_3, C_4, X, and θ are constants determined by the initial conditions.   Equation (4-99) is the simplest form of the solution, and it is plotted in Fig. 4-14.

[1] The roots are of the form $A + jB$ and $A - jB$ and are called *conjugate complex* roots.   Since complex numbers result from taking the square root of a negative number, complex roots will always be found in conjugate pairs.

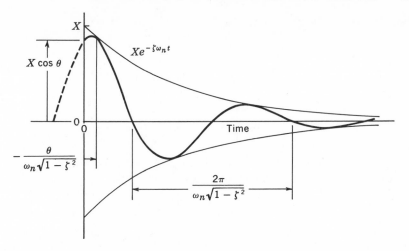

Fig. 4-14   *Free vibration of an underdamped system.*

It should be noted that the frequency of free vibration with damping is given by

$$\omega_{nd} = \omega_n \sqrt{1 - \zeta^2} \qquad (4\text{-}100)$$

and the period is given by

$$T = \frac{2\pi}{\omega_{nd}} = \frac{2\pi}{\omega_n \sqrt{1 - \zeta^2}} \qquad (4\text{-}101)$$

As can be seen, the damped natural frequency is always less than the undamped value.   However, the difference is small when ζ is small and it can be (and is) ignored in the great majority of real problems where the damping is both small and *unknown* in magnitude.

As shown in Fig. 4-14, the exponential decay term $Xe^{-\zeta\omega_n t}$ in Eq. (4-99) represents the envelope of the decaying vibration.   The actual curve and the envelope will be tangent whenever

$$x = Xe^{-\zeta\omega_n t} \qquad (4\text{-}102)$$

which will be true whenever

$$\cos(\omega_n \sqrt{1 - \zeta^2}\, t + \theta) = 1.0 \qquad (4\text{-}103)$$

As can be seen in the figure, the maximum displacements occur slightly before the points of tangency.   However, the difference is usually quite small and the ratio of consecutive maximum displacements will be practically the same as the ratio of displacements at consecutive points of tangency.   Therefore, since, from Eq. (4-101), the time interval between consecutive cycles is

$$T = \frac{2\pi}{\omega_n \sqrt{1 - \zeta^2}} \tag{4-104}$$

the ratio of consecutive maximum displacements is

$$\frac{x_i}{x_{i+1}} = \frac{Xe^{-\zeta\omega_n t}}{X \exp\left[-\zeta\omega_n\left(t + \dfrac{2\pi}{\omega_n \sqrt{1 - \zeta^2}}\right)\right]} \tag{4-105}$$

where x_i is the amplitude of any cycle and x_{i+1} is the amplitude of the next following cycle. Equation (4-105) can be simplified to give

$$\frac{x_i}{x_{i+1}} = \exp\frac{2\pi\zeta}{\sqrt{1 - \zeta^2}} = e^\delta \tag{4-106}$$

where δ, the *logarithmic decrement*, is defined as

$$\delta = \ln\frac{x_i}{x_{i+1}} = \frac{2\pi\zeta}{\sqrt{1 - \zeta^2}} \tag{4-107}$$

The logarithmic decrement is most useful in calculating the damping ratio from experimental data taken (recorded) during free vibrations of an actual system. For this purpose Eq. (4-107) is more convenient when solved for ζ as a function of δ. Thus,

$$\zeta = \frac{\delta}{\sqrt{\delta^2 + 4\pi^2}} \tag{4-108}$$

Applying Eq. (4-106) to successive amplitudes gives

$$\frac{x_i}{x_{i+n}} = \frac{x_i}{x_{i+1}}\frac{x_{i+1}}{x_{i+2}} \cdots \frac{x_{i+n-1}}{x_{i+n}} = e^{n\delta} \tag{4-109}$$

from which

$$\delta = \frac{1}{n}\ln\frac{x_i}{x_{i+n}} \tag{4-110}$$

Equation (4-110) is particularly useful when the damping is small and it is difficult to make accurate measurements of consecutive amplitudes.

In Eq. (4-107) it can be seen that when the damping ratio is small, for example, $\zeta < 0.3$, the square-root term $\cong 1$ and

$$\delta \cong 2\pi\zeta \tag{4-111}$$

Substituting from Eq. (4-111) into Eq. (4-110) and rearranging to solve for the damping ratio results in

$$\zeta = \frac{1}{2\pi n}\ln\frac{x_i}{x_{i+n}} \tag{4-112}$$

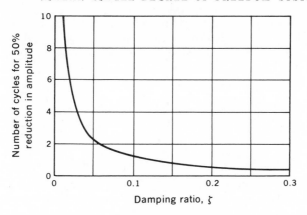

Fig. 4-15   *Number of cycles for* 50 *percent reduction in amplitude of free vibration as a function of the damping ratio.*

When there is little damping, it is often convenient to count the number of cycles for the amplitude to decrease to one-half its initial value.   Solving Eq. (4-112) for this case results in

$$n = \frac{1}{2\pi\zeta} \ln 2 = \frac{0.110}{\zeta} \tag{4-113}$$

The relationship in Eq. (4-113) is plotted in Fig. 4-15.

4-9. Overdamped Free Vibrations   When $\zeta > 1$, the roots, Eqs. (4-92) and (4-93), of the characteristic equation are real and negative and the complete solution of Eq. (4-90) becomes

$$x = C_1 \exp\left[(-\zeta\omega_n + \omega_n \sqrt{\zeta^2 - 1})t\right]$$
$$+ C_2 \exp\left[(-\zeta\omega_n - \omega_n \sqrt{\zeta^2 - 1})t\right] \tag{4-114}$$

where C_1 and C_2 are constants determined by the initial condition.   Since the roots are real, there will be no oscillation; and since both roots are negative, both terms decrease (decay) as time increases.

4-10. Critically Damped Free Vibrations   When $\zeta = 1$, the roots, Eqs. (4-92) and (4-93), of the characteristic equation are real, negative, and equal and the complete solution of Eq. (4-90) can be written as

$$x = C_1 e^{-\omega_n t} + C_2 t e^{-\omega_n t} \tag{4-115}$$

Critically damped motion represents the dividing line between oscillatory motion when $\zeta < 1$ and nonoscillatory motion when $\zeta > 1$; since both roots are real and negative, it also is nonoscillatory.

From a physical viewpoint, critical damping can be considered as that value of damping which permits the quickest free return of a deflected body to its equilibrium position *without overshooting*.   Critically damped motion is often called "dead-beat motion," and it is particularly useful in meters, artillery recoil systems, automobile suspension, and similar situations where it is desired to have the effect of a sudden change in position die out in the shortest possible time.

Example 4-3 A mass-spring-damper system with an undamped natural frequency of 3 cps is deflected 1.0 in. from its equilibrium position and released. Determine the equations of motion for the following cases: (a) when $\zeta = 0.2$, (b) when $\zeta = 2.0$, and (c) when $\zeta = 1.0$.

Solution. For all cases,

$$\omega_n = 2\pi f_n = 2\pi \times 3 = 18.85 \text{ rad/sec}$$

and the initial conditions are $x = 1.0$ in. and $Dx = 0$ when $t = 0$.

Case a. $\zeta = 0.2$. This is underdamped motion, and the appropriate equation, Eq. (4-99), is

$$x = Xe^{-\zeta\omega_n t} \cos (\omega_n \sqrt{1 - \zeta^2}\, t + \theta) \tag{4-116}$$

Substituting $\zeta = 0.2$ and $\omega_n = 18.85$ rad/sec into Eq. (4-116) gives

$$\begin{aligned} x &= Xe^{-0.2\times 18.85t} \cos (18.85 \sqrt{1 - 0.2^2}\, t + \theta) \\ &= Xe^{-3.77t} \cos (18.47t + \theta) \end{aligned} \tag{4-117}$$

Substituting the initial conditions into Eq. (4-117) leads to

$$x \Big|_0 = 1.0 = X \cos \theta \tag{4-118}$$

and
$$Dx \Big|_0 = 0 = -18.47X \sin \theta - 3.77X \cos \theta \tag{4-119}$$

From Eq. (4-119),

$$\frac{\sin \theta}{\cos \theta} = \tan \theta = -\frac{3.77}{18.47} = -0.204$$

from which $\theta = -11.5°$. Substituting for θ in Eq. (4-118) and solving for X gives

$$X = \frac{1.0}{\cos \theta} = \frac{1.0}{\cos (-11.5°)} = \frac{1.0}{\cos 11.5°} = 1.021$$

Thus, the answer for Case *a* is

$$x = 1.021e^{-3.77t} \cos (18.47t - 11.5°) \tag{4-120}$$

Case b. $\zeta = 2.0$. This is overdamped motion, and the appropriate equation, Eq. (4-114), is

$$x = C_1 \exp [(-\zeta\omega_n + \omega_n \sqrt{\zeta^2 - 1})t] + C_2 \exp [(-\zeta\omega_n - \omega_n \sqrt{\zeta^2 - 1})t] \tag{4-121}$$

Substituting $\zeta = 2.0$ and $\omega_n = 18.85$ rad/sec into Eq. (4-121) and simplifying gives

$$x = C_1 e^{-5.05t} + C_2 e^{-70.4t} \tag{4-122}$$

Substituting the initial conditions into Eq. (4-122) leads to

$$x \Big|_0 = 1.0 = C_1 + C_2 \tag{4-123}$$

and
$$Dx \Big|_0 = 0 = -5.05C_1 - 70.4C_2 \tag{4-124}$$

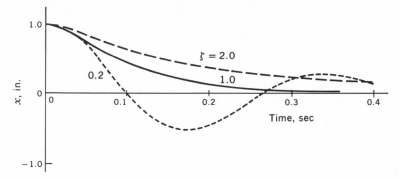

Fig. 4-16 *Free vibration of underdamped, critically damped, and overdamped systems.*

Solving Eqs. (4-123) and (4-124) simultaneously gives

$$C_1 = 1.077 \quad \text{and} \quad C_2 = -0.0773$$

Therefore, the answer for Case b is

$$x = 1.077e^{-5.05t} - 0.0773e^{-70.4t} \tag{4-125}$$

Case c. $\zeta = 1.0$. This is critically damped motion, and the appropriate equation, Eq. (4-115), is

$$x = C_1e^{-\omega_n t} + C_2 t e^{-\omega_n t} \tag{4-126}$$

which, upon substituting $\omega_n = 18.85$ rad/sec and rearranging, becomes

$$x = (C_1 + C_2 t)e^{-18.85t} \tag{4-127}$$

Substituting the initial conditions into Eq. (4-127) leads to

$$x \Big|_0 = 1.0 = C_1 \tag{4-128}$$

and

$$Dx \Big|_0 = 0 = -18.85C_1 + C_2 \tag{4-129}$$

Substituting $C_1 = 1.0$ into Eq. (4-127) gives

$$C_2 = 18.85$$

Therefore, the solution for Case c is

$$x = (1.0 + 18.85t)e^{-18.85t} \tag{4-130}$$

The three curves are plotted in Fig. 4-16.

4-11. Steady-state Forced Vibrations As discussed in Sec. 4-2, the term steady-state forced vibration applies to the motion of a system subjected to a periodic forcing function that continues after the transient vibration, introduced by the initial application of the forcing function, has died out. In general, forced vibrations are the result of a periodic force acting on the mass, as in Fig. 4-17a, or a periodic motion of the support, as in Fig. 4-17b.

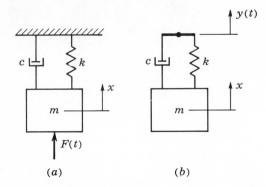

Fig. 4-17 *Forced vibrations. (a) Force disturbance; (b) motion disturbance.*

The periodic disturbing force $F(t)$ can result from the operational function of machines and/or the inertia of unbalanced rotating and reciprocating masses. The disturbing motion $y(t)$ is most often associated with objects, such as instruments and communications equipment, mounted on machines or vehicles of much larger mass that are themselves vibrating as a spring-mass system. It will also be associated with an automobile on a washboard road, but, in general, suspension systems will be more concerned with shock or the transients resulting from bumps in the road surface than with steady-state vibration due to a periodic variation in the profile of the road.

As shown in most books on advanced calculus or advanced mathematics,[1] every periodic function can be represented by a series of sine and cosine terms. Such a series is known as a Fourier series, and the general expression for the forcing function shown in Fig. 4-17a can be written as

$$F(t) = \sum_{n=0}^{\infty} (A_n \sin n\omega t + B_n \cos n\omega t) \tag{4-131}$$

or

$$F(t) = A_1 \sin \omega t + A_2 \sin 2\omega t + A_3 \sin 3\omega t + \cdots$$
$$+ B_0 + B_1 \cos \omega t + B_2 \cos 2\omega t + B_3 \cos 3\omega t + \cdots \tag{4-132}$$

As discussed in Sec. 4-5, sine and cosine terms of the same frequency can be combined into a single term involving a phase angle. Thus, Eq. (4-132) can be written as

$$F(t) = B_0 + C_1 \sin (\omega t + \theta_1) + C_2 \sin (2\omega t + \theta_2)$$
$$+ C_3 \sin (3\omega t + \theta_3) + \cdots \tag{4-133}$$

[1] Wilfred Kaplan, "Advanced Calculus," chap. 7, Addison-Wesley Publishing Company, Inc., Reading, Mass., 1952.
I. S. Sokolnikoff and R. M. Redheffer, "Mathematics of Physics and Modern Engineering," pp. 175–211, McGraw-Hill Book Company, New York, 1958.

where B_0 = constant term
  $C_1 \sin (\omega t + \theta_1)$ = fundamental, or first harmonic
  $C_2 \sin (2\omega t + \theta_2)$ = second harmonic
  $C_3 \sin (3\omega t + \theta_3)$ = third harmonic
                                    etc.

When only the fundamental, or first-harmonic, term is present, we have, as the reader will recognize, *simple harmonic motion*.

One important property of a linear system is that known as superposition. That is, the response of the system to any periodic input will be identical with the sum of the responses to each of the terms in the series expansion for the input. Thus, the basic forcing function is simply

$$F(t) = F_0 \sin \omega t \qquad (4\text{-}134)$$

where F_0 is the amplitude of the sinusoidally varying force. Similarly, for the case in Fig. 4-17*b* the basic forcing function is

$$y = Y \sin \omega t \qquad (4\text{-}135)$$

where Y is the amplitude of the sinusoidally varying displacement of the frame or base.

The systems in Fig. 4-17 have much in common, but since the equations of motion, and thus the solutions, are different, they will be considered separately in the following sections.

4-12. Sinusoidal Response—Force Disturbance   When

$$F(t) = F_0 \sin \omega t$$

the equation of motion for the system in Fig. 4-17*a*, Eq. (4-15), becomes

$$m D^2 x + c D x + k x = F_0 \sin \omega t \qquad (4\text{-}136)$$

or, in terms of complex algebra,

$$m D^2 x + c D x + k x = F_0 e^{j\omega t} \qquad (4\text{-}137)$$

Either expression can be used satisfactorily, but in most respects the complex-number representation is the more useful. However, the "classical" method of solution emphasizes several important physical relationships more readily and, therefore, it will be used in the initial solution.

Classical solution   Dividing both sides of Eq. (4-136) by k results in

$$\frac{m}{k} D^2 x + \frac{c}{k} D x + x = \frac{F_0}{k} \sin \omega t \qquad (4\text{-}138)$$

Substituting $1/\omega_n^2$ for m/k and $2\zeta/\omega_n$ for c/k, from Eq. (4-89), Eq. (4-138) becomes

$$\frac{1}{\omega_n^2} D^2 x + \frac{2\zeta}{\omega_n} D x + x = \frac{F_0}{k} \sin \omega t \qquad (4\text{-}139)$$

The complete solution of Eq. (4-139) will be made up of the sum of the complementary solution and a particular solution. The complementary solution will be the same as the solutions for free vibration: Eq. (4-99) for $\zeta < 1$, Eq. (4-114) for $\zeta > 1$, and Eq. (4-115) for $\zeta = 1$. Since for any damping—no matter how small—the free or transient vibration soon dies out, the complementary solution can be ignored and only the particular solution must be found when considering steady-state forced vibrations.

As discussed in texts on the calculus and differential equations, the particular solution can be made by assuming

$$x = A \sin \omega t + B \cos \omega t \qquad (4\text{-}140)$$

and solving for the constants A and B by substituting from Eq. (4-140) and its derivatives for x, Dx, and D^2x in Eq. (4-139). However, in vibrations it is more useful to combine the sine and cosine terms at the start, as in Sec. 4-5, into a single term as

$$x = X \sin (\omega t + \phi) \qquad (4\text{-}141)$$

from which

$$Dx = \omega X \cos (\omega t + \phi) \qquad (4\text{-}142)$$

and

$$D^2x = -\omega^2 X \sin (\omega t + \phi) \qquad (4\text{-}143)$$

Substituting from Eqs. (4-141) through (4-143) into Eq. (4-139) gives

$$-\frac{\omega^2}{\omega_n^2} X \sin (\omega t + \phi) + 2\zeta \frac{\omega}{\omega_n} X \cos (\omega t + \phi)$$
$$+ X \sin (\omega t + \phi) = \frac{F_0}{k} \sin \omega t \qquad (4\text{-}144)$$

It will be noted that each term is the projection of a rotating vector (Sec 4-5), and Eq. (4-144) can be solved most readily for X and ϕ by using a graphical representation for the equation. Using the relationships

$$- \sin (\omega t + \phi) = \sin (\omega t + \phi + \pi) \qquad (4\text{-}145)$$

and

$$\cos (\omega t + \phi) = \sin \left(\omega t + \phi + \frac{\pi}{2} \right) \qquad (4\text{-}146)$$

Eq. (4-144) can be written as

$$\left(\frac{\omega}{\omega_n} \right)^2 X \sin (\omega t + \phi + \pi) + 2\zeta \frac{\omega}{\omega_n} X \sin \left(\omega t + \phi + \frac{\pi}{2} \right)$$
$$+ X \sin (\omega t + \phi) = \frac{F_0}{k} \sin \omega t \qquad (4\text{-}147)$$

Considering each as a vector, the terms on the left-hand side of Eq. (4-147) will be as shown in Fig. 4-18*a*. Their sum and, therefore, the solution of Eq. (4-147) are shown in Fig. 4-18*b*.

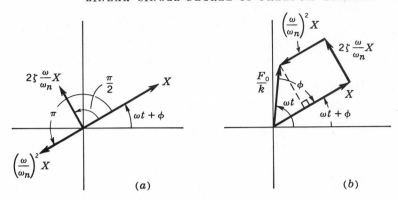

Fig. 4-18 *Rotating-vector representations. (a) Displacement, velocity, and acceleration terms; (b) solution for steady-state response to a sinusoidal force disturbance.*

From the right triangle in Fig. 4-18*b*,

$$X \sqrt{\left[1 - \left(\frac{\omega}{\omega_n}\right)^2\right]^2 + \left(2\zeta \frac{\omega}{\omega_n}\right)^2} = \frac{F_0}{k} \qquad (4\text{-}148)$$

or

$$X = \frac{F_0/k}{\sqrt{[1 - (\omega/\omega_n)^2]^2 + (2\zeta\omega/\omega_n)^2}} \qquad (4\text{-}149)$$

Substituting from Eq. (4-149) into Eq. (4-141) gives for the steady-
state solution of Eq. (4-139)

$$x = \frac{F_0/k}{\sqrt{[1 - (\omega/\omega_n)^2]^2 + (2\zeta\omega/\omega_n)^2}} \sin (\omega t + \phi) \qquad (4\text{-}150)$$

where, also from the right triangle in Fig. 4-18*b*,

$$\phi = - \tan^{-1} \frac{2\zeta\omega/\omega_n}{1 - (\omega/\omega_n)^2} \qquad (4\text{-}151)$$

As shown in Eq. (4-151), the phase angle of the displacement relative
to the forcing function is negative.   A negative phase angle is called a *lag*
angle.   From consideration of the relation between mass and force, it can
be seen that *all real systems must have negative phase angles*—that is, an
object can have a displacement only after it is acted upon by a force.

Complex-algebra solution   Dividing both sides of Eq. (4-137) by k
and then substituting $1/\omega_n{}^2$ for m/k and $2\zeta/\omega_n$ for c/k, as above, results in

$$\frac{1}{\omega_n{}^2} D^2x + \frac{2\zeta}{\omega_n} Dx + x = \frac{F_0}{k} e^{j\omega t} \qquad (4\text{-}152)$$

For the particular (steady-state) solution we assume

$$x = X e^{j(\omega t + \phi)} \qquad (4\text{-}153)$$

which is the complex equivalent of Eq. (4-141).

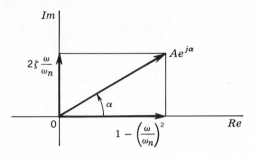

Fig. 4-19 *Complex-plane representation of $Ae^{j\alpha} = 1 - (\omega/\omega_n)^2 + j2\zeta\omega/\omega_n$.*

From Eq. (4-153),

$$Dx = j\omega X e^{j(\omega t + \phi)} \tag{4-154}$$

and

$$D^2x = -\omega^2 X e^{j(\omega t + \phi)} \tag{4-155}$$

Substituting from Eqs. (4-153), (4-154), and (4-155) into Eq. (4-152) results in

$$\left[-\left(\frac{\omega}{\omega_n}\right)^2 + j2\zeta\frac{\omega}{\omega_n} + 1 \right] X e^{j(\omega t + \phi)} = \frac{F_0}{k} e^{j\omega t} \tag{4-156}$$

The coefficient of $Xe^{j(\omega t + \phi)}$ is seen to contain both real and imaginary terms. As discussed in Sec. 4-6, multiplication by j results in a 90° rotation of a vector in the positive direction. Thus, if the real part of the coefficient, $1 - (\omega/\omega_n)^2$, is plotted along the positive real axis in the complex plane, the imaginary part, $j2\zeta\omega/\omega_n$, will lie on the positive imaginary axis, as shown in Fig. 4-19. The resultant vector is

$$Ae^{j\alpha} = \overset{Re}{\left[1 - \left(\frac{\omega}{\omega_n}\right)^2 \right]} + \overset{Im}{j2\zeta\frac{\omega}{\omega_n}} \tag{4-157}$$

where A is the magnitude, absolute value, or *modulus* of the vector and α is the *argument* of the vector.[1] From the right triangle,

$$A = \sqrt{\left[1 - \left(\frac{\omega}{\omega_n}\right)^2 \right]^2 + \left(2\zeta\frac{\omega}{\omega_n} \right)^2} \tag{4-158}$$

and

$$\alpha = \tan^{-1}\frac{2\zeta\omega/\omega_n}{1 - (\omega/\omega_n)^2} \tag{4-159}$$

Substituting from Eq. (4-157) into Eq. (4-156) gives

$$Ae^{j\alpha}Xe^{j(\omega t + \phi)} = \frac{F_0}{k} e^{j\omega t} \tag{4-160}$$

---

[1] R. V. Churchill, "Complex Variables and Applications," 2d ed., McGraw-Hill Book Company, Inc., New York, 1960.
Kaplan, *op. cit.*, p. 2.

which can be simplified to

$$A X e^{j(\alpha+\phi)} = \frac{F_0}{k} \tag{4-161}$$

Since A, X, and F_0/k are positive real numbers (absolute values), it follows that

$$e^{j(\alpha+\phi)} = 1 \tag{4-162}$$

Therefore,

$$\phi = -\alpha = -\tan^{-1} \frac{2\zeta\omega/\omega_n}{1 - (\omega/\omega_n)^2} \tag{4-163}$$

and

$$X = \frac{F_0/k}{A} = \frac{F_0/k}{\sqrt{[1 - (\omega/\omega_n)^2]^2 + (2\zeta\omega/\omega_n)^2}} \tag{4-164}$$

Thus, the solution to Eq. (4-152) is

$$x = \frac{F_0/k}{\sqrt{[1 - (\omega/\omega_n)^2]^2 + (2\zeta\omega/\omega_n)^2}} e^{j(\omega t+\phi)} \tag{4-165}$$

where

$$\phi = -\tan^{-1} \frac{2\zeta\omega/\omega_n}{1 - (\omega/\omega_n)^2} \tag{4-166}$$

Comparison of Eqs. (4-165) and (4-166) with Eqs. (4-150) and (4-151) shows that the solutions are identical. The major advantage of the method of complex algebra is the relative simplicity in determining the magnitude and phase angles without having to draw the vectors. This becomes more important as the complexity of the systems increases.

4-13. Magnification Factor In Eq. (4-164), F_0/k is the deflection that would be given if F_0 were applied as a static load, i.e., the amplitude when the frequency is zero cycles per unit time, and X is the amplitude at any frequency. Thus $1/A$ is the factor by which the zero-frequency amplitude[1] is multiplied to give the actual amplitude at any combination of frequency and damping ratios. Although somewhat misleading, $1/A$ has been termed the *magnification factor*. Thus,

$$\text{Magnification factor} = \frac{X}{F_0/k} = \frac{1}{\sqrt{[1 - (\omega/\omega_n)^2]^2 + (2\zeta\omega/\omega_n)^2}} \tag{4-167}$$

Both the magnification factor, Eq. (4-167), and the phase angle, Eq. (4-166), are nondimensional and are seen to be functions of only the non-dimensional ratios ω/ω_n and ζ. The response of any linear single-degree-of-freedom system to a periodic force disturbance can be determined by use of the curves in Fig. 4-20.

It should be noted that the effect of damping is small when ω/ω_n is either very small or very large. Actually, when $\omega/\omega_n \ll 1.0$, both the

[1] It should be noted that the zero-frequency amplitude is *not* the same as the static deflection of the system.

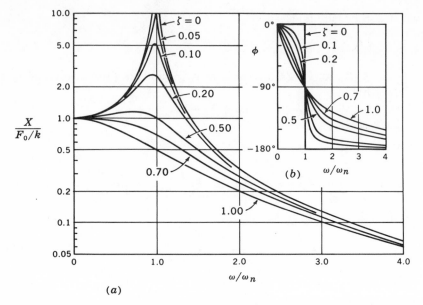

Fig. 4-20 *Steady-state response to a sinusoidal force disturbance.* (a) *Magnification factor;* (b) *phase angle of response relative to disturbance.*

damping and inertia forces are small and the impressed force goes largely into deflection of the spring. When $\omega/\omega_n \gg 1.0$, the inertia becomes the most important factor and the impressed force goes largely into accelerating the mass, as in the case of rigid-body dynamics.

From Eqs. (4-166) and (4-165) it can be seen that, at $\omega = \omega_n$, $\phi = -90°$ for all damping ratios and

$$X \bigg|_{\omega_n} = \frac{F_0/k}{2\zeta} \qquad (4\text{-}168)$$

Therefore, at $\omega = \omega_n$ the amplitude is limited solely by the damping. When $\zeta = 0$, the amplitude is infinite. For finite damping, the amplitude is also finite, although it may be destructively great for low values of damping ratio.

4-14. Resonance and Critical Frequency The term *resonance* is usually applied in physics to the situation "in which a body is set into vigorous vibration by an agitation having a period that corresponds to one of the natural periods of the body."[1] In many of the applications in physics damping is negligibly small and the definition is adequate and

[1] R. A. Millikan, D. Roller, and E. C. Watson, "Mechanics, Molecular Physics, Heat, and Sound," paperback ed., pp. 393, 394, The M.I.T. Press, Cambridge, Mass., 1965.

appropriate.   In the case of a vibrating system with damping, we found that the damped natural frequency, Eq. (4-100), was less than the undamped natural frequency.   Logically, we might extend the concept of resonance to mean that it exists whenever the frequency of the forcing function coincides with the natural frequency, damped or undamped as the case may be, of the system.   However, we are more generally interested in the conditions under which the amplitude of the vibration is a maximum.

Applying the usual method for determining the maximum to Eq. (4-167) leads to

$$\omega \Big|_{\max \frac{X}{F_0/k}} = \omega_n \sqrt{1 - 2\zeta^2} \tag{4-169}$$

Comparing Eq. (4-169) with Eq. (4-100) leads to the conclusion that in this case the maximum amplitude occurs at a frequency below the damped natural frequency and that the elementary concept of resonance no longer is adequate.   Furthermore, we shall find in following sections that the frequency at which the maximum response occurs is also a function of the type of forcing function.

Consequently, *we shall limit the term resonance to situations where damping is negligibly small and use the more general term critical frequency to describe the frequency at which there is a maximum response to a particular input.*   Thus, the critical frequency for a sinusoidal force disturbance is given by Eq. (4-169).[1]

From a practical viewpoint, we can conclude from Eq. (4-169) that for small values of damping ratio the critical frequency will be very close to the natural frequency.

4-15. Sinusoidal Response—Inertia-force Excitation   As discussed in earlier chapters, the inertia or shaking forces resulting from the operation of linkages with oscillating or reciprocating members and unbalanced rotating members become disturbing forces on the frame.   If the motions are periodic, as would be the case in many machines, the inertia forces are also periodic and they can be expressed by a Fourier series.   Thus, each term of the series can be considered as a separate input and the resultant output will then be the summation of the responses to the individual terms of the series.   From this viewpoint, systems excited by inertia forces can be treated as in Secs. 4-11 and 4-12.   However, since the amplitude of the inertia force is not constant, but varies as the square of the frequency, and since this is one of the most frequently encountered cases in machines, it is useful to consider it in a little more detail.

[1] It should be noted that Eq. (4-169) is valid only for $\zeta \leq 0.707$.   For $\zeta > 0.707$, the maximum value of the magnification factor is 1.0 at $\omega/\omega_n = 0$.

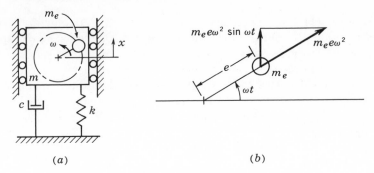

Fig. 4-21   *System with inertia-force excitation.*

Figure 4-21a shows a machine with a total mass m that contains an unbalanced rotating member with a mass m_e. The center of gravity of the rotating member is located a distance e (the eccentricity) from the center of rotation. The machine is constrained to move in the vertical direction to make this a single-degree-of-freedom problem and thus keep it within the scope of this section.

The inertia force resulting from the vertical component of the acceleration of the small mass is

$$F_i = m_e(e\omega^2 \sin \omega t - D^2 x) \tag{4-170}$$

If we consider F_i as the disturbing force acting on the net mass $m - m_e$, we find the equation of motion is

$$(m - m_e) D^2 x + c\,Dx + kx = m_e(e\omega^2 \sin \omega t - D^2 x) \tag{4-171}$$

from which

$$m\,D^2 x + c\,Dx + kx = m_e e\omega^2 \sin \omega t$$

or

$$m\,D^2 x + c\,Dx + kx = m_e e\omega^2 e^{j\omega t} \tag{4-172}$$

It should be noted from Eq. (4-172) that in effect the inertia-force disturbance is simply $m_e e\omega^2 \sin \omega t$, as shown in Fig. 4-21$b$, acting on the total mass m.

Equation (4-171) will be identical with Eq. (4-136) if in the latter F_0 is replaced by $m_e e\omega^2$. Thus, the solution obtained by substituting $m_e e\omega^2$ for F_0 in Eq. (4-150) is

$$x = \frac{m_e e\omega^2 / k}{\sqrt{[1 - (\omega/\omega_n)^2]^2 + (2\zeta\omega/\omega_n)^2}} \sin (\omega t + \phi) \tag{4-173}$$

where, as before,

$$\phi = -\tan^{-1} \frac{2\zeta\omega/\omega_n}{1 - (\omega/\omega_n)^2} \tag{4-174}$$

In terms of amplitudes, Eq. (4-173) can be nondimensionalized by multiplying both sides of the equation by m and then substituting $1/\omega_n^2$ for m/k to give

$$\frac{mX}{m_e e} = \frac{(\omega/\omega_n)^2}{\sqrt{[1 - (\omega/\omega_n)^2]^2 + (2\zeta\omega/\omega_n)^2}} \tag{4-175}$$

Curves showing the relationships in Eqs. (4-174) and (4-175) are presented in Fig. 4-22. Since the inertia force is zero when $\omega = 0$, it is to be expected that $X = 0$ and, therefore, $mX/m_e e = 0$ when $\omega/\omega_n = 0$. However, at first glance, it is not so obvious that $mX/m_e e = 1.0$ and $\phi = -180°$ when ω/ω_n becomes very large.

If we consider the case when $\zeta = 0$ and $\omega/\omega_n = \infty$, we find that $mX/m_e e = 1.0$ and $\phi = -180°$. Physically, this results because for a finite value of ω and $\omega/\omega_n = \infty$, k must equal zero because $\omega_n = 0$, and we have a system that cannot be acted upon by any external force. According to Newton's first law, in the absence of an external force there can be no change in momentum of the system and therefore its center of gravity must remain in a fixed position. Thus, as the small mass m_e moves upward, the rest of the mass $m - m_e$ must move downward so that $mX = m_e e$.

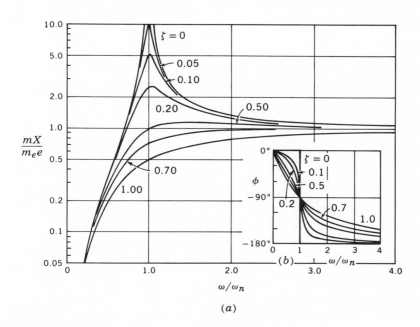

(a)

Fig. 4-22   *Steady-state response to the inertia-force excitation of a rotating unbalanced body.   (a)* $mX/m_e e$; *(b) phase angle of response relative to excitation.*

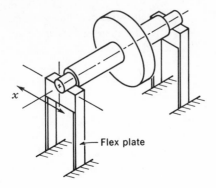

Flex plate

Fig. 4-23   *Simple frame for balancing machine.*

This behavior is used to advantage in most commercial balancing machines.   For example, consider the system shown in Fig. 4-23.   The rotor is supported in bearings that are mounted on flex plates (flat springs that are very soft in one direction and very stiff in the other directions).   As shown, the system has a low natural frequency in the horizontal plane and has practically zero damping.   In this case the rotor mass is the only mass, that is, $m = m_e$, and the curves in Fig. 4-22 apply to the ratio X/e.   Consequently, when rotating at a sufficiently high speed ($\omega/\omega_n > 10$), a measurement of X gives e directly and the angular position of e, the high side, is exactly opposite to (out of phase with) the point on the rotor that is coincident with the extreme positions of the horizontal motion of the bearing supports.

The unbalance $m_e e$ is usually spoken of in terms of ounce-inches, and the magnitude of e is often measured in millionths of an inch.   Highly refined electromechanical devices, called displacement transducers, and circuits are used in measuring the amplitudes and angles.   When dynamic unbalance exists, the motion at each end is the result of the unbalances that we consider to exist in two planes (see Sec. 2-3).   The circuitry in most balancing machines permits the separation of the effects of the two planes so that only one correction in each plane needs to be made.

When a rotor cannot be easily removed from a machine and put in a balancing machine, *field balancing*, or balancing under normal service operating conditions, is often required.   Field balancing is considerably more complicated because the speed of rotation is usually not much greater than the natural frequency of the system.   Consequently, the phase angle is not exactly $-180°$ and it must be determined for each case.   Actually, the phase angle will normally be close to $0°$ because most machines operate well below the natural frequency and there is very little damping.   Separation of the effects of the unbalances in the two arbitrarily chosen balancing planes required in the case of a long rotor is somewhat complicated,[1] and only the case of a short rotor, i.e., static unbalance, will be considered here.

Figure 4-24*a* shows a narrow rotor that is unbalanced. The machine has been operated, and point A has been found to be the high spot. A known trial weight W_t has been added at radius R_t, and the new high spot is located at B.

The amplitudes of the vibration before and after the trial weight was added are laid out to scale in Fig. 4-24*b*, where *oa* has the direction of OA and the magnitude of the vibration amplitude before the addition of the trial weight and *ob* has the direction of OB and the magnitude of the vibration amplitude after the addition of the trial weight. The vector *ab* is the change in vibration amplitude caused by W_t. The phase angle ϕ, by which the effect lags the weight, is determined as indicated in Fig. 4-24*b*.

Since the frequency remains constant, the amplitude of the vibration will be directly proportional to the magnitude of the unbalance. Therefore, since the trial balance $W_t R_t$ results in the vibration amplitude *ab*, the original unbalance $W_O R_O$ must be

$$W_O R_O = \frac{oa}{ab} W_t R_t \qquad (4\text{-}176)$$

and it must be located $\phi°$ ahead of OA, as shown in Fig. 4-24*c*. The actual correction can be made by removing weight at any convenient radius O to $W_O R_O$, in Fig. 4-24*c*, such that $W_R R_R = W_O R_O$ or by adding a similar amount at a point diametrically opposite $W_O R_O$.

As a matter of practicality, good results can be achieved rather quickly by simply using some form of vibration meter (see Sec. 4-22) to measure

[1] R. P. Kroon, Balancing of Rotating Apparatus—II, *Trans. ASME*, vol. 66, pp. A47–50, 1944; also, "Design Data and Methods," pp. 169–172, American Society of Mechanical Engineers, New York, 1953.

W. T. Thomson, "Vibration Theory and Applications," pp. 89–91, Prentice-Hall, Inc., Englewood Cliffs, N.J., 1965.

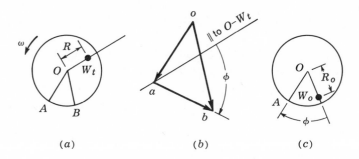

| (*a*) | (*b*) | (*c*) |

Fig. 4-24   *Field balancing of a narrow rotor.*

the vibration of the bearing supports and then applying correction weights by trial and error until the vibration amplitudes have decreased to the level desired.

4-16. Transmissibility—Isolation In many situations flexible members, such as metallic springs, air springs, rubber, cork, and felt, are placed between a machine and the floor or foundation to "isolate" the force disturbance in the machine from its surroundings.[1] For the system in Fig. 4-17a, the force transmitted to the foundation is the sum of the spring and damping forces. Thus,

$$f_{tr} = kx + c \, Dx \tag{4-177}$$

where f_{tr} is the instantaneous value of the transmitted force.

The ratio of the amplitude of the transmitted force to the amplitude of the impressed or exciting force is called *transmissibility ratio* TR or just *transmissibility*. Thus,

$$\text{TR} = \frac{F_{tr}}{F_0} \tag{4-178}$$

where F_{tr} is the amplitude of the transmitted force.

In Sec. 4-12 it was shown that when $F = F_0 e^{j\omega t}$,

$$x = X e^{j(\omega t + \phi)} \tag{4-179}$$

and
$$Dx = j\omega X e^{j(\omega t + \phi)} \tag{4-180}$$

where
$$\phi = -\tan^{-1} \frac{2\zeta\omega/\omega_n}{1 - (\omega/\omega_n)^2} \tag{4-181}$$

Substituting from Eqs. (4-179) and (4-180) into Eq. (4-177) gives

$$f_{tr} = kX e^{j(\omega t + \phi)} + jc\omega X e^{j(\omega t + \phi)}$$

or
$$f_{tr} = (k + jc\omega)X e^{j(\omega t + \phi)} \tag{4-182}$$

The complex coefficient can be rewritten in terms of an amplitude and an angle as

$$k + jc\omega = \sqrt{k^2 + (c\omega)^2}\, e^{j\beta} \tag{4-183}$$

where
$$\beta = \tan^{-1} \frac{c\omega}{k} \tag{4-184}$$

Equations (4-183) and (4-184) can be rearranged by use of the relationships in Eq. (4-89) to give, respectively,

$$k + jc\omega = k \sqrt{1 + \left(2\zeta\,\frac{\omega}{\omega_n}\right)^2}\, e^{j\beta} \tag{4-185}$$

and
$$\beta = \tan^{-1} 2\zeta\,\frac{\omega}{\omega_n} \tag{4-186}$$

[1] See also Example 5-5 and Secs. 6-6 and 6-7.

Substituting from Eqs. (4-164) and (4-185) into Eq. (4-182) and simplifying results in

$$f_{tr} = \frac{\sqrt{1 + (2\zeta\omega/\omega_n)^2}}{\sqrt{[1 - (\omega/\omega_n)^2]^2 + (2\zeta\omega/\omega_n)^2}} F_0 e^{j(\omega t + \phi + \beta)} \qquad (4\text{-}187)$$

or
$$f_{tr} = \frac{\sqrt{1 + (2\zeta\omega/\omega_n)^2}}{\sqrt{[1 - (\omega/\omega_n)^2]^2 + (2\zeta\omega/\omega_n)^2}} F_0 e^{j(\omega t + \psi)} \qquad (4\text{-}188)$$

where $\psi = \phi + \beta$ = phase angle of the transmitted force with respect to the exciting force.

Thus, the transmissibility ratio is

$$\text{TR} = \frac{F_{tr}}{F_0} = \frac{\sqrt{1 + (2\zeta\omega/\omega_n)^2}}{\sqrt{[1 - (\omega/\omega_n)^2]^2 + (2\zeta\omega/\omega_n)^2}} \qquad (4\text{-}189)$$

The transmitted-force phase angle is found by use of functions of multiple angles, from which

$$\tan \psi = \tan (\phi + \beta) = \frac{\tan \phi + \tan \beta}{1 - \tan \phi \tan \beta} \qquad (4\text{-}190)$$

Substituting from Eqs. (4-166) and (4-186) into Eq. (4-190) results in

$$\psi = -\tan^{-1} \frac{2\zeta(\omega/\omega_n)^3}{1 - (1 - 4\zeta^2)(\omega/\omega_n)^2} \qquad (4\text{-}191)$$

Curves showing transmissibility and the phase angle between the transmitted and exciting forces as functions of the frequency and damping ratios are presented in Fig. 4-25. Comparison of the curves in Fig. 4-25 with those in Fig. 4-20 shows that, except for zero damping, the curves are quite different. As shown in the figures, or by letting $\zeta = 0$ in Eqs. (4-166), (4-167), (4-189), and (4-191), the magnification factor and transmissibility, as well as the respective phase angles, are identical when there is no damping. For example, Eqs. (4-167) and (4-189) become

$$\frac{X}{F_0/k} = \text{TR} = \frac{1}{\sqrt{[1 - (\omega/\omega_n)^2]^2}} = \frac{1}{|1 - (\omega/\omega_n)^2|} \qquad (4\text{-}192)$$

If the absolute-value signs are not used, one must remember that the ratios are positive for $\omega/\omega_n < 1$ and negative for $\omega/\omega_n > 1$.

It should be noted that all curves for transmissibility pass through 1.0 at $\omega/\omega_n = 0$ and $\omega/\omega_n = \sqrt{2}$ and that between these frequency ratios TR > 1.0. Thus, when flexible members are used to isolate a disturbing force from its surroundings, no benefit results unless the frequency of the disturbing force is greater than $\sqrt{2}$ times the undamped natural frequency of the system.   Below this value, springs only make matters worse.   In general, isolation systems are designed to give $\omega/\omega_n = 3.5$, or higher if practical.

As a matter of interest, it should be noted that the phase angle between the transmitted and exciting forces approaches $-90°$ for $\omega/\omega_n \gg 1$. Physically, this means that the transmitted force is due almost entirely to damping.   That is, the spring force kX is extremely small because X is small, but the damping force $c\omega X$ is still significant because ω is very large.

From the above, and as shown by the curves in Fig. 4-25, it can be seen that the most effective isolation will be given when damping is zero.   In most practical situations, machines must start and stop; consequently, some damping is desirable to absorb energy and prevent excessive buildup of the amplitude when passing through the critical frequency.   The lower the rate of acceleration or deceleration of the machine, the greater the damping required for satisfactory operation.

4-17. Sinusoidal Response—Motion Disturbance   The system in Fig. 4-26a is excited by a motion disturbance $y = Y \sin \omega t$.   The free-body diagram showing the forces acting on the mass is drawn in Fig. 4-26b for the situation in which the displacement y and velocity Dy of the frame are greater than the corresponding values for the mass.   By

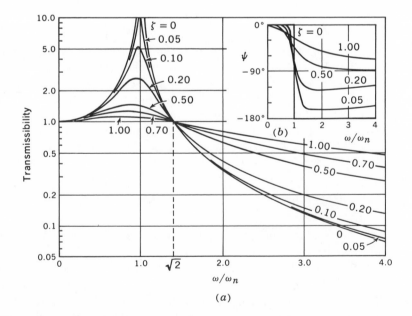

Fig. 4-25   (a) *Transmissibility;* (b) *phase angle of transmitted quantity relative to disturbance.*

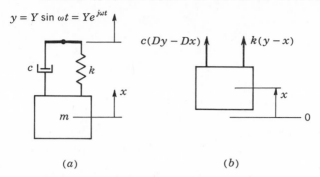

$$y = Y \sin \omega t = Y e^{j\omega t}$$

(a)

(b)

Fig. 4-26   (a) *System with sinusoidal motion disturbance;* (b) *free-body diagram of the mass.*

application of Newton's second law,

$$m\,D^2x = \Sigma F = c(Dy - Dx) + k(y - x) \tag{4-193}$$

which can be rearranged to give

$$m\,D^2x + c\,Dx + kx - c\,Dy - ky = 0 \tag{4-194}$$

or
$$m\,D^2x + c\,Dx + kx = c\,Dy + ky \tag{4-195}$$

In most respects Eq. (4-195) is the more useful form because the forcing function terms are now on the right-hand side of the equation where we are used to seeing them.

Dividing Eq. (4-195) through by k and substituting $1/\omega_n{}^2$ for m/k and $2\zeta/\omega_n$ for c/k gives

$$\frac{1}{\omega_n{}^2}\,D^2x + \frac{2\zeta}{\omega_n}\,Dx + x = \frac{2\zeta}{\omega_n}\,Dy + y \tag{4-196}$$

For this case, the input is

$$y = Y e^{j\omega t} \tag{4-197}$$

and the output or response will be

$$x = X e^{j(\omega t + \gamma)} \tag{4-198}$$

Substituting from Eqs. (4-197) and (4-198) and their derivatives into Eq. (4-196) and collecting like terms results in

$$\left[-\left(\frac{\omega}{\omega_n}\right)^2 + j2\zeta\,\frac{\omega}{\omega_n} + 1\right] X e^{j(\omega t + \gamma)} = \left[j2\zeta\,\frac{\omega}{\omega_n} + 1\right] Y e^{j\omega t} \tag{4-199}$$

The terms within brackets can be written as, respectively,

$$\sqrt{\left[1 - \left(\frac{\omega}{\omega_n}\right)^2\right]^2 + \left(2\zeta\,\frac{\omega}{\omega_n}\right)^2}\; e^{j\beta} \tag{4-200}$$

where
$$\beta = \tan^{-1} \frac{2\zeta\omega/\omega_n}{1 - (\omega/\omega_n)^2} \tag{4-201}$$

and
$$\sqrt{1 + \left(2\zeta\,\frac{\omega}{\omega_n}\right)^2}\; e^{j\lambda} \tag{4-202}$$

where
$$\lambda = \tan^{-1} 2\zeta\,\frac{\omega}{\omega_n} \tag{4-203}$$

Substituting from Eqs. (4-200) and (4-202) into Eq. (4-199) and solving for the dimensionless ratio of the amplitudes of output to input results in

$$\frac{X}{Y} = \frac{\sqrt{1 + (2\zeta\omega/\omega_n)^2}}{\sqrt{[1 - (\omega/\omega_n)^2]^2 + (2\zeta\omega/\omega_n)^2}}\,\frac{e^{j\lambda}}{e^{j(\gamma+\beta)}} \tag{4-204}$$

Since, by definition, both X and Y are positive real numbers, i.e., amplitudes with magnitude only, the resultant of the complex terms must also be real. Therefore,

$$\lambda - (\gamma + \beta) = 0 \tag{4-205}$$

from which

$$\gamma = \lambda - \beta \tag{4-206}$$

From functions of multiple angles,

$$\tan \gamma = \tan (\lambda - \beta) = \frac{\tan \lambda - \tan \beta}{1 + \tan \lambda \tan \beta} \tag{4-207}$$

Substituting from Eqs. (4-201) and (4-203) into Eq. (4-207) and simplifying results in

$$\gamma = -\tan^{-1} \frac{2\zeta(\omega/\omega_n)^3}{1 - (1 - 4\zeta^2)(\omega/\omega_n)^2} \tag{4-208}$$

The ratio of amplitudes, Eq. (4-204), is the same as the transmissibility ratio, Eq. (4-189), and γ, Eq. (4-208), is identical with ψ, Eq. (4-191). In addition, velocity and acceleration amplitudes are directly proportional to displacement amplitudes at a given frequency. Therefore, transmissibility for a single-degree-of-freedom linear system is simply the ratio of the output to input amplitudes of *similar quantities*. In other words,

$$\text{TR} = \frac{F_{tr}}{F_0} = \frac{X}{Y} = \frac{|Dx|}{|Dy|} = \frac{|D^2x|}{|D^2y|} \tag{4-209}$$

and Eqs. (4-189) and (4-192) or the curves in Fig. 4-25 can be used in all such cases.

Example 4-4   A piece of electronic apparatus weighing 26 lb must be mounted on a machine that contains a member that reciprocates at frequencies from 400 to 4,800 cpm.   An investigation of the vibrational sensitivity of the electronic apparatus has been made by mounting the apparatus on the table of an electrodynamic shaker and noting the performance while the frequency was varied.   The apparatus was found to be particularly sensitive to vibrations only when the frequency was in the vicinity of 39 cps, at which time the effect was serious.

Since the shaking force cannot be eliminated, it has been decided to isolate the apparatus from the machine, and we are asked to specify the spring rate for each of four springs through which the apparatus will be attached to the machine.

Solution.   Although this is really a problem with more than one degree of freedom (Secs. 6-6 and 6-7), the reciprocating inertia force will act in one direction and we shall assume that the location of the springs will be such that only the vertical mode of vibration will be excited.

Since no information is available about magnitudes, we shall be conservative and select the springs so that at 39 cps less than 5 percent of the machine vibration will be transmitted to the case of the electronic apparatus.   In general, the stiffest possible mounting is to be preferred.   Thus, if possible, metallic springs with very little damping will be used.

From Fig. 4-25 or Eq. (4-192), for TR < 0.05, $\omega/\omega_n = f/f_n > 4.58$ and for $f = 39$ cps,

$$f_n < \frac{39}{4.58} = 8.52 \text{ cps, or 511 cpm}$$

This would be satisfactory from the viewpoint of disturbances around 39 cps, but it cannot be used because it will coincide frequently with the disturbing frequency as the machine operates between 400 and 4,800 cpm.   On the basis of the vibration test results it can be concluded that if TR ≤ 1.0 at all frequencies between 400 and 4,800 cpm and < 0.05 at 39 cps (2,340 cpm), everything will be fine.   In fact, TR > 1.0 might be acceptable over much of the range of frequencies, but insufficient data are available to justify using a higher value.

Since TR $= 1.0$ at $\omega/\omega_n = f/f_n = \sqrt{2}$, we find for $f = 400$ cpm

$$f_n = \frac{f}{\sqrt{2}} = \frac{400}{\sqrt{2}} = 282.8 \text{ cpm, or 4.71 cps}$$

Solving Eq. (4-49) for k, we find

$$k = m(2\pi f_n)^2 = \frac{W}{g} 4\pi^2 f_n^2$$

Since four springs will be used, $W = {}^{26}\!/_4 = 6.5$ lb, and

$$k = \frac{6.5}{386} 4\pi^2 \times 4.71^2 = 14.7 \text{ lb/in.}$$

Thus, we shall specify that each of the springs is to have a spring rate of 14.7 lb/in.

Note.   Operation at 39 cps will now be improved because the frequency ratio is $39/4.71 = 8.28$ and TR < 0.015.

4-18. Energy Methods   As is evident from the preceding sections, the most important observation about steady-state forced vibrations is that one wants to avoid operating near the critical frequency.   Thus, in many cases it would be sufficient to ignore values of magnification factor and transmissibility and to keep out of trouble by minimizing the dis-

turbances by balancing parts, etc., *and* ensuring that the critical frequency is well below the operating frequency. But, in most cases, the damping is unknown and it is impossible to calculate the critical frequency. However, the damping is usually quite small and, since the difference between the critical and natural frequencies is negligible when the damping is small, Sec. 4-14, it is often sufficient to calculate only the natural frequency and to ensure operation is not close to it. For example, keeping below $\omega/\omega_n = 0.5$ or above $\omega/\omega_n = 1.4$ will usually be satisfactory as far as the operation of the internal parts of the machine is concerned. *Thus, the natural frequency is the most important characteristic of a vibrating system.*

As shown in Sec. 4-4, the free vibrations of an undamped, single-degree-of-freedom linear system give what is called simple harmonic motion, and for the system in Fig. 4-27 the motion can be described by

$$x = X \cos \omega_n t \tag{4-210}$$

where X is the amplitude of the vibration and ω_n is the natural frequency. In most cases the spring and the mass can readily be identified and the natural frequency can be computed as

$$\omega_n = \sqrt{\frac{k}{m}} \tag{4-211}$$

but in many other situations it is simpler to work with energy.

When there is no damping, we have what is called a *conservative system:* that is, energy is conserved because none is dissipated. The energy within the system remains constant and, since a vibrating system is characterized by a continuous changing back and forth between kinetic and potential energy, the energy relationships for conservative systems can be expressed in the following ways:

$$E_K + E_P = \text{const} \tag{4-212}$$

and
$$E_{K,\max} = E_{P,\max} \tag{4-213}$$

where E_K = kinetic energy at a given instant
E_P = potential energy at a given instant
$E_{K,\max}$ = amplitude of kinetic-energy variation during a cycle
$E_{P,\max}$ = amplitude of potential-energy variation during a cycle

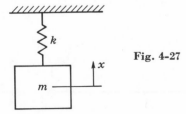

Fig. 4-27

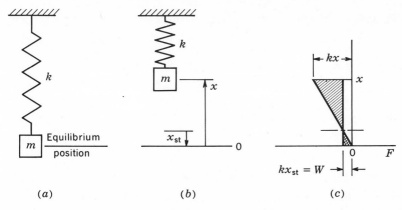

Fig. 4-28 *Potential-energy relationships for mass-spring system with vertical displacement.*

The second expression is the more useful in vibrations and only it will be considered further in this text.

Figure 4-28a shows a mass-spring system in its equilibrium position, and b and c show the system and the spring force-displacement diagram when the mass is raised a distance x above the equilibrium position.   The potential energy in the system is increased by raising the weight and by compressing the spring.   However, energy was stored in the spring when statically deflected, and this energy is lost and represents a decrease. The net change in potential energy is

$$E_P = Wx + \tfrac{1}{2}k(x - x_{st})^2 - \tfrac{1}{2}kx_{st}{}^2 \qquad (4\text{-}214)$$

Simplifying Eq. (4-214) gives

$$E_P = Wx + \tfrac{1}{2}kx^2 - kxx_{st} \qquad (4\text{-}215)$$

But, $kx_{st} = W$ and therefore

$$E_P = \tfrac{1}{2}kx^2 \qquad (4\text{-}216)$$

It should be noted that Eq. (4-216) is simply the energy stored in a spring compressed (or extended) a distance x and, therefore, is exactly the same as will be found for a horizontal system with no change in potential energy due to elevation.   This result was to be expected because, as shown in Sec. 4-1, gravity has no effect on the vibrations of a linear mass-spring vibrating system.   From Eq. (4-210), the maximum displacement will be X and the maximum change in potential energy will be

$$E_{P,\max} = \tfrac{1}{2}kX^2 \qquad (4\text{-}217)$$

By definition, the kinetic energy of the system in Fig. 4-28 will be

$$E_K = \tfrac{1}{2}m(Dx)^2 \qquad (4\text{-}218)$$

Differentiation of Eq. (4-210) with respect to time leads to

$$Dx = -\omega_n X \sin \omega_n t \qquad (4\text{-}219)$$

which has a maximum value of

$$Dx \Big|_{\max} = \omega_n X \qquad (4\text{-}220)$$

Substituting from Eq. (4-220) into Eq. (4-218) gives

$$E_{K,\max} = \tfrac{1}{2}m(\omega_n X)^2 \qquad (4\text{-}221)$$

Then, substituting from Eqs. (4-217) and (4-221) into Eq. (4-213) gives

$$\tfrac{1}{2}m(\omega_n X)^2 = \tfrac{1}{2}kX^2 \qquad (4\text{-}222)$$

which, after simplifying and rearranging, becomes

$$\omega_n{}^2 = \frac{k}{m}$$

or
$$\omega_n = \sqrt{\frac{k}{m}} \qquad (4\text{-}223)$$

Thus, by equating the maximum kinetic energy to the maximum potential energy, we are able to determine the natural frequency of the vibration without actually writing and solving the differential equation. This method offers little in the case of simple systems, such as in Fig. 4-28, where k and m are readily recognized and we know that $\omega_n = \sqrt{k/m}$ applies. However, as will be shown in Sec. 4-19, it becomes a powerful tool in solving many practical problems in which it is difficult or impossible to separate the system into a massless spring and a rigid mass.

4-19. Lateral Vibrations of Beams and Shafts—Rayleigh's Method Many practical problems in vibrations of mechanical components involve shafts, beams, and other members in which the mass and elasticity are distributed in such a manner that it is impossible to designate a simple massless spring and a simple rigid mass. For example, in Fig. 4-29a and b, the shaft and the beam are both mass and spring. In Fig. 4-29c, d, and e, the springs can often be considered massless when the concentrated masses are much greater than the mass of the spring element. But, when this is not the case, the error introduced by ignoring the mass of the spring may be serious.[1]

---

[1] In Sec. 1-9 it was shown that when one end of a helical spring does not move, the equivalent mass at the moving end is one-third the mass of the spring.

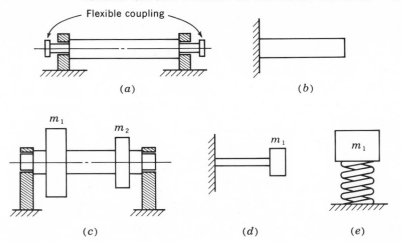

Fig. 4-29   *Typical systems with distributed mass where energy methods can be particularly useful.*

In all cases involving distributed mass, the system has an infinite number of degrees of freedom and an infinite number of natural frequencies, and the discussion in Secs. 6-9 through 6-12 will be useful in calculating those of interest.   However, in most applications the designer can make the system stiff enough to ensure that the lowest natural frequency (fundamental) is well above the highest operating or disturbing frequency.   Under these conditions, the method proposed by Lord Rayleigh for finding an approximate value for the lowest natural frequency of a complex system can be of great use to the designer.

Rayleigh's method is basically an energy method in which the maximum kinetic and potential energies are equated, as in Sec. 4-18.   In general, the method is most useful when the system is conservative and linear or can be assumed to be so without introducing serious error. There is so little damping inherent in most beams and shafts that it can be safely neglected as far as natural-frequency calculations are concerned. The linearity requirement is also not a serious limitation, but the method becomes complicated if the vibration cannot be assumed to be simple harmonic motion.

From strength of materials we know that the potential (strain) energy stored in the elastic bending of a beam is

$$E_P = \tfrac{1}{2}\textstyle\int M \, d\theta \tag{4-224}$$

where M is the moment and θ is the slope of the elastic curve.   For small deflections, as will usually be the case for lateral vibrations, we can also write

$$\theta = \frac{dy}{dx} \tag{4-225}$$

and

$$\frac{d\theta}{dx} = \frac{1}{R} = \frac{d^2y}{dx^2} \tag{4-226}$$

and

$$\frac{1}{R} = \frac{M}{EI} \tag{4-227}$$

where y = deflection

x = distance along neutral axis

R = radius of curvature

E = modulus of elasticity

I = rectangular moment of inertia of the beam cross section

As written, Eqs. (4-224) through (4-226) apply to the static case where the deflection is not a function of time. When the beam is vibrating, we must write

$$y = f(x,t) \tag{4-228}$$

and Eqs. (4-225) and (4-226) become, respectively,

$$\theta = \frac{\partial y}{\partial x} \tag{4-229}$$

and

$$\frac{\partial \theta}{\partial x} = \frac{1}{R} = \frac{\partial^2 y}{\partial x^2} \tag{4-230}$$

From Eqs. (4-227), (4-229), and (4-230), we can write Eq. (4-224) in terms of maximum values as

$$E_{P,\max} = \tfrac{1}{2} \int_0^L \frac{M_{\max}^2}{EI} \, dx \tag{4-231}$$

and

$$E_{P,\max} = \tfrac{1}{2} \int_0^L EI \left(\frac{\partial^2 y}{\partial x^2}\right)_{\max}^2 dx \tag{4-232}$$

For systems without lumped masses, as in Fig. 4-29a and b, the kinetic energy is also distributed along the beam and the maximum value can be found from

$$E_{K,\text{beam},\max} = \frac{1}{2g} \int_0^L \gamma A \left(\frac{\partial y}{\partial t}\right)_{\max}^2 dx \tag{4-233}$$

where γ = weight density of the material

A = area of the beam cross section

In Sec. 4-4 we found that the free vibration of an undamped single-degree-of-freedom system was simple harmonic motion. Thus, Eq. (4-228) can be written as

$$y = Y \cos \omega_n t \tag{4-234}$$

where $Y = Y(x)$ = deflection curve of the beam in its position of maximum deflection.

From Eq. (4-234),

$$\left.\frac{\partial^2 y}{\partial x^2}\right|_{\text{max}} = \frac{\partial^2 Y}{\partial x^2} = \frac{d^2 Y}{dx^2} \tag{4-235}$$

and

$$\left.\frac{\partial y}{\partial t}\right|_{\text{max}} = \omega_n Y \tag{4-236}$$

Substituting from Eq. (4-235) into Eq. (4-232) leads to

$$E_{P,\text{max}} = \frac{1}{2} \int_0^L EI \left(\frac{d^2 Y}{dx^2}\right)^2 dx \tag{4-237}$$

and substituting from Eq. (4-236) into Eq. (4-233) leads to

$$E_{K,\text{beam,max}} = \frac{\omega_n^2}{2g} \int_0^L \gamma A Y^2 \, dx \tag{4-238}$$

When the system has both distributed mass and one or more lumped masses, as in Fig. 4-29c, d, and e, the strain energy is still the strain energy of the beam but the kinetic energy of the masses must be added to that of the beam or shaft by itself. If the maximum value of the deflection at mass 1 is Y_1, at mass 2 is Y_2, etc., the kinetic energy of n masses can be expressed as

$$E_{K,\text{masses,max}} = \frac{\omega_n^2}{2} \sum_{i=1 \text{ to } n} m_i Y_i^2 = \frac{\omega_n^2}{2g} \sum_{i=1 \text{ to } n} W_i Y_i^2 \tag{4-239}$$

Thus, the total kinetic energy of a beam or shaft with n lumped masses is the sum of Eqs. (4-238) and (4-239),

$$E_{K,\text{max}} = \frac{\omega_n^2}{2g} \left(\int_0^L \gamma A Y^2 \, dx + \sum W_i Y_i^2 \right) \tag{4-240}$$

Equating the potential and kinetic energies, Eqs. (4-231) and (4-237), with Eq. (4-240) and solving for ω_n^2 gives, respectively,

$$\omega_n^2 = \frac{g \int_0^L (M_{\text{max}}^2/EI) \, dx}{\int_0^L \gamma A Y^2 \, dx + \sum W_i Y_i^2} \tag{4-241}$$

and

$$\omega_n^2 = \frac{g \int_0^L EI(d^2 Y/dx^2)^2 \, dx}{\int_0^L \gamma A Y^2 \, dx + \sum W_i Y_i^2} \tag{4-242}$$

Neither Eq. (4-241) nor Eq. (4-242) can be used directly to design a beam or shaft to give a specified natural frequency; but they can be used to analyze a given combination of materials and dimensions, *provided* the *moment distribution* or the *curvature and the deflection* of the beam at its maximum amplitude of vibration are known. Unfortunately, none of these is generally known but, fortunately, Lord Rayleigh has shown that the assumption of any reasonable deflection curve will lead to answers that are within several percent of the correct answer. The frequency calculated on basis of an assumed deflection curve can never be lower than the true value. Physically this results from the "stiffening effect" of introducing additional imaginary constraints to make the deflection coincide with the assumed curve.

Equations (4-241) and (4-242) are not necessarily limited to calculating only the lowest, or fundamental, natural frequency; but for practical purposes they are, because it is almost impossible to make a reasonable assumption for the deflection curves for frequencies higher than the lowest one.[1]

Examination of Eqs. (4-241) and (4-242) shows that the moment, the curvature, and the amplitudes enter as squared terms. Consequently, the major contributions to the integrals come from the regions of the maximums, and deviations from the exact curve away from the maximums are not very important. Thus, in general, the simplest curve that provides for maximum curvature and deflection in the correct location will be the one to use. Usually the choice is between a sine (or cosine), a parabola, and the static-deflection curve.[2]

When the beam is uniform, E, I, A, and γ are constant and Eq. (4-242) can often be solved directly in terms of the equation for the static deflection or by assuming a sine or cosine function for the deflection curve. The effect of lumped masses can be included by using superposition; but changes in diameter, as in Fig. 4-29c, result in changes in I and A, and it is then usually more convenient to use Eq. (4-241) with the moments and deflections associated with the static-deflection curve and perform the integrations by use of numerical and/or graphical methods. In fact, the deflection curve itself will usually be determined by use of numerical and/or graphical methods.[3]

When the mass of the beam or shaft is small relative to the lumped masses or if the beam or shaft is broken down into a series of lumped masses, the procedure can be simplified somewhat by using the static-deflection curve and assuming that the strain energy is equal to the work done in moving the masses to their positions of maximum static deflection. Thus, considering the beam or shaft as a linear spring, the potential

---

[1] Stodola's method, which involves using an iterative process to determine the true deflection curve corresponding to the inertial loading, can be extended to determining higher-order natural frequencies. See, for example, J. P. Den Hartog, "Mechanical Vibrations," 4th ed., pp. 162, 163, McGraw-Hill Book Company, New York, 1956.

[2] The bending formulas in Appendix B will be useful in many cases. For formulas covering many more situations, see R. J. Roark, "Formulas for Stress and Strain," 4th ed., McGraw-Hill Book Company, 1965.

[3] R. M. Phelan, "Fundamentals of Mechanical Design," 2d ed., pp. 108–113, McGraw-Hill Book Company, New York, 1962.

energy stored in a system containing n masses will be

$$E_P = \frac{1}{2} \sum_{i=1 \text{ to } n} W_i \delta_{st,i} \qquad (4\text{-}243)$$

Substituting δ_{st} for Y in Eq. (4-239) gives for the kinetic energy, when vibrating with the maximum amplitudes corresponding to the static-deflection curve,

$$E_K = \frac{\omega_n{}^2}{2g} \sum_{i=1 \text{ to } n} W_i \delta_{st,i}^2 \qquad (4\text{-}244)$$

Equating Eqs. (4-243) and (4-244) and solving for ω_n leads to

$$\omega_n = \sqrt{g \frac{\Sigma W_i \delta_{st,i}}{\Sigma W_i \delta_{st,i}^2}} \qquad (4\text{-}245)$$

Example 4-5 The steel shaft in Fig. 4-30*a* is carrying a rotor weighing 26.7 lb. We are asked to determine the lowest natural frequency for lateral vibrations.

Solution. The changes in diameter and the lengths of shaft involved in the bearings are relatively small and will be neglected. Bearing clearances in journal bearings are large enough to permit considerable rotation (due to deflection), and the shaft will be considered to be simply supported.

The appropriate equation is Eq. (4-242), which is

$$\omega_n{}^2 = \frac{g \int_0^L EI (d^2Y/dx^2)^2 \, dx}{\int_0^L \gamma A Y^2 \, dx + \sum W_i Y_i{}^2}$$

Since the materials and dimensions are uniform throughout the length of the shaft and since only one lumped mass is involved, the above equation can be rewritten as

$$\omega_n{}^2 = \frac{gEI \int_0^L (d^2Y/dx^2)^2 \, dx}{\gamma A \int_0^L Y^2 \, dx + W_1 Y_1{}^2} \qquad (4\text{-}246)$$

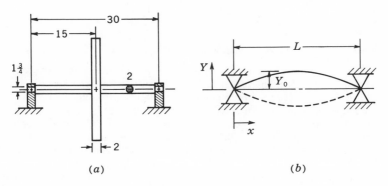

(a)   (b)

Fig. 4-30

The remaining decision to make before Eq. (4-246) can be solved is the choice of deflection curve. A rough sketch of the type of deflection curve expected is given in Fig. 4-30b. The curvature at the ends will be zero, and the curve is symmetrical about the midpoint. These conditions agree with both a sine curve and the static-deflection curve. Either will give a usable answer, and the choice of the sine curve is based purely upon its relative simplicity. For the coordinate system shown in Fig. 4-30b,

$$Y = Y_0 \sin \frac{\pi x}{L} \tag{4-247}$$

From Eq. (4-247),

$$\frac{d^2 Y}{dx^2} = -\left(\frac{\pi}{L}\right)^2 Y_0 \sin \frac{\pi x}{L} \tag{4-248}$$

Also,

$$Y_1 = Y_0 \tag{4-249}$$

Substituting from Eqs. (4-247), (4-248), and (4-249) into Eq. (4-246) gives

$$\omega_n{}^2 = \frac{gEI \displaystyle\int_0^L (\pi/L)^4 Y_0{}^2 \sin^2 (\pi x/L)\, dx}{\gamma A \displaystyle\int_0^L Y_0{}^2 \sin^2 (\pi x/L)\, dx + W_1 Y_0{}^2} \tag{4-250}$$

Integrating Eq. (4-250), substituting the limits, and simplifying leads to

$$\omega_n{}^2 = \frac{gEI(\pi/L)^4 L/2}{\gamma A L/2 + W_1} \tag{4-251}$$

For the dimensions and materials given we know the following:

$g = 386$ in./sec
$E = 30 \times 10^6$ psi
$I = \pi d^4/64 = \pi \times 2^4/64 = 0.785$ in.^{4}
$L = 30$ in.
$\gamma = 0.283$ lb/in.^{3}
$A = \pi d^2/4 = \pi \times 2^2/4 = 3.14$ in.^{2}
$W_1 = 26.7$ lb

Substituting the above values into Eq. (4-251) and solving for ω_n gives

$$\omega_n = 642 \text{ rad/sec}$$

or

$$f_n = \frac{\omega_n}{2\pi} = \frac{642}{2\pi} = 102.2 \text{ cps}$$

Closure. Several additional points are worth noting.

1 The strain energy was integrated over the entire length even though the rotor would prevent bending of part of it. The reasons for ignoring this were that the effect is minor and hard to predict and the error introduced is in the direction of indicating a lower natural frequency than will actually be found. This tends to counteract the error introduced by the sine curve not being exact (in this case); thus the calculated value will be higher than the actual value.

2 The term $\gamma AL/2$ in the denominator of Eq. (4-251) is one-half the weight of the shaft. Thus, the mass of a uniform shaft can be accounted for by using an equivalent mass with half the shaft weight at the midpoint between the bearings. It is interesting to note that if the deflection curve is assumed to be that given

by a concentrated load at the midspan, Eq. (4-251) becomes

$$\omega_n{}^2 = \frac{48gEI}{(0.486 \ \gamma AL + W_1)L^3} \tag{4-252}$$

and $\omega_n = 638 \text{ rad/sec}$

Since $638 < 642$, the deflection curve for the concentrated load is closer to the exact curve than is the sine curve.   For engineering purposes the difference is insignificant.

3   If the rotor is removed from the shaft, Eq. (4-251) becomes

$$\omega_n{}^2 = \frac{gEI}{\gamma A} \left(\frac{\pi}{L}\right)^4$$

or $\omega_n = \left(\frac{\pi}{L}\right)^2 \sqrt{\frac{gEI}{\gamma A}} \tag{4-253}$

As shown in Example 6-2, Eq. (4-253) is the exact equation for the fundamental frequency of a simply supported uniform beam.   Thus, the sine deflection curve is the exact curve for this case *without the rotor*.

4-20. Shaft Whirl   The dynamics of unbalanced *rigid* rotating bodies was considered in Secs. 2-2 and 2-3.   In many cases, where the rotors are stiff or disks are mounted on shafts that are stiff, with the bearings located close to the disks, the assumption of a rigid system leads to satisfactory results.   But when the rotor or shaft becomes relatively flexible, the designer can no longer consider the system to be rigid and only a problem in static or dynamic unbalance.

In general, the inertia force of an unbalanced rotor or shaft causes a deflection and the rotor or shaft no longer rotates about its geometric or design axis of rotation, but, rather, "whirls" about some other axis. Although the rotating inertia force of a whirling shaft will be a disturbing force, as for any unbalanced rotating member, this is not really a vibrations problem, because there is no interchange between kinetic and potential energy.   However, as will be seen, the solution has so much in common with the lateral vibration of a shaft that this is the logical place for discussing it.

Figure 4-31*a* shows an unbalanced disk on a flexible shaft.   The disk is assumed to be mounted at the point of zero slope (to eliminate gyroscopic effects) and the mass of the shaft is so small in comparison with the mass of the disk that it can be neglected.   Thus, the shaft acts as a spring and its effective spring constant at the location of the mass is

$$k = \frac{F}{y} = \frac{W}{\delta_{st}} \tag{4-254}$$

where F = applied force
     y = deflection under force F
     W = weight of rotor
     δ_{st} = static deflection at rotor

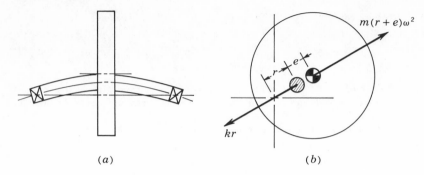

Fig. 4-31   (a) *Shaft whirl;* (b) *free-body diagram of rotor.*

The inertia force of an unbalanced mass rotating about a fixed axis of rotation is

$$F_R = me\omega^2 \tag{4-255}$$

where m is the mass of the disk and e is the eccentricity of the center of gravity with respect to the center of rotation. In reality the reaction force must come from the shaft and the shaft will deflect under the inertia load, which then increases the radius from the center of rotation to the center of gravity of the disk.

Neglecting friction, the forces acting on the disk will be as shown in Fig. 4-31b. For equilibrium,

$$kr - m(r + e)\omega^2 = 0 \tag{4-256}$$

where r = deflection of the shaft from the at-rest equilibrium position.

Solving Eq. (4-256) for r gives

$$r = e\, \frac{\omega^2}{k/m - \omega^2} \tag{4-257}$$

In Eq. (4-257) it can be seen that r becomes infinite when $\omega^2 = k/m$. Thus, $\sqrt{k/m}$ is the critical frequency for shaft whirl when there is no damping. Noting that $\sqrt{k/m}$ is also the natural frequency, or critical frequency for zero damping, for lateral vibration leads to the conclusion that whirling and lateral vibration become serious at the same frequencies. The main difference is that lateral vibration will not exist without an external source of excitation, whereas shaft whirl can become serious in every case because it is impossible, in any practical sense, to make $e = 0$.

Equation (4-257) can be made more convenient to use by substituting ω_n^2 for k/m and rearranging it in nondimensional form as

$$\frac{r}{e} = \frac{(\omega/\omega_n)^2}{1 - (\omega/\omega_n)^2} \tag{4-258}$$

The relationship expressed by Eq. (4-258) is shown in Fig. 4-32. It should be noted that, when $\omega/\omega_n < 1$, the ratio r/e is positive and that, when $\omega/\omega_n > 1$, the ratio is negative. This is the same as saying that r and e are in phase below $\omega/\omega_n = 1$ and are out of phase above $\omega/\omega_n = 1$.

In reality, there is little point in trying to include damping in this situation, since it is extremely difficult to introduce significant damping, and thereby dissipate energy in a shaft that is whirling, because of the absence of appreciable relative motion. However, damping can be included;[1] and if it is, the phase angle will have to be found separately and the complete curves will be found to be identical with those in Fig. 4-22. In fact, Eqs. (4-174) and (4-175) apply directly to this case. It should be noted that for the shaft $m_e = m$ and X becomes r.

The aspects of the shaft whirl most important to the designer are those related to (1) possible direct damage to the shaft itself or adjacent parts and (2) the increased bearing loads and the excitation of vibrations in supporting or nearby members by the rotating inertia-force vector. The first aspect is concerned with overstressing (yielding) the shaft and with interference between parts that result from excessive deflections. From Fig. 4-32 it can be seen that, in general, the best operating region will be with the stiffest possible shaft, i.e., with $\omega/\omega_n < 0.7$. However, when

[1] F. S. Tse, I. E. Morse, and R. T. Hinkle, "Mechanical Vibrations," pp. 69–71, Allyn and Bacon, Inc., Boston, 1963.

W. T. Thomson, "Vibration Theory and Applications," pp. 80–84, Prentice-Hall, Inc., Englewood Cliffs, N.J., 1965.

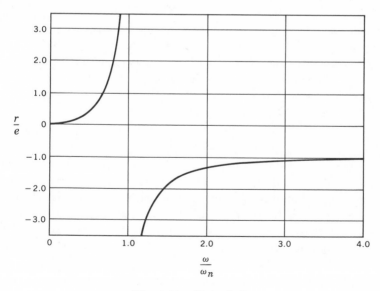

Fig. 4-32 *Shaft whirl.*

size and/or weight limitations and a high operating speed make this impractical, the conditions become good again when $\omega/\omega_n > 1.5$. In this case the problem is to ensure that the shaft accelerates rapidly through the critical frequency so that there is not enough time for the deflection to build up to a destructive level.[1] This is not normally an insurmountable problem, and many high-speed machines, such as centrifuges, turbines, and supercharger drives, have been designed to operate at speeds much greater than the natural frequency.

As far as the forces are concerned, the rotating inertia force will be

$$F_R = m(r + e)\omega^2 \qquad (4\text{-}259)$$

which, in terms of Eq. (4-258), becomes

$$F_R = \frac{1}{|1 - (\omega/\omega_n)^2|} \, me\omega^2 \qquad (4\text{-}260)$$

When the shaft is very stiff and the speed is low, $\omega/\omega_n \ll 1$ and Eq. (4-260) becomes

$$F_R \approx me\omega^2 \qquad (4\text{-}261)$$

As is to be expected, Eq. (4-261) is the inertia force associated with a rotating unbalanced rigid body for which $\omega/\omega_n = 0$.

When the shaft is very flexible and the speed is high, $\omega/\omega_n \gg 1$, $r \to -e$, and the disk rotates about its center of gravity, the result is relatively smooth operation with the force becoming

$$F_R \approx me\omega_n{}^2 \qquad (4\text{-}262)$$

It is interesting to note that, even when the disk rotates about its center of gravity, the inertia force will not be zero unless the disk is perfectly balanced, that is, $e = 0$. The physical explanation is that the shaft must deflect to make $r = -e$ and the force required to give the deflection must also act through the bearings to the support. Substituting k/m for $\omega_n{}^2$ in Eq. (4-262) gives

$$F_R \approx ke \qquad (4\text{-}263)$$

as the minimum inertia force for operating at $\omega/\omega_n > 1.0$.

Very few machinery shafts are as simple as the one in Fig. 4-31. In practically all cases the shaft will be stepped; in most cases the mass of the shaft itself cannot be ignored; and in many cases there may be more than one disk. Under these conditions, the best approach is to use the fact that the critical frequency for lateral vibration is the same as that for whirl and apply one of the methods in Sec. 4-19 to find the *lowest* critical frequency for shaft whirl.

[1] Dominick Macchia, Acceleration of an Unbalanced Rotor through the Critical Speed, ASME Paper 63-WA-9, 1963.

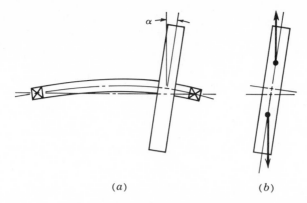

Fig. 4-33 *Stiffening effect when rotor is not located at point of zero slope of the shaft.*

When, as in Fig. 4-33a, the disk is not located at a point of zero slope of the shaft-deflection curve, the disk will not be rotating about an axis of symmetry and a couple will act, as shown in Fig. 4-33b, to straighten the shaft. In effect, this increases the stiffness of the shaft and thus raises the critical frequency. In most cases the angle α is very small and the effect is negligible.[1]

4-21. Transient Response The transient vibrations of most interest to the designer are those associated with the starting up of a machine and those associated with shock loading. As discussed in earlier sections, the transient due to starting up soon dies out and the remaining steady-state vibration is normally of major interest. An exception to this is when a machine must accelerate slowly through a critical frequency and damaging amplitudes may be reached before steady-state operation can be achieved.

Most problems in shock loading are related to transportation in some manner. For example, the designer may be interested in minimizing the discomfort of passengers, minimizing damage to cargo, ensuring adequate protection for instruments and radios, and determining the response of other members of the elastic systems to shock encountered by automobiles, railroad cars, airplanes, missiles, ships, etc.

There are many sources of shock loading, but the two most important are (1) the sudden application of a force, as shown in Fig. 4-34a, and (2) a sudden change in displacement, as shown in Fig. 4-34b. These figures have been encountered in earlier sections where $F(t)$ and $y(t)$ were periodic, sine or cosine, functions of time. The equations of motion will be recognized as, respectively,

$$m\,D^2x + c\,Dx + kx = F(t) \qquad (4\text{-}264)$$

[1] J. P. Den Hartog, "Mechanical Vibrations," 4th ed., pp. 253–265, McGraw-Hill Book Company, New York, 1956.

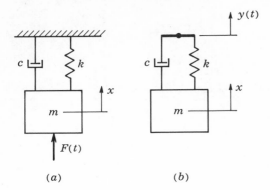

Fig. 4-34 (*a*) *System with force disturbance;* (*b*) *system with motion disturbance.*

$$\text{and} \qquad m\,D^2x + c(Dx - Dy) + k(x - y) = 0$$
$$\text{or} \qquad m\,D^2x + c\,Dx + kx = c\,Dy + ky \qquad (4\text{-}265)$$
$$\text{where} \qquad y = f(t) \qquad \text{and} \qquad Dy = \frac{df(t)}{dt}$$

The four methods most often used in solving Eqs. (4-264) and (4-265) are as follows:

1 Classical method of differential equations
2 Operational mathematics (Laplace transform)
3 Numerical methods (digital computer)
4 Analog computer

When the forcing function can be expressed as, or made up from, reasonably simple functions of time, as in Fig. 4-35, all of the methods can be used with relative ease. But, if the forcing function is more complicated, as will be true in most real situations, the use of numerical methods, with or without a digital computer, or the use of analog computer will not only be convenient but will usually be necessary. The analog computer, or electronic differential analyzer, is a particularly useful tool in solving complex problems involving ordinary differential equations and will be considered in some detail in Appendix A. Although somewhat less useful, numerical methods in combination with high-speed digital com-

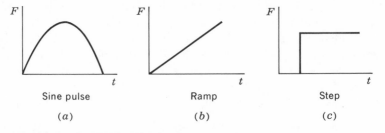

Fig. 4-35 *Simple shock pulses.*

puters should not be ignored, and the reader is referred to the literature[1] for further information.

Electrical engineers were the first to become deeply concerned with transients, such as those due to lightning, opening or closing of switches, and electrical pulses in television circuitry. The differential equations involved in many cases are difficult, if not impossible, to solve by classical methods, and considerable attention has been given to the application of the Laplace transformation to this type of problem. However, the Laplace transformation offers little advantage over classical methods in the simpler cases, rapidly becomes laborious as the complexity of the system or input increases, and cannot be used for nonlinear problems. Consequently, the development and widespread availability of modern computers have decreased the importance of the Laplace transformation to the point where it will not be considered at all in this book in relation to vibrations. It does have some degree of usefulness in the study of control systems and will be discussed briefly in Sec. 7-3.

A considerable amount of information is available in the literature[2] for use by the designer when the forcing function is, or can be approximated by, a simple pulse, such as a rectangle, a half-sine, a triangle, a trapezoid, etc. A typical set of curves for several types of pulses acting on an undamped system is given in Fig. 4-36. The ratios are left in terms of excitation and response because the curves apply to all situations as long as the excitation and response are similar quantities, e.g., force, acceleration, velocity, or displacement. This is directly analogous to transmissibility, as discussed in Secs. 4-16 and 4-17. In general, if the excitation is a force, it will be applied to the mass, as in Fig. 4-34a, and the response will be the force acting on the ground. If the excitation is acceleration, velocity, or displacement, we shall have a motion disturbance, as in Fig. 4-34b, and the response will be in terms of motion of the mass.

In Fig. 4-36, R and E are the maximum values of the response and excitation, respectively, τ is the duration of the pulse, and T is the natural period $(T = 2\pi/\omega_n)$ of the system. As can be seen, the rectangular pulse is the worst case. This will be true in general because the area under this pulse is a maximum for a given magnitude and duration.

---

[1] D. D. McCracken and W. S. Dorn, "Numerical Methods and Fortran Programming," John Wiley & Sons, Inc., New York, 1964.

K. S. Kunz, "Numerical Analysis," McGraw-Hill Book Company, New York, 1957.

F. B. Hildebrand, "Introduction to Numerical Analysis," McGraw-Hill Book Company, New York, 1956.

[2] L. S. Jacobsen and R. S. Ayre, "Engineering Vibrations," chap. 4, McGraw-Hill Book Company, New York, 1958.

C. W. Harris and C. E. Crede (eds.), "Shock and Vibration Handbook," vol. I, pp. 8-14–8-54, McGraw-Hill Book Company, New York, 1961.

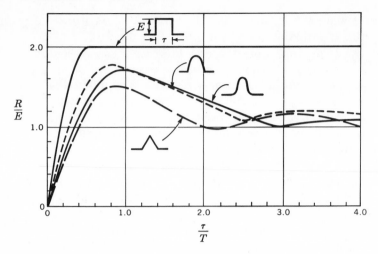

Fig. 4-36 . *Ratio of the maximum value of the response to the maximum value of the disturbance as a function of the ratio of the duration of the pulse to the natural period of the system for square, triangular, half-sine, and versed-sine pulses.* (After L. S. Jacobsen and R. S. Ayre, "Engineering Vibrations," chap. 4, McGraw-Hill Book Company, New York, 1958.)

When $\tau/T \ll \frac{1}{2}$, the system is very soft and the ratio of response to excitation is small; and when $\tau/T \gg \frac{1}{2}$, the system is very stiff and the response equals (nearly so) the excitation. The curve in Fig. 4-36 for the rectangular pulse does not appear to agree with the latter part of this statement, and the reader may find it worthwhile to give some thought to this point.

When a system is excited by a steady-state sinusoidal disturbance, the main functions of damping are to limit the amplitude when operating at the critical frequency to some reasonable value and to dissipate the transient introduced when starting up or changing the speed. In the case of a pulse disturbance, damping is found to have, relatively, a much smaller effect on the peak value of the response than on the amplitude at resonance.[1] Therefore, its main function is to destroy the residual free vibration (after the pulse is over) in the shortest possible time without unduly increasing the magnitude of force transmitted through the combined spring and damper. In many cases, such as automotive suspension systems and artillery recoil systems, the damping ratio is close to 1.0. That is, the systems are close to being critically damped.

Example 4-6 During normal operation, a machine weighing 8,600 lb is subjected to force pulses in the vertical direction. The pulses occur at intervals of several seconds, and each can be approximated by a half-sine pulse with an amplitude of 900 lb and a duration of 0.05 sec. The plant engineer has specified that no periodic

[1] Harris and Crede, *op. cit.*, p. 8-48.

force greater than 200 lb can be applied to the floor in the area in which the machine is to be installed.

We are asked to determine whether or not it will be feasible to use helical compression springs to isolate the machine from the floor.

Solution. Since the excitation is a force, the system in Fig. 4-34*a* will apply and the response will be the force exerted by the spring on the floor. Thus,

$$R = 4kX$$

where k = spring rate of each of four springs, lb/in.

X = maximum deflection of the springs, in.

From the problem specifications,

$$\frac{R}{E} \leq \frac{200}{900} = 0.222$$

Neglecting damping, from Fig. 4-36 for $R/E = 0.222$, we find $\tau/T = 0.08$, and, since τ is given as 0.05 sec,

$$T \geq \frac{0.05}{0.08} = 0.625 \text{ sec}$$

Therefore,

$$\omega_n \leq \frac{2\pi}{T} = \frac{2\pi}{0.625} = 10.05 \text{ rad/sec}$$

or

$$f_n \leq \frac{1}{T} = \frac{1}{0.625} = 1.4 \text{ cps}$$

Since $\omega_n = \sqrt{4k/m}$,

$$k \leq \frac{m}{4} \omega_n{}^2 = \frac{W}{4g} \omega_n{}^2 = \frac{8,600}{4 \times 386} \times 10.05^2 = 563 \text{ lb/in.}$$

The total maximum deflection of the spring will be the sum of the static deflection and the deflection resulting from the pulse.

For $k = 563$ lb/in. for each spring, the static deflection will be

$$\delta_{st} = \frac{W}{4k} = \frac{8,600}{4 \times 563} = 3.82 \text{ in.}$$

The pulse deflection will be

$$X = \frac{R}{4k} = \frac{200}{4 \times 563} = 0.089 \text{ in.}$$

Thus, the total deflection will be

$$\delta_{\text{total}} = \delta_{st} + X = 3.82 + 0.09 = 3.91 \text{ in.}$$

Each spring must be designed to carry a maximum force of $(8,600 + 200)/4 = 2,200$ lb, to have a spring rate of 563 lb/in., and to have a solid deflection of $1.15 \times 3.91 = 4.50$ in. (including a clash allowance of 15 percent). These specifications can easily be met. Therefore, our answer is that it will be feasible to use helical springs in the isolation system.

4-22. Vibration Measurements The measurement of motion-time characteristics of components and complete machines is a field of study in itself. The major reason for requiring measurements of the behavior of actual parts under normal and/or simulated operating conditions is that the predictions based upon mathematical solutions often differ from observed behavior. Usually the differences arise from the necessity for using a reasonably simple model so that a mathematical solution can be obtained; in the simplification significant terms are often ignored or left out because they are not recognized as being important. The major complexities result from nonlinearities in elastic action and in the damping and the almost certain error in neglecting or trying to guess at the degree to which damping is present. Another major reason for a difference between theory and reality is incomplete or incorrect knowledge of the type, magnitude, and, in many cases, even the source of the excitation. In general, the deviations become greater where a maximum level of performance must be combined with a minimum of weight, as in the aerospace industry.

The widespread—almost universal in military and aerospace fields— use of vibration and shock testing in development and as proof of performance before equipment will be accepted by the purchaser is further evidence of the disparity that is often found between purely mathematical solutions and reality. It should be observed that the monitoring of vibration and shock tests also requires the measurement of both the excitation and response.

Vibration measurements may be broken down into two major classifications:

1   Direct measurements of the motion from a fixed reference point
2   Indirect measurements by use of a seismic system

The major advantage of direct measurement is that the calibration can be both simple and accurate. Its major disadvantage is that only values of amplitude or, in some cases, the actual displacement as a function of time can be measured or recorded. If the motion is simple-harmonic, a knowledge of amplitude and frequency is sufficient because the velocity and acceleration can be readily calculated. However, if the motion is not simple-harmonic, the displacement-time data will have to be differentiated once to obtain velocity and twice to obtain acceleration values. Differentiation of test data by graphical,[1] numerical, or electrical[2] means is at best an undesirable operation, and if the data do not plot as an extremely smooth curve, the answers may bear little relation to the true story.

Of the numerous methods available, the simplest, and by far the cheapest, method is the use of a stroboscope and a microscope, as shown in Fig. 4-37a. In general, it can be used only with periodic motion and, without auxiliary equipment for adjusting the phase between the motion and

---

[1] Phelan, *op. cit.*, pp. 50–52.
[2] M. Hetenyi (ed.), "Handbook of Experimental Stress Analysis," pp. 382–386, John Wiley & Sons, Inc., New York, 1949.

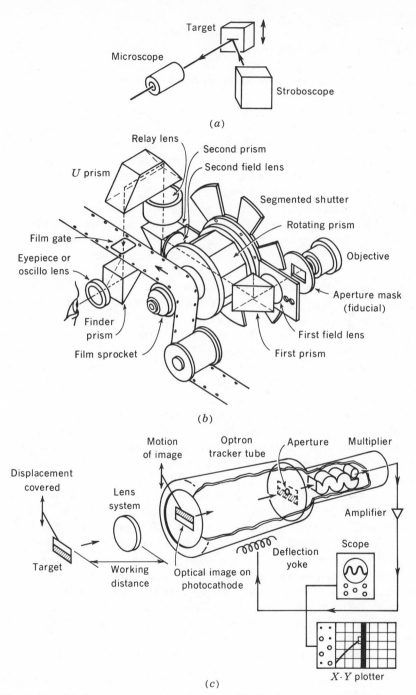

Fig. 4-37 *Displacement measuring systems.* (*a*) *Stroboscope and microscope;* (*b*) *Hycam high-speed motion-picture camera.* (Red Lake Laboratories, Inc.) (*c*) *The Optron Tracker.* (Optron Corporation.)

light flashing in known increments, it is limited to determining the total excursion or peak-to-peak travel. This scheme finds considerable use as a standard for calibrating other types of instruments under steady-state sinusoidal excitation.

The frequency of the flashing·light can be adjusted until the motion stops,. and the frequency is read from the calibrated dial. Since the stroboscopic action of the light can make the motion appear to stop at a large number of frequencies, care must be exercised to be sure the correct one is being read. In general, the correct frequency is the highest at which only a single image can be seen.

Although it is possible to stop the motion in order to read the displacement when the object is at the extreme limits of travel, better results will be obtained by adjusting the frequency so that the object appears to move very slowly and reading the displacement as it reaches the limits of travel.

High-speed movie cameras can take pictures at rates up to 44,000 frames per second. The dynamic forces involved are so great at these rates that it is no longer possible to use intermittent-motion devices to stop the film, open the shutter, and then close the shutter for each frame—as in home and standard movie cameras with their normal rates of 16 and 24 frames per second, respectively. The film moves continuously in most high-speed motion-picture cameras, and framing of each individual picture is accomplished by the combined action of a rotating segmented shutter and a rotating compensating prism, as shown in Fig. 4-37b.

Higher rates, up to 67 million frames per second, have been obtained in framing cameras in which an electronic shutter controls the admission of the light, which is directed by a rotating mirror to a stationary strip of film.[1] Still higher speeds, up to 200 million frames per second, can be provided by image-converter cameras in which the image for each frame is reproduced electronically on a separate image-converter tube (much as in television) and then photographed by use of an ordinary still camera.[2]

The analysis of hundreds of feet of film can be tedious, but there is no better way for observing transients than by running the developed film at normal speeds and seeing the vibration in slow motion. For quantitative results, the displacement must be read from the film (usually from enlarged pictures), and time information is obtained from the spacing of timing marks on the film (from a flashing lamp) or from the speed of the film.

The Optron Tracker, Fig. 4-37c, is a device that utilizes a feedback

---

[1] W. C. Goss, Kerr Cell Framing Camera, *Proc. Fifth Intern. Congr. High-speed Phot.*, pp. 135–137, New York, 1962.

[2] For example, the image-converter cameras manufactured by Beckman and Whitley, Inc., San Carlos, Calif.

control circuit that acts through the deflection yoke to make the image formed on the photocathode pass through the aperture. The photomultiplier tube senses the amount of light passing through the aperture, and the control circuit acts to maintain this at a constant level by deflecting the image. The voltage supplied to the deflection yoke varies directly with the displacement of the target; and when this voltage is applied to an oscilloscope, or any other type of recording oscillograph, the displacement can be observed or recorded as a function of time. The main advantage here is that the output is electrical, and it can be observed and analyzed in a relatively simple manner.

Seismic is defined[1] as "of, pertaining to, of the nature of, subject to, or caused by, an earthquake," and the instruments developed to measure and record the shocks and motions of earthquakes are called seismographs. Seismographs are simply single-degree-of-freedom systems having an extremely low natural frequency, and the term *seismic system* is now used to describe any device utilizing the dynamic behavior of a mass-spring-damper system in measuring any of the displacement-time characteristics of motion.

The basic system shown in Fig. 4-38 is identical with that shown in Fig. 4-17*b* and discussed in Sec. 4-17. However, we are no longer interested in the response x to the excitation $y(t)$, but, rather, in the relative response $z = x - y$, because this is now the only quantity that is available for measurement. Many different methods have been used to measure and record $z(t)$, and some of them will be discussed briefly in this section, but for further information the reader should refer to manufacturers' bulletins and other literature.[2]

Rearranging Eq. (4-193) and writing it in terms of the undamped natural frequency and the damping ratio can lead to

$$\frac{1}{\omega_n{}^2} D^2 x + \frac{2\zeta}{\omega_n} (Dx - Dy) + (x - y) = 0 \qquad (4\text{-}266)$$

[1] "Webster's New Collegiate Dictionary," G. & C. Merriam Company, Springfield, Mass., 1959.

[2] Harris and Crede, *op. cit.*, vol. I, chaps. 12–17.
Hetenyi, *op. cit.*, chap. 8.

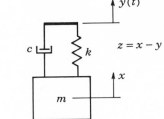

Fig. 4-38   *Seismic system.*

Since $z = x - y$, we can write

$$x - y = z \tag{4-267}$$
and
$$Dx - Dy = D(x - y) = Dz \tag{4-268}$$
and
$$D^2x = D^2z + D^2y \tag{4-269}$$

Substituting from Eqs. (4-267), (4-268), and (4-269) into Eq. (4-266) and rearranging terms gives

$$\frac{1}{\omega_n{}^2} D^2z + \frac{2\zeta}{\omega_n} Dz + z = -\frac{1}{\omega_n{}^2} Dy \tag{4-270}$$

For a steady-state sinusoidal input, we can write

$$y = Ye^{j\omega t} \tag{4-271}$$

In Sec. 4-12 we found that the steady-state response to a sinusoidal input is also sinusoidal with the same frequency as the input and a lagging phase angle. Thus, we can write

$$z = Ze^{j(\omega t + \phi)} \tag{4-272}$$

Differentiating Eqs. (4-271) and (4-272) the necessary number of times, substituting the proper values into Eq. (4-270), and collecting like terms gives

$$\left[-\left(\frac{\omega}{\omega_n} \right)^2 + j2\zeta \frac{\omega}{\omega_n} + 1 \right] Ze^{j(\omega t + \phi)} = \left(\frac{\omega}{\omega_n} \right)^2 Ye^{j\omega t} \tag{4-273}$$

or

$$\sqrt{\left[1 - \left(\frac{\omega}{\omega_n} \right)^2 \right]^2 + \left(2\zeta \frac{\omega}{\omega_n} \right)^2} \, Ze^{j(\omega t + \phi + \alpha)} = \left(\frac{\omega}{\omega_n} \right)^2 Ye^{j\omega t} \tag{4-274}$$

where

$$\alpha = \tan^{-1} \frac{2\zeta\omega/\omega_n}{1 - (\omega/\omega_n)^2} \tag{4-275}$$

Solving Eq. (4-274) for Z/Y leads to

$$\frac{Z}{Y} = \frac{(\omega/\omega_n)^2}{\sqrt{[1 - (\omega/\omega_n)^2]^2 + (2\zeta\omega/\omega_n)^2}} \, e^{-j(\phi + \alpha)} \tag{4-276}$$

Now, since Z and Y have magnitude only, $e^{-j(\phi + \alpha)}$ must equal 1.0. Therefore,

$$\frac{Z}{Y} = \frac{(\omega/\omega_n)^2}{\sqrt{[1 - (\omega/\omega_n)^2]^2 + (2\zeta\omega/\omega_n)^2}} \tag{4-277}$$

and

$$\phi = -\alpha = -\tan^{-1} \frac{2\zeta\omega/\omega_n}{1 - (\omega/\omega_n)^2} \tag{4-278}$$

If we replace Z/Y with $mX/m_e e$, we find Eqs. (4-277) and (4-278) are identical with Eqs. (4-175) and (4-174), respectively. Thus, the curves in Fig. 4-22 also give the relationship of Z to Y and ϕ as functions of

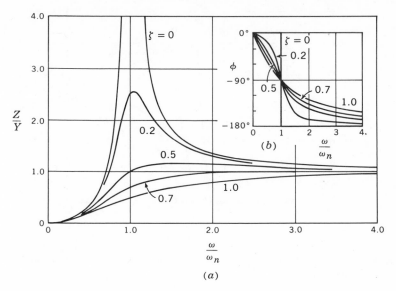

Fig. 4-39 *Steady-state response of a seismic system to a sinusoidal motion disturbance. (a) Ratio of the amplitude of spring deflection to the amplitude of disturbance; (b) phase angle of the response relative to the disturbance.*

ω/ω_n and ζ. However, we are not particularly interested in Z itself, but in how Z is related to the amplitude of displacement, velocity, and acceleration of $y(t)$. For this purpose, the shapes of the curves become important, and they are more readily recognized if plotted, as in Fig. 4-39, with linear scales for all axes.

For $y = Ye^{j\omega t}$, the amplitude is

$$|y| = Y \tag{4-279}$$

and for perfect correlation of Z with Y,

$$Z = KY \tag{4-280}$$

or

$$\frac{Z}{Y} = K \tag{4-281}$$

where K is an arbitrary constant. In Fig. 4-39 it can be seen that for $\omega/\omega_n \gg 1$ the condition in Eq. (4-281) can be met to a high degree of accuracy if $K = 1.0$.

To take a closer look at the error involved, let us consider the amplitude error given if Z is measured and called Y.

$$\text{Amplitude error} = \frac{Z - Y}{Y} 100 \quad \text{percent} \tag{4-282}$$

or

$$\text{Amplitude error} = \left(\frac{Z}{Y} - 1\right) 100 \quad \text{percent} \tag{4-283}$$

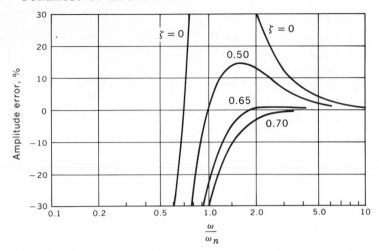

Fig. 4-40 *Percent error in measuring the amplitude of a sinusoidal motion disturbance by use of a seismic system.*

The error can be calculated by taking values from the curves in Fig. 4-39 or by substituting Eq. (4-277) in Eq. (4-283). Curves for the amplitude error as functions of ω/ω_n and ζ are given in Fig. 4-40.

On the basis of the curves shown, it would appear that $\zeta \approx 0.65$ is optimum because it has the smallest error over the widest range of frequencies. For example, the error will be less than 5 percent for $\omega/\omega_n > 1.4$. If the motion is pure simple harmonic motion, the above conclusion would be correct. However, if the motion is periodic, but not simple-harmonic, a Fourier series will have to be used to represent the motion and the measurement of recorded response will be correct only if the sum of the responses to the individual terms of the series agrees exactly with the excitation. In general, this will be true only when every term in the response is exactly in phase *or* is exactly out of phase with the corresponding term in the excitation *or* when every term in the response lags the corresponding term in the excitation *by the same time interval*. The first two conditions require the phase angle to be 0 or $-180°$, respectively, and the third condition requires that the phase angle vary directly with the frequency. Thus, for one-to-one correspondence between response and excitation we must have

$$\phi = 0° \tag{4-284}$$
$$\phi = -180° \tag{4-285}$$

or
$$\phi = K\omega \tag{4-286}$$

where K is an arbitrary constant.

As can be seen in Fig. 4-39b, $\phi = 0°$ when $\zeta = 0$ and $\omega/\omega_n < 1.0$ and $\phi = -180°$ when $\zeta = 0$ and $\omega/\omega_n > 1.0$. Equation (4-286) describes a straight line passing through the origin, and the reader can see that the longest line possible in Fig. 4-39b will be when ζ is slightly less than 0.7 and that the agreement will be quite good for $\omega/\omega_n \leq 1.0$. Since it was concluded above that for displacement measurements $\omega/\omega_n > 1.0$, the phase-angle condition that should be met is for $\phi = -180°$, which in turn requires $\zeta = 0$.

Thus, *a seismic system to be used to measure displacement should have minimum damping and the lowest possible natural frequency.* Physically this means that the spring is so soft that the mass hardly moves. In a seismograph, the suspension system is so soft that during an earthquake the mass literally continues in its path in space while the earth shakes under it.

Instruments for measuring the displacement of vibrations are often called vibrometers, and many different devices have been manufactured. Most of them are hand-held, with the hand and arm becoming part of the mass of the system. Dial indicators, moving light beams, ink pens, etc., have been used as readout and/or recording devices. For frequency determination, time is often introduced by having the light beam fall on a moving strip of photosensitive paper.

For velocity measurements with $y = Ye^{j\omega t}$, the velocity amplitude is

$$|Dy| = \omega Y \tag{4-287}$$

and for perfect correlation of Z with ωY,

$$\frac{Z}{Y} = K\omega \tag{4-288}$$

Equation (4-288) describes a family of straight lines passing through the origin, and reference to Fig. 4-39a will show that the only curve with which there can be any agreement is the one for $\zeta = 1.0$. Even here the range of values of ω/ω_n over which there is reasonable agreement is too small to be of any practical use. Consequently, it appears that a seismic mass is not particularly useful in measuring the velocity of a steady-state vibration.[1] This is true as far as the seismic system by itself is concerned; but when the output is an electrical signal generated by relative motion of a magnet and a coil, as in Fig. 4-41, the voltage is proportional

---

[1] A velocity meter that utilizes a highly overdamped system ($\zeta \approx 100$) to achieve a useful frequency range of from 0.5 to 500 cps is described by E. G. Chilton and T. D. Witherly, Transient-velocity Gage for Structures, *Exptl. Mechanics*, vol. 6, no. 6, pp. 306–312, 1966.

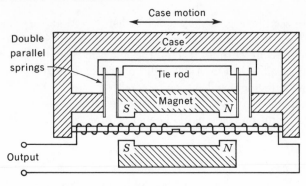

Fig. 4-41 *Schematic diagram of a seismic velocity pickup.* (Consolidated Electrodynamics Corporation.)

to the rate at which the magnetic lines of flux are cut. Thus,

$$E = KZ\omega \qquad (4\text{-}289)$$

where E is the amplitude of output voltage and K is an arbitrary constant.

If the same operating conditions are met as for displacement measurements, that is, $\omega/\omega_n \gg 1$ and $\zeta \approx 0$, $Z \approx Y$ and Eq. (4-289) becomes

$$E \approx KY\omega \qquad (4\text{-}290)$$

or $$E \approx K|Dy| \qquad (4\text{-}291)$$

We now have a device that has an electrical output proportional to the velocity of a steady-state vibration. The velocity can be read from a calibrated meter, and the output can be integrated electrically to give displacement. Some commercial velocity meters use a damping ratio of 0.65. As discussed above, this decreases its usefulness for all except simple harmonic motions, but most steady-state vibrations are essentially simple-harmonic and advantages of an increased frequency range and the avoidance of any buildup of amplitude near $\omega/\omega_n = 1.0$ outweigh the loss in fidelity with non-simple-harmonic motion.

It should be noted that none of the methods discussed above for using seismic systems to measure displacement and velocity can be applied to constant values, i.e., at zero frequency. In fact, they cannot be used for $\omega < \omega_n$ and they work best when $\omega \gg \omega_n$. Considerable ingenuity[1] has been exercised in developing systems with extremely low natural frequencies for use in measuring the amplitude of low-frequency vibrations associated with the vibrations and motions of ships, bridges, buildings, etc.

For acceleration measurements with $y = Ye^{j\omega t}$,

$$|D^2y| = \omega^2 Y \qquad (4\text{-}292)$$

[1] Hetenyi, *op. cit.*, pp. 313–341.

and for perfect correlation of Z with $\omega^2 Y$,

$$\frac{Z}{Y} = K\omega^2 \qquad (4\text{-}293)$$

where K is an arbitrary constant.

Equation (4-293) describes a family of parabolas. Examination of the curves in Fig. 4-39, or study of Eq. (4-277), will show that for values of $\omega/\omega_n \ll 1$, all curves, irrespective of the amount of damping, are good approximations to parabolas. It can also be noted that when $\omega/\omega_n > 1$, no curve even remotely resembles a parabola. Thus, seismic systems used to measure accelerations will be "stiff" systems with high natural frequencies. Such devices are called *accelerometers*.

Although the degree of damping does not appear to be critical with respect to amplitude, the question of phase angle still must be answered. As before, valid results for motions that are not simple-harmonic will be given only when the phase angle is $0°$, is $-180°$, or varies directly with the frequency. Reference to Fig. 4-39b will show that, for $\omega/\omega_n < 1$, $\phi = 0°$ for $\zeta = 0$, and the curve for ζ slightly less than 0.7 is almost a straight line. Therefore, the only amplitude-ratio curves we need to consider further are those for $\zeta = 0$ and $\zeta \approx 0.65$.

The acceleration error will be

$$\text{Acceleration error} = \frac{Z - K\omega^2 Y}{K\omega^2 Y} 100 \qquad \text{percent} \qquad (4\text{-}294)$$

After substituting from Eq. (4-277) and simplifying, Eq. (4-294) becomes

Acceleration error

$$= \left\{ \frac{1/\omega_n{}^2}{K \sqrt{[1 - (\omega/\omega_n)^2]^2 + (2\zeta\omega/\omega_n)^2}} - 1 \right\} 100 \qquad \text{percent} \qquad (4\text{-}295)$$

The determination of the optimum value of K will depend upon how one defines optimum with respect to an error that is a function of frequency. However, K will not vary greatly from $1/\omega_n{}^2$, and that value will be used in the following discussion.

For $K = 1/\omega_n{}^2$, Eq. (4-295) becomes

Acceleration error

$$= \left\{ \frac{1}{\sqrt{[1 - (\omega/\omega_n)^2]^2 + (2\zeta\omega/\omega_n)^2}} - 1 \right\} 100 \qquad \text{percent} \qquad (4\text{-}296)$$

and the curves for $\zeta = 0$ and $\zeta = 0.65$ are given in Fig. 4-42. As can be seen, if a 5 percent error can be tolerated, an accelerometer with zero damping will be good up to a value of $\omega/\omega_n \approx 0.22$, while if the damping ratio is 0.65, the range is extended to $\omega/\omega_n \approx 0.74$. Thus, for a given

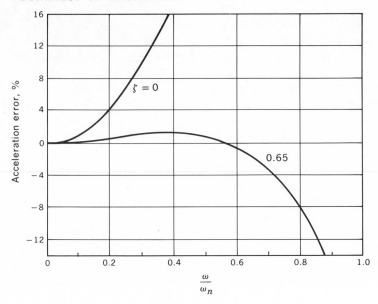

Fig. 4-42 *Percent error in measuring the amplitude of the acceleration of a sinusoidal motion disturbance by use of a seismic system.*

seismic system, i.e., a given natural frequency, the system with $\zeta = 0.65$ will have over three times the useful range of an undamped system. The damped system will have two other advantages over the undamped system in that (1) its response falls off rapidly as ω/ω_n increases and there will be no violent, and erroneous, response to frequencies near ω_n and (2) the free vibrations, called *ringing*, introduced by transients will not be as great and will quickly die out.

Consequently, one might expect all accelerometers to be made with $\zeta = 0.65$. Many are, but introducing sufficient damping in a system with a high natural frequency is not a simple matter, and most present-day accelerometers for use at frequencies $> 1,000$ cps are undamped.

In general, the output from the accelerometer is an electrical signal and filters are used to shape the output signal to extend the linear range and to minimize the effect of resonance with higher harmonics, i.e., to minimize the ringing. It should be noted that the filter circuits affect only the electrical signal and do not change the behavior of the seismic system.

The measurement of motion-time characteristics of short-duration pulses by use of seismic systems is most important in the general area of transportation—particularly where ordnance and aerospace equipment is concerned. As discussed in Sec. 4-21, there are no curves for general use in transient cases as there are for steady-state sinusoidal excitations. However, some qualitative statements of general usefulness can be made.

For example, if the displacement is the desired quantity, the system should be as soft, i.e., have as low a natural frequency, as possible. As before, this would mean that the mass would remain almost motionless and the relative motion would be a good representation of the absolute motion.

If acceleration is the desired quantity, then the system should be as stiff, i.e., have as high a natural frequency, as possible. Physically this means that the mass will follow closely the motion of the base, with the spring force—and thus the deflection—being that required to provide the acceleration.

Another way to look at this is to consider the frequency content of the pulse. A thorough discussion of this subject is beyond the scope of this book, but it can be said that rapid changes result in high frequencies, in much the same way that a Fourier series representation of a square wave requires more terms than for a sine wave. Since the error in the output from an undamped accelerometer becomes large when $\omega/\omega_n > 0.3$, it becomes apparent that ω_n must be very high when ω itself is high.

This may also be related to the ratio of the pulse duration to the natural period of the system, as in Sec. 4-21. For example, Fig. 4-36 shows the ratio of response amplitude to excitation amplitude. Thus, in terms of amplitudes of accelerations, reasonable responses to the half-sine, triangular, and versed-sine pulses would be found for $\tau/T > 3$. Accurate acceleration-time records would require greater ratios, but in many cases the maximum value is the most important information. It should be noted that if the pulse is rectangular, τ/T would have to be infinite for the ratio R/E to equal 1. This case is somewhat academic, because it is physically impossible to have a perfect rectangular pulse.

Most accelerometers use either the phenomenon of piezoelectricity or the change in resistance of bonded or unbonded strain gages to provide an electrical signal that is proportional to acceleration. Crystals of piezoelectric materials, such as rochelle salt, quartz, barium titanate, and lead zirconate titanate, develop a charge, or voltage, on opposing faces when the crystal is deformed. Crystals in the form of plates, where the crystal is both the mass and the spring, are used for low-frequency accelerometers ($<1,200$ cps), but in the great majority of accelerometers the crystal is in compression at all times. Figure 4-43a illustrates the principle of the compression-type piezoelectric accelerometer. The spring is relatively soft in comparison with the crystal, and the force exerted by the spring is sufficient to keep the mass in contact with the crystal for accelerations, within the rated capacity, in either direction along the longitudinal axis of the accelerometer. The compressive force

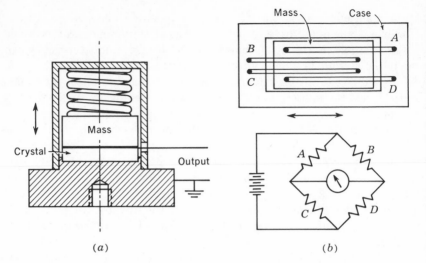

Fig. 4-43 *Accelerometers.* (*a*) *Compression-type piezoelectric;* (*b*) *unbonded strain gage.*

on, and therefore the deformation of, the crystal varies with the acceleration. Piezoelectric accelerometers are self-generating and cannot be used to measure zero-frequency (constant-level) accelerations, although the frequency response can be extended to "almost dc" by use of a charge amplifier. Compression-type accelerometers are available with natural frequencies up to 135 kilohertz (135,000 cps), with usable frequencies from 0.2 cps to over 12 kilohertz, and with the capability of measuring accelerations with magnitudes up to 40,000g.

Figure 4-43b illustrates the use of unbonded strain gages in accelerometers manufactured by Statham Instruments, Inc. The resistance of the wires in the loops A, B, C, and D changes with strain and the loops are connected together in the form of a Wheatstone bridge, as shown. The loops are wound under initial tension so that motion of the mass relative to the case increases the tension in two loops and decreases the tension in the other two loops. This unbalances the bridge, and the meter reading becomes a measure of the relative displacement. The flex plates (not shown) between the mass and the case offer little restraint to relative motion in the direction indicated, and the wires provide most of the spring effect. The bridge must be supplied with a source of power, and this unit can be used to measure zero-frequency accelerations. Damping is about 0.65 critical, and units are available with natural frequencies ranging from 5 to 675 cps, with usable frequencies from 0 to 500 cps, and with maximum acceleration levels from 0.5 to 200g.

An electrical signal corresponding to velocity is obtained by integrating electrically the voltage output of an accelerometer, and an electrical signal corresponding to displacement is obtained by subsequently integrating the velocity signal.

Figure 4-44 shows several types of devices that have been used in measuring acceleration during shock loading. The crushing of the ball in Fig. 4-44a will be a measure of the peak acceleration of the case in the x direction—provided the peak value is sustained long enough for the deformation to be completed. The devices in Fig. 4-44b and c are both of the "no-go" or "go" type in that the contacts open and the necked section breaks only when the acceleration exceeds some specified level. Some of these find use where electrical circuits cannot be used, but they have been relegated to minor usage by the development of high-performance piezoelectric accelerometers. For further information the reader should consult the literature.[1]

4-23. Stability Stability is usually defined in terms of the response of the system in either of the following two ways:

1 A stable system is one whose response remains finite for all finite inputs.

2 A stable system is one that returns to its equilibrium position after being disturbed.

Both definitions are correct, but the second is easier to use because only the free-vibration part of the transient response to a disturbance needs to be considered. As discussed in Secs. 4-3 through 4-10, the characteristics of free vibration of a *linear* system are determined completely by the roots of the characteristic equation, with the response in the form

$$x = C_1 e^{s_1 t} + C_2 e^{s_2 t} \tag{4-297}$$

[1] Hetenyi, *op. cit.*, pp. 368–374.
"Shock and Vibration Instrumentation," American Society of Mechanical Engineers, New York, 1952.

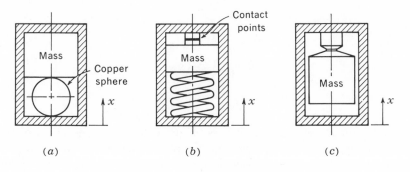

(a)   (b)   (c)

Fig. 4-44 *Accelerometers for shock loading.*

For example, when, in Sec. 4-9, the roots were both negative real numbers, the system was overdamped and, obviously, stable. When, in Sec. 4-8, the roots were complex conjugates with negative real parts, the system was underdamped and again was obviously stable. In both cases, the negative real parts resulted in exponentially decaying terms that rapidly restore the system to its equilibrium position.

Although not illustrated, it should be apparent that whenever a root contains a positive real part, that term will increase exponentially with time instead of decaying to zero and, by definition, the system will be unstable. Therefore, *any system for which any root of the characteristic equation contains a positive real part will be unstable.*

The remaining case to be considered is when the roots are complex and contain no real parts. This case corresponds to a system with zero damping, and, as shown in Sec. 4-4, the motion resulting from a disturbance will continue indefinitely without an increase or decrease in amplitude. Strictly speaking, by either definition, this case is unstable. However, the amplitude does not increase and it is not unstable in the same sense as for the case with a positive real root. For most purposes it is convenient to think of roots with zero real parts as being neutral between positive and negative and, thus, to consider the system to be *neutrally stable.*

The problem of stability is much more important in relation to control systems, where an outside source of energy is controlled by the dynamic behavior of the system, than it is in relation to vibrations, where the forcing function is unaffected by the response of the system. The apparent exception to the latter is the case of a self-excited vibration—which in reality is a special feedback control system.

In Sec. 4-3 the characteristic equation for a linear single-degree-of-freedom system, such as shown in Fig. 4-1, was found to be

$$s^2 + \frac{c}{m} s + \frac{k}{m} = 0 \tag{4-298}$$

and the roots of Eq. (4-298) are

$$s_1 = -\frac{c}{2m} + \sqrt{\left(\frac{c}{2m}\right)^2 - \frac{k}{m}} \tag{4-299}$$

and
$$s_2 = -\frac{c}{2m} - \sqrt{\left(\frac{c}{2m}\right)^2 - \frac{k}{m}} \tag{4-300}$$

Considering only positive mass as being significant, the parts of the roots that must be real, i.e., outside the radical, can and will be positive when the damping coefficient is negative. This requires that the damping force assist rather than oppose the motion and does not agree with physical reality. There are cases, such as the humming of a wire, the vibration of a bridge when the wind is blowing, the vibration or chatter of a cutting tool in a machining operation, and oil-film whirl in lightly loaded journal bearings, where the forcing function is a function of the velocity and can be considered as negative damping.[1] The result is a *self-excited* vibration. In general, self-excited vibrations are nonlinear and stability analysis becomes considerably more complicated because it is no longer a simple matter of whether the real parts of the roots of the auxiliary equation are negative or positive.

When damping is positive, s_2, Eq. (4-300), can never become positive, although s_1, Eq. (4-299), will be positive if the spring has a negative spring rate. Such a spring is possible,[2] but it would not be practical in a suspension system because it would automatically collapse. Thus, *real linear* vibration systems will be stable and the designer need not make a stability analysis in each case.

As mentioned above, stability considerations are much more important in relation to feedback control systems. In fact, stability is one of the major problems in control systems and will be discussed in detail in Chaps. 9 and 10.

[1] J. P. Den Hartog, "Mechanical Vibrations," 4th ed., chap. 7, McGraw-Hill Book Company, New York, 1956.

[2] For example, the Neg'ator spring manufactured by the Hunter Spring Company, Lansdale, Pa.

chapter 5

NONLINEAR
SINGLE-DEGREE-OF-FREEDOM
SYSTEMS

The systems discussed in Chap. 4 were all called linear. In reality, and if one wants to look closely enough, no system is perfectly linear. However, Chap. 4 has not been a wasted exercise in mathematics, because in most cases the effect of the deviations from linearity is negligible for all practical purposes. This chapter will be concerned with some of the most frequently encountered situations where the nonlinearities cannot be ignored.

Since, in general, nonlinear differential equations cannot be solved in closed form, recourse must be had to special techniques, approximations, or to the use of numerical methods and a digital computer or an analog computer to obtain useful information. Of these various methods, the analog computer is probably the most useful, with the numerical methods and digital computer being a close second.

5-1. Classifications of Systems   Depending upon the type of nonlinearity and the manner in which it affects the solution, nonlinear systems can be divided into three groups, as follows:

1   Systems that can be considered to be linear over a small region about the equilibrium position
2   Systems that are piecewise linear
3   Systems that must be considered nonlinear at all times

5-2. Small-amplitude Linear Systems   Since it is not possible to design and manufacture springs that have a linear force-deflection curve for unlimited deflections, it can be argued that this classification actually includes all so-called linear systems. This is true, but the behavior of

142

many real systems agrees closely with that predicted by using a linear model until the amplitude of vibration has increased to the point where damage occurs.   This section will be concerned with systems that appear to be, and are, nonlinear, but under certain conditions can be treated as linear in order to obtain useful information.

The great majority of systems for which linearization over a small range of amplitudes is possible are those in which the potential energy is stored in a nonlinear element.   The most common situations are related to gravity pendulums and nonlinear springs.   In some cases, particularly where trigonometric functions are involved, it is convenient to expand the function in a Maclaurin's series[1] and discard the nonlinear terms.   In most cases, linearization can most readily be accomplished by studying the situation and arriving at a reasonable linear approximation to the exact equation.

Example 5-1   Derive the equation for the natural frequency of small oscillations of the simple gravity pendulum in Fig. 5-1a.   (The term simple pendulum means that the cable has no mass and that the mass m is concentrated at a point.)

Solution.   The free-body diagram in Fig. 5-1b shows the forces acting on the mass at the instant of release after the pendulum has been displaced through the angle θ. Designating the direction along the cable as radial and the direction perpendicular to the string as tangential, we find that in accordance with Newton's second law, we have, radially,

$$\Sigma F_r = mA_r$$

and $$\Sigma F_r = F_T - W \cos \theta = 0 \tag{5-2}$$

[1] As will be found in all books on the calculus, Maclaurin's series is

$$f(x) = f(0) + xf'(0) + \frac{x^2 f''(0)}{2!} + \frac{x^3 f'''(0)}{3!} + \cdots \tag{5-1}$$

where $f(0)$ = value of the function at $x = 0$
   $f'(0)$ = value of the first derivative with respect to x at $x = 0$
   etc.

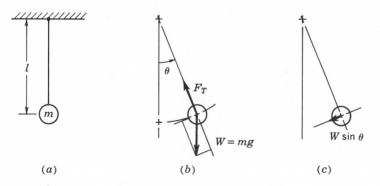

(a)　　　　　　　(b)　　　　　　　(c)

Fig. 5-1   *Simple pendulum.*

where F_T = tension force in the string, and, tangentially,

$$\Sigma F_t = mA_t$$

and $$\Sigma F_t = -W \sin \theta \qquad (5\text{-}3)$$

Thus, as shown in Fig. 5-1c, only a single unbalanced force acts on the mass. From Eq. (5-2) we see that $A_r = 0$, and from Eq. (5-3)

$$-W \sin \theta = mA_t \qquad (5\text{-}4)$$

Since m moves on a circular path with radius l, its distance x along the path from the equilibrium position is

$$x = l\theta$$

and thus, $$A_t = D^2 x = l\, D^2\theta \qquad (5\text{-}5)$$

Substituting from Eq. (5-5) into Eq. (5-4) results in

$$-W \sin \theta = ml\, D^2\theta = \frac{W}{g}\, l\, D^2\theta$$

or $$\frac{l}{g} D^2\theta + \sin \theta = 0 \qquad (5\text{-}6)$$

where $\sin \theta$ is a nonlinear function of θ.

Since we are interested in linearizing $\sin \theta$ about $\theta = 0$, we write Maclaurin's series, Eq. (5-1), as

$$f(\theta) = f(0) + \theta f'(0) + \frac{\theta^2 f''(0)}{2!} + \frac{\theta^3 f'''(0)}{3!} + \cdots \qquad (5\text{-}7)$$

But $f(\theta) = \sin \theta$. Therefore,

$$\begin{aligned} f(0) &= \sin 0 = 0 \\ f'(0) &= \cos 0 = 1 \\ f''(0) &= -\sin 0 = 0 \\ f'''(0) &= -\cos 0 = -1 \\ &\text{etc.} \end{aligned} \qquad (5\text{-}8)$$

Substituting from Eqs. (5-8) into Eq. (5-7) and simplifying results in

$$\sin \theta = \theta - \frac{\theta^3}{6} + \frac{\theta^5}{120} - \cdots \qquad (5\text{-}9)$$

The linear term is θ; therefore, for small angles,

$$\sin \theta = \theta \qquad (5\text{-}10)$$

Substituting from Eq. (5-10) into Eq. (5-6) and rearranging terms gives

$$D^2\theta + \frac{g}{l}\, \theta = 0 \qquad (5\text{-}11)$$

Equation (5-11) is a second-order linear differential equation, and from its similarity with Eq. (4-34) we recognize that it can be written as

$$D^2\theta + \omega_n{}^2\theta = 0 \qquad (5\text{-}12)$$

where $$\omega_n{}^2 = \frac{g}{l} \qquad (5\text{-}13)$$

Thus, from Eq. (5-13),

$$\omega_n = \sqrt{\frac{g}{l}} \tag{5-14}$$

which is the equation derived, in a less rigorous manner, in high school physics textbooks.

Note. Equation (5-9) can be used to answer the question, "When does the oscillation cease to be small?" For example, when $\theta = 0.1$ rad, or $5.73°$,

$$\sin \theta = 0.1 - 0.0001667 + 0.00000008 - \cdots$$

and the error introduced by neglecting the second term is less than 0.2 percent. When $\theta = 0.3$ rad ($17.19°$), the error increases only to 1.5 percent, which is still negligible for most engineering purposes.

Example 5-2 Derive the equation for the natural frequency of the compound pendulum in Fig. 5-2*a*. The mass moment of inertia of the pendulum about its center of gravity is I, and the weight of the pendulum is W.

Solution. Figure 5-2*b* is a free-body diagram showing the forces that would appear to act if only conditions for static equilibrium were considered. Applying Newton's second law to the vertical and horizontal components of the forces, we find, respectively,

$$\Sigma F_v = mA_v$$
$$\Sigma F_v = F_0 - W = 0 \tag{5-15}$$

and

$$\Sigma F_h = mA_h$$
$$\Sigma F_h = 0 \tag{5-16}$$

But, although $\Sigma F_v = 0$, the vertical forces are not collinear and there is a couple acting on the body. Applying Newton's second law to the moment about O_2, we find

$$\Sigma T = I\alpha$$

and

$$\Sigma T = -W\bar{l} \sin \theta \tag{5-17}$$

Thus, from Eq. (5-17),

$$I_o D^2\theta = -W\bar{l} \sin \theta \tag{5-18}$$

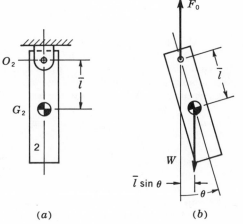

Fig. 5-2 *Compound pendulum.*

(*a*)   (*b*)

where I_o is the mass moment of inertia of the pendulum about O_2. Rewriting Eq. (5-18) results in

$$I_o \, D^2\theta + W\bar{l} \sin \theta = 0 \qquad (5\text{-}19)$$

Equation (5-19) is a second-order nonlinear differential equation. But, as in Example 5-1, it can be linearized for small amplitudes by substituting θ for $\sin \theta$ to give

$$D^2\theta + \frac{W\bar{l}}{I_o} \theta = 0 \qquad (5\text{-}20)$$

or

$$D^2\theta + \omega_n{}^2\theta = 0 \qquad (5\text{-}21)$$

where

$$\omega_n{}^2 = \frac{W\bar{l}}{I_o} \qquad (5\text{-}22)$$

Therefore, from Eq. (5-22),

$$\omega_n = \sqrt{\frac{W\bar{l}}{I_o}} \qquad (5\text{-}23)$$

In accordance with the parallel-axis theorem,

$$I_o = I + m\bar{l}^2 = I + \frac{W}{g}\bar{l}^2 \qquad (5\text{-}24)$$

Substituting from Eq. (5-24) into Eq. (5-23) gives

$$\omega_n = \sqrt{\frac{W\bar{l}}{I + (W/g)\bar{l}^2}} \qquad (5\text{-}25)$$

Note 1.   Equation (5-25) is the basis for the experimental determination of the moments of inertia of complex members and bodies, such as connecting rods and complete automobiles, where computations based on dimensions and materials cannot readily be made.   Since the period of oscillation is

$$T = \frac{2\pi}{\omega_n} \qquad (5\text{-}26)$$

we can substitute $2\pi/T$ for ω_n in Eq. (5-25) and solve for I.   Thus,

$$I = W\bar{l}\left(\frac{T^2}{4\pi^2} - \frac{\bar{l}}{g}\right) \qquad (5\text{-}27)$$

and all that is necessary is to weigh the member, determine the position of the center of gravity, and determine the period of oscillation by using a stopwatch or other timing device.   It should be noted that maximum accuracy, for a given accuracy in determining T, will be had when $\bar{l}$ is a minimum.   Thus, it is desirable to swing the member about a point as close to the center of gravity as possible.

Note 2.   The simple pendulum that is kinetically equivalent to the compound pendulum is of interest because the distributed mass is replaced by a point mass. Since the inertia force on a point mass acts through the point, as in Fig. 5-1c, there will not be a couple developed when a horizontal force acts to accelerate the mass, i.e., the pendulum.   Since there will not be a couple, there will not be a reaction at the pivot.   The point at which the mass of a compound pendulum can be considered to be concentrated is known as the *center of percussion*.

Example 5-3   The use of the compound-pendulum theory in determining the moment of inertia of a complex member was discussed in Example 5-2.   The method

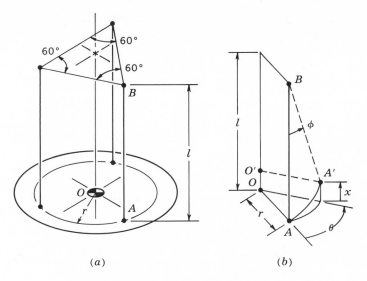

(a) (b)

Fig. 5-3   *Trifilar gravity torsional pendulum.*

is particularly useful when the member is a connecting rod or other link that looks like a compound pendulum, but when the body is more or less cubical, supporting it as a compound pendulum with the axis of rotation just above the center of gravity often becomes a difficult problem.   In such a situation the gravity torsional pendulum, as shown in Fig. 5-3a, becomes quite useful.

We are asked to derive an equation relating the moment of inertia of the body to the natural period of oscillation of the gravity torsional pendulum.

Solution.   The first step will be to consider the importance of symmetry.   For example, if the cables are all the same length, the platform will remain horizontal at all times.   Also, if the cables are attached in the form of an equilateral triangle and if the centers of gravity of the platform and the object are positioned to be equidistant from the cables, each cable will carry one-third of the total weight and the pendulum can oscillate about a vertical axis through the centers of gravity without developing any other motion.   (This system actually has three degrees of freedom and we wish to use only one in this case; this is called *decoupling* of modes of vibration and will be discussed in some detail in Secs. 6-5 through 6-7.)

When the platform is rotated about the center line through the center of gravity, it remains horizontal but rises.   Figure 5-3b shows the motion of the point of attachment of cable AB to the platform when the platform has been rotated through the angle θ.   As shown, the platform has risen a distance x and the cable makes an angle ϕ with the vertical.

Let us use an energy method (Sec. 4-18) to determine the relationship between the natural frequency of the torsional oscillations and the moment of inertia of the platform.   Assuming simple harmonic motion, we shall have

$$\theta = \Theta \sin \omega_n t \tag{5-28}$$

where Θ is the amplitude of the oscillation, and

$$x = X|\sin \omega_n t| \tag{5-29}$$

where X is the maximum value of the displacement and the absolute value of $\sin \omega_n t$ must be used because x is always positive.

The maximum value of the kinetic energy will be

$$E_{K,\max} = \tfrac{1}{2}I(D\theta)^2_{\max} = \tfrac{1}{2}I\omega_n{}^2\Theta^2 \qquad (5\text{-}30)$$

and the maximum value of the potential energy will be

$$E_{P,\max} = Wx_{\max} = WX \qquad (5\text{-}31)$$

Equating the maximum kinetic energy to the maximum potential energy, we find

$$E_{K,\max} = E_{P,\max}$$
$$\tfrac{1}{2}I\omega_n{}^2\Theta^2 = WX \qquad (5\text{-}32)$$

We must now eliminate either Θ or X. To do this, let us find X as a function of Θ. From Fig. 5-3b,

$$x = l(1 - \cos \phi) \qquad (5\text{-}33)$$

and we must now find ϕ as a function of θ.

The path from A to A' is a curved path in space, and this system is nonlinear in three dimensions. Maclaurin's series is not directly applicable here, but we can linearize part of the problem by noting that for small angles the arc approximates the chord. Thus, we can write

$$AA' = r\theta = l\phi \qquad (5\text{-}34)$$

from which

$$\phi = \frac{r}{l}\,\theta \qquad (5\text{-}35)$$

Substituting from Eq. (5-35) into Eq. (5-33), we find

$$x = l\left(1 - \cos\frac{r}{l}\,\theta\right) \qquad (5\text{-}36)$$

and in terms of maximum values

$$X = l\left(1 - \cos\frac{r}{l}\,\Theta\right) \qquad (5\text{-}37)$$

Substituting from Eq. (5-37) into Eq. (5-32) gives

$$\tfrac{1}{2}\,I\omega_n{}^2\Theta^2 = Wl\left(1 - \cos\frac{r}{l}\,\Theta\right) \qquad (5\text{-}38)$$

There is still a problem: the Θ's do not cancel as in Sec. 4-18. The reason for this is that the energy method can be used to determine natural frequencies for only linear systems and Eq. (5-38) still contains a nonlinear term, $1 - \cos (r\Theta/l)$. However, since the Maclaurin's series expansion of the function was found to be worthwhile in converting the nonlinear system in Example 5-1 into a linear one, we can expect it to be of use in this case, too. Since the variable is $(r/l)\Theta$, we can write Maclaurin's series, Eq. (5-1), as

$$f\left(\frac{r}{l}\,\Theta\right) = f(0) + \frac{r}{l}\,\Theta\,f'(0) + \left(\frac{r}{l}\,\Theta\right)^2\frac{f''(0)}{2!} + \left(\frac{r}{l}\,\Theta\right)^3\frac{f'''(0)}{3!} + \cdots \qquad (5\text{-}39)$$

But,

$$f(0) = 1 - \cos 0 = 1 - 1 = 0$$
$$f'(0) = \sin 0 = 0$$
$$f''(0) = \cos 0 = 1 \tag{5-40}$$
$$f'''(0) = - \sin 0 = 0$$
etc.

Substituting from Eqs. (5-40) into Eqs. (5-39) gives

$$1 - \cos\frac{r}{l}\Theta = 0 + 0 + \left(\frac{r}{l}\Theta\right)^2 \times \tfrac{1}{2} + 0 - \left(\frac{r}{l}\Theta\right)^4 \times \tfrac{1}{24} + \cdots \tag{5-41}$$

Then, substituting the first nonzero term, the only one significant for $\Theta \ll 1$, from Eq. (5-41) into Eq. (5-38) and simplifying, we find

$$I\omega_n{}^2 = \frac{Wr^2}{l} \tag{5-42}$$

which can be solved for ω_n. Therefore, series expansion is useful in linearizing this system, too. Since we are interested in finding I by timing the torsional oscillations of the system, Eq. (5-42) can be put in more convenient form by substituting $2\pi/T$ for ω_n and solving for I. Doing this, we find

$$I = \frac{Wr^2T^2}{4\pi^2l} \tag{5-43}$$

where T is the period of small-amplitude natural oscillation.

Note. One major advantage of using a torsional pendulum rather than a compound pendulum is that the moment of inertia about the center of gravity can be determined directly without use of the parallel-axis theorem.

Example 5-4   Four nonlinear springs, each with a force-deflection characteristic defined by

$$F = 50x + 400x^2 \tag{5-44}$$

will support a mass weighing 800 lb. We are asked to determine the natural frequency of the system for small oscillations about the equilibrium point, i.e., about the static-deflection position.

Solution. The force-deflection curve is shown in Fig. 5-4a. Since the force and displacement involved in the vibration are about the static-deflection position, it will

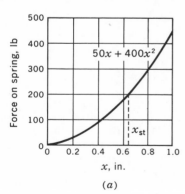

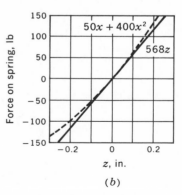

(a)   (b)

Fig. 5-4   *Force-deflection curves.   (a) Actual; (b) linearized for small-amplitude vibrations about position of static deflection.*

be helpful if we introduce a new variable

$$z = x - x_{st} \tag{5-45}$$

so that the force will now be a function of z. In fact, we want the force to be a linear function of z, such as

$$F = kz \tag{5-46}$$

where k is the spring rate of a linear spring. The spring rate is defined as the slope of the force-deflection curve, and for small displacements the slope at $x = x_{st}$ or $z = 0$ will be the best possible approximation.

From Eq. (5-44),

$$\frac{dF}{dx} = 50 + 800x \tag{5-47}$$

Considering that each spring carries one-quarter of the total weight, the static deflection can be found by substituting 200 lb for F in Eq. (5-44) and solving for x. Thus,

$$200 = 50x_{st} + 400x_{st}{}^2$$

from which

$$x_{st} = 0.647 \text{ in.}$$

At $x = x_{st} = 0.647$ in., by use of Eq. (5-47),

$$k = \frac{dF}{dx}\bigg|_{x_{st}} = 50 + 800(0.647) = 568 \text{ lb/in.}$$

and, therefore, from Eq. (5-46),

$$F = 568z \tag{5-48}$$

The linear approximation and the true curves are shown in Fig. 5-4*b*. As can be seen, the approximation is quite good for deflections up to 0.1 in. either side of the equilibrium position.

For free vibrations the equation of motion becomes

$$m\,D^2z + kz = 0 \tag{5-49}$$

or

$$D^2z + \frac{k}{m}\,z = 0 \tag{5-50}$$

From Sec. 4-4 we know that for Eq. (5-50) we can write

$$\omega_n = \sqrt{\frac{k}{m}} = \sqrt{\frac{kg}{W}} \tag{5-51}$$

On a per-spring basis, Eq. (5-51) becomes

$$\omega_n = \sqrt{\frac{568 \times 386}{200}} = 33.1 \text{ rad/sec}$$

or

$$f_n = \frac{\omega_n}{2\pi} = \frac{33.1}{2\pi} = 5.27 \text{ cps}$$

Example 5-5 A foundation for an electrodynamic shaker is to be isolated from its surroundings. The total weight of the table and foundation will not exceed

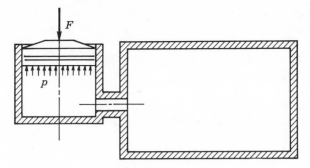

Fig. 5-5   *Air spring.*

150 tons, and the natural frequency of the system is to be less than 0.5 cps.   Preliminary calculations, by use of Eq. (4-53), indicated that if linear springs are used, the static deflection will have to be about 39.2 in.   Since this is not readily accomplished, other methods have been investigated and the use of air springs appears to be the best.

As shown in Fig. 5-5, the air spring is essentially an air cylinder connected to a large reservoir containing air, or other gas, under pressure.   The supply of air is automatically controlled so that the equilibrium (static) position of the piston is always the same, regardless of the mass being supported.   (This self-leveling feature of air springs is a major advantage and is one of the reasons for their widespread use in buses, where there is a great difference between the loaded and unloaded weights.)

We are asked to determine the specifications for the suspension system on the basis of using 12 springs.   Air at a pressure of 150 psi (gauge) is already available in the building.

Solution.   To allow for possible overload, we shall use an air pressure of about 140 psi in calculating the diameter of the springs.   Assuming that the weight is distributed equally over the 12 springs,

$$W = 12 p_s A \tag{5-52}$$

where W = total weight = $150 \times 2,000$ = 300,000 lb
$\quad p_s$ = supply air pressure, psig
$\quad A$ = area of bore of spring = $\pi d^2 / 4$
Thus,

$$d = \sqrt{\frac{4W}{12 p_s \pi}} = \sqrt{\frac{4 \times 300,000}{12 \times 140 \times \pi}} = 15.08 \text{ in.}$$

Rounding off d to 15 in., we find that the pressure will have to be

$$p_s = \frac{4W}{12 \pi d^2} = \frac{4 \times 300,000}{12 \pi \times 15^2} = 141.5 \text{ psig}$$

which should be satisfactory.

We must now derive an expression for the spring rate of an air spring.   By definition,

$$k = \frac{dF}{dx} \tag{5-53}$$

where k = spring rate, lb/in.

F = force, lb

x = displacement, in.

From conditions of equilibrium of forces and assuming that ambient pressure is 14.7 psig,

$$F = (p - 14.7)A \qquad (5\text{-}54)$$

where p = absolute pressure, psia.

Differentiating Eq. (5-54) with respect to x gives

$$\frac{dF}{dx} = A \frac{dp}{dx} = k \qquad (5\text{-}55)$$

Assuming isentropic compression and expansion, we know, from elementary thermo-dynamics, that for air

$$pv^{1.4} = C \qquad (5\text{-}56)$$

where C is a constant. Solving Eq. (5-56) for p, we find

$$p = Cv^{-1.4}$$

from which

$$\frac{dp}{dx} = -1.4Cv^{-2.4} \frac{dv}{dx} \qquad (5\text{-}57)$$

Letting the equilibrium pressure and volumes be p_0 and v_0, respectively, we can write

$$C = p_0v_0^{1.4} \qquad (5\text{-}58)$$

and, from Fig. 5-5,

$$v = v_0 - Ax \qquad (5\text{-}59)$$

Hence,

$$\frac{dv}{dx} = -A \qquad (5\text{-}60)$$

Substituting from Eqs. (5-58), (5-59), and (5-60) into Eq. (5-57) gives

$$\frac{dp}{dx} = 1.4p_0v_0^{1.4}(v_0 - Ax)^{-2.4}A \qquad (5\text{-}61)$$

Substituting from Eq. (5-61) into Eq. (5-55) gives

$$k = 1.4p_0v_0^{1.4}(v_0 - Ax)^{-2.4}A^2 \qquad (5\text{-}62)$$

As can be seen, k varies with x and the spring is nonlinear. However, v_0 is usually a large value and, thus, for small displacements, i.e., small values of x, Ax can be neglected. Under these conditions, k can be considered to be constant and Eq. (5-62) becomes

$$k = \frac{1.4p_0A^2}{v_0} \qquad (5\text{-}63)$$

From Eq. (4-49),

$$k = m(2\pi f_n)^2 = \frac{W}{g} 4\pi^2 f_n^2$$

In this case $f_n = 0.5$ cps and W is one-twelfth the total weight.   Thus,

$$k = \frac{300,000}{12 \times 386} 4\pi^2(0.5)^2 = 639 \text{ lb/in.}$$

We also know that, by definition,

$$p_0 = p_s + p_A$$

where p_A is the ambient (atmospheric) pressure.   We can now write

$$p_0 = 141.5 + 14.7 = 156.2 \text{ psia}$$

and

$$A = \frac{\pi d^2}{4} = \frac{\pi}{4} 15^2 = 176.7 \text{ in.}^2$$

Therefore, Eq. (5-63) can be solved for v_0 to give

$$v_0 = \frac{1.4 p_0 A^2}{k} = \frac{1.4 \times 156.2 \times 176.7^2}{639} = 10,700 \text{ in.}^3$$

Summary   Each of the 12 air springs should have an internal diameter of 15.0 in.; the volume of the cylinder plus reservoir should be 10,700 in.^{3}; and the normal operating air pressure will be 141.5 psig.

Note.   Since p_0 and A are determined by the ambient pressure, the weight to be carried, and the pressure desired, or allowable, in the cylinder, the reservoir volume is the major variable with which the designer can work.   It should also be noted that damping, although also nonlinear, can be introduced by putting a restriction or orifice in the line between the cylinder and the reservoir.

5-3. Piecewise Linear Systems

The systems in this classification contain spring and/or damping elements whose force-displacement and/or force-velocity curves are made up of segments of straight lines.   The discontinuity in the spring force-deflection curve is usually the result of backlash or play in parts, Fig. 5-6, or the use of a precompressed spring, Fig. 4-44b, and the discontinuity in the damping force is almost always due to the presence of dry friction or Coulomb (rather than viscous) damping, as discussed in Sec. 3-4 and illustrated in Fig. 3-6.

As shown in Fig. 5-6b, the relationship between the spring force and the displacement for the system in Fig. 5-6a is linear in three pieces, in the regions $x < -a$, $-a < x < a$, and $x > a$.   Thus, for free vibrations we

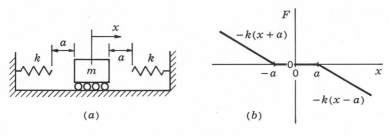

(a)   (b)

Fig. 5-6

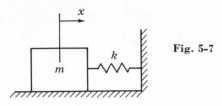

Fig. 5-7

have the following three equations of motion:

For $x < -a$

$$m\,D^2x + k(x + a) = 0 \qquad (5\text{-}64)$$

For $-a < x < a$

$$m\,D^2x = 0 \qquad (5\text{-}65)$$

For $x > a$

$$m\,D^2x + k(x - a) = 0 \qquad (5\text{-}66)$$

The solution for a given set of initial conditions requires solving each equation in turn, with the final value of velocity when leaving a region becoming the initial condition for the next region. When the system is conservative, as in this case, the solution for free vibrations is not particularly time-consuming. However, when damping is present or if there is a forcing function, the time required becomes much greater and the use of an analog or digital computer is to be recommended.

The most important piecewise linear vibration system is that in which the damping can be considered to be Coulomb damping. Replacing μN by F_μ in Eq. (3-20), we find that the damping force *acting on the mass* can be expressed as

$$F = -F_\mu \frac{Dx}{|Dx|} \qquad (5\text{-}67)$$

where F_μ is the magnitude of the friction force.

For the system in Fig. 5-7 the equation of motion becomes

$$m\,D^2x + F_\mu \frac{Dx}{|Dx|} + kx = 0 \qquad (5\text{-}68)$$

which is nonlinear. However, Eq. (5-68) is piecewise linear in that the damping term is constant as long as the velocity remains in one direction. Thus, we can write two linear equations, one applying to the region Dx positive (> 0) and the other to the region Dx negative (< 0), as follows:

For Dx positive

$$\Sigma F = m\,D^2x = -kx - F_\mu$$

from which

$$m\,D^2x + kx = -F_\mu \qquad (5\text{-}69)$$

For Dx negative

$$F = m\, D^2x = -kx + F_\mu$$

from which

$$m\, D^2x + kx = F_\mu \tag{5-70}$$

Dividing through by m and recognizing that $k/m = \omega_n^2$ (from Sec. 4-4), we can rewrite Eqs. (5-69) and (5-70) as, respectively,

$$D^2x + \omega_n^2 x = -\frac{F_\mu}{m} \tag{5-71}$$

and

$$D^2x + \omega_n^2 x = \frac{F_\mu}{m} \tag{5-72}$$

Two points can be noted immediately: (1) the damping force acts as a constant-magnitude forcing function (a force step function) and (2) the natural frequency with Coulomb damping is the same as the undamped natural frequency.

Considering Eq. (5-71) first, we find that, from Sec. 4-4, the complementary solution can be written as

$$x = C_1 \sin \omega_n t + C_2 \cos \omega_n t \tag{5-73}$$

To find the particular solution, we let

$$x = C_3 \tag{5-74}$$

and, substituting from Eq. (5-74) into Eq. (5-71) and solving for C_3, we find

$$C_3 = -\frac{F_\mu}{\omega_n^2 m} = -\frac{F_\mu}{k} \tag{5-75}$$

The general solution to Eq. (5-71) is the sum of Eqs. (5-73) and (5-75), which is

$$x = C_1 \sin \omega_n t + C_2 \cos \omega_n t - \frac{F_\mu}{k} \tag{5-76}$$

Now if we consider the case when the mass is displaced to $-X$, so that Dx will be positive, we have as initial conditions at $t = 0$, $x = -X$ and $Dx = 0$. Solving for C_1 and C_2 for these initial conditions, we find that Eq. (5-76) becomes

$$x = -\left(X - \frac{F_\mu}{k}\right) \cos \omega_n t - \frac{F_\mu}{k} \tag{5-77}$$

The velocity will change from positive to negative when $t = \pi/\omega_n$. Solving Eq. (5-77) for the amplitude at this time, we find

$$x \bigg|_{t=\pi/\omega_n} = X - 2\frac{F_\mu}{k} \tag{5-78}$$

and we see that in one half cycle the amplitude has decreased by $2F_\mu/k$.

To continue the solution for the next half cycle when Dx is negative,

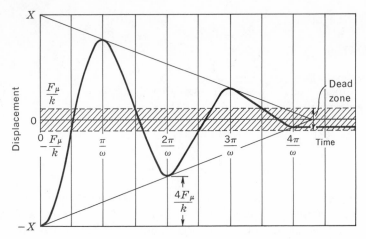

Fig. 5-8   *Free vibration of a system with Coulomb damping.*

we have as initial conditions, when $t = \pi/\omega_n$, $x = X - 2F_\mu/k$ and $Dx = 0$. Solving Eq. (5-72) for these conditions will result in

$$x = -\left(X - 3\frac{F_\mu}{k}\right)\cos\omega_n t + \frac{F_\mu}{k} \tag{5-79}$$

At the end of the half cycle, $t = 2\pi/\omega_n$ and, from Eq. (5-79),

$$x\Big|_{t=2\pi/\omega_n} = -X + 4\frac{F_\mu}{k} = -\left(X - 4\frac{F_\mu}{k}\right) \tag{5-80}$$

Thus, the amplitude has decreased again by $2F_\mu/k$. The reader should see the pattern that has developed, which is that the amplitude for a free vibration with Coulomb damping will decrease at the constant rate of $4F_\mu/k$ per cycle, as shown in Fig. 5-8.

The physical significance of F_μ/k is that this is the spring deflection required to overcome the friction force. Since there will be no acceleration unless the spring force is greater than the friction force, it can be seen that the vibration will cease whenever the mass comes to rest within the region $-F_\mu/k < x < F_\mu/k$. This region is called the *dead zone,* and the reader has undoubtedly encountered this phenomenon in the laboratory where the pointers of meters have been observed to fail to move for a slight change in the value of the quantity being measured. Coulomb friction and the resulting dead zone become a serious problem when a feedback control system is being used to control position and a high degree of accuracy is required, as in tracer-controlled machine tools and automatic pilots for aerospace applications. This type of control system is called a *servomechanism* and will be discussed in Chaps. 7 through 10.

The effect of the difference in mechanism of energy dissipation has just been illustrated by showing that a free vibration decays at a constant rate for Coulomb damping in comparison with the logarithmic decrement (Sec. 4-8) for viscous damping. From a physical viewpoint the difference arises from the fact that the Coulomb damping force has a constant magnitude, independent of amplitude and frequency, whereas the magnitude of the viscous damping force is proportional to the velocity and is, therefore, proportional to the product of the amplitude and frequency. The result is that Coulomb damping is the more effective when low amplitudes and low frequencies are involved, and viscous damping becomes more effective when large amplitudes and high frequencies are involved. Although of more academic than practical value, it can be shown[1] that in the case of forced vibration an infinite amplitude can be reached at resonance when only Coulomb damping is present whereas, as shown in Sec. 4-13, the amplitude will remain finite at resonance with viscous damping, no matter how little it may be.

5-4. General Nonlinear Systems   Whenever (1) a system is not piecewise linear in a relatively simple manner, (2) the amplitude of the motion or the degree of nonlinearity is so great that linearization for small amplitudes will not give satisfactory results, or (3) the response to a periodic or pulse forcing function is desired, the designer will usually find it desirable to use an analog computer or to resort to numerical methods and a digital computer in arriving at a solution to the problem. However, there is one technique, the phase-plane method, that can be applied to the free vibrations of every single-degree-of-freedom system and, thus, can often be used to advantage in obtaining some information about the behavior of a grossly nonlinear system when a computer is not immediately available. A brief discussion of the method will be given below. Any mention of its application to linear single-degree-of-freedom systems will be purely incidental, because more information is already available, in Secs. 4-3 through 4-10, about their free-vibration behavior than would be presented in the phase plane.

The term phase plane is applied to a plot of one quantity as a function of another when knowledge of both completely fixes the state of the system. The pertinent quantities in vibrations are velocity and displacement and, in this discussion, the term *phase plane will always mean a plot of the velocity as a function of the displacement.*[2]

---

[1] J. P. Den Hartog, "Mechanical Vibrations," 4th ed., p. 375, McGraw-Hill Book Company, New York, 1956.

[2] In some books on vibrations the phase plane is a plot of displacement as a function of velocity. However, the application to vibrations is relatively recent, and the phase plane shows up in the literature far more frequently in relation to nonlinear electric circuits and feedback control systems where there is almost universal use of velocity as a function of displacement. In the interest of minimizing confusion, this text will follow the coordinate system the reader will encounter most often in further studies in theory of nonlinear oscillations.

Applying Newton's second law to the general single-degree-of-freedom system, we find for free vibrations

$$m\,D^2x = f(Dx,x) \tag{5-81}$$

which can be rewritten as

$$D^2x = \frac{f(Dx,x)}{m} \tag{5-82}$$

Using v for velocity, to minimize confusion, we can write Eq. (5-82) in terms of the velocity as

$$D^2x = \frac{dv}{dt} = \frac{f(v,x)}{m} \tag{5-83}$$

By definition,

$$\frac{dx}{dt} = v \tag{5-84}$$

Dividing Eq. (5-83) by Eq. (5-84), we find

$$\frac{dv/dt}{dx/dt} = \frac{dv}{dx} = \frac{f(v,x)}{mv} \tag{5-85}$$

The independent variables are now v and x (time has been eliminated), and Eq. (5-85) is the basis for the phase-plane method. As can be seen, the slope of the curve or trajectory in the phase plane can be calculated for every point (x,v). Therefore, if values of dv/dx are calculated for enough combinations of x and v, a graphical plot, using arrows showing the slope at each point, of the information will look something like Fig. 5-9a.

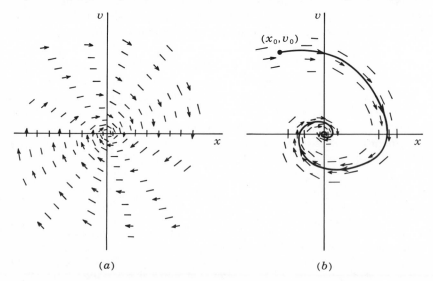

(a)   (b)

Fig. 5-9   *The phase plane.   (a) Phase portrait;   (b) trajectory.*

A pattern should be discernible, and one needs only to start at the point representing the initial conditions and then sketch in a curve that follows the arrows. This procedure is illustrated in Fig. 5-9*b*, where the initial conditions are $x = x_0$ and $v = v_0$.

It is useful to note, from Eq. (5-85) or deduced from previous knowledge of the behavior of vibrating systems, that the slope of the trajectory must be either $+\infty$ or $-\infty$ whenever $v = 0$. Thus, the trajectory will always be vertical when crossing the x axis.

On the basis of the trajectory in Fig. 5-9*b* we can draw several conclusions: (1) The system is stable because the amplitude decreases to zero. (2) Although the type of damping may be complicated and nonlinear, the effective damping appears to be considerably less than critical (because the displacement goes from positive to negative a number of times), but appreciable (because each time the trajectory cuts the positive x axis the amplitude is only about 40 percent of the amplitude of the preceding cycle—this would indicate $\zeta = 0.14$ if the damping were purely viscous). (3) Coulomb damping is probably zero or negligible because the system comes to rest at zero displacement.[1]

The only information missing in the phase-plane plot is that pertaining to time. This is usually of major importance because of the possibility of resonance with a disturbance. Since, for a nonlinear system, the apparent natural frequency, damped or undamped, is not a constant but varies with amplitude, the phenomenon of resonance can differ greatly from that for a linear system as discussed in Secs. 4-14 through 4-19. The subject of forced vibrations of nonlinear systems is beyond the scope of this book,[2] but it should be noted that except for Coulomb damping and a particular spring characteristic, very little can be done analytically without recourse to a computer.

To bring time back into the solution, we shall use Eq. (5-84), which can be rewritten as

$$dt = \frac{1}{v}\,dx \qquad (5\text{-}86)$$

Integrating both sides of Eq. (5-86) gives

$$\int_{t_1}^{t_2} dt = t_2 - t_1 = \int_{x_1}^{x_2} \frac{1}{v}\,dx \qquad (5\text{-}87)$$

In most cases, integration of the right-hand term must be accomplished by using numerical methods or by plotting $1/v$ as a function of x and measuring the area by use of a planimeter.

---

[1] See Sec. 5-3 for discussion of effect of Coulomb damping on system behavior.

[2] Den Hartog, *op. cit.*, pp. 370–379.

Example 5-6 A recoil system consisting of a mass, a linear spring, and a damper is being designed. Preliminary calculations have determined that (a) when the forcing function ceases to act, the mass is displaced a distance of 2.75 in. from the equilibrium position and is moving away from the equilibrium position at a speed of 2.1 fps, (b) the mass will weigh 600 lb, and (c) the spring rate will be 150 lb/in. It has been proposed that a combination of viscous and Coulomb damping be used, with the viscous damping coefficient being such that the damping ratio would be 0.5 if only it were to be used and the Coulomb damping being a constant force of 40 lb.

We are asked to determine (1) the maximum deflection of the spring and (2) the number of oscillations before the mass comes to rest.

Solution. The Coulomb damping force will change sign when the velocity changes sign. Therefore, two equations of motion are required:

For Dx positive

$$m\,D^2x = -c\,Dx - kx - F_\mu \tag{5-88}$$

For Dx negative

$$m\,D^2x = -c\,Dx - kx + F_\mu \tag{5-89}$$

It should be noted that Eqs. (5-88) and (5-89) are linear. Thus, this system is actually piecewise linear, and it can be solved in a piecewise manner, as indicated in Sec. 5-3. However, for the purpose of illustration, we shall use the phase-plane method in arriving at our answers.

From the problem statement we know

$$m = \frac{W}{g} = \frac{600}{386} = 1.55 \text{ lb-sec}^2/\text{in.}$$

and

$$k = 150 \text{ lb/in.}$$

and

$$F_\mu = 40 \text{ lb}$$

At this point it should be noted that the Coulomb damping will result in a dead zone and the system will remain at rest once the velocity becomes zero within the zone. From Sec. 5-3, the dead zone will be the region $-F_\mu/k < x < F_\mu/k$.

Since

$$\frac{F_\mu}{k} = \frac{40}{150} = 0.267 \text{ in.}$$

the dead zone will lie between $x > -0.267$ in. and $x < +0.267$ in.

To calculate c, we can use Eq. (4-87) and the given condition that $c = 0.5c_c$. Thus,

$$c = 0.5c_c = 0.5 \times 2\sqrt{mk} = 0.5 \times 2\sqrt{1.55 \times 150} = 15.25 \text{ lb/in./sec}$$

Substituting the above values in Eqs. (5-88) and (5-89) gives, respectively:

For Dx positive

$$1.55\,D^2x = -15.25\,Dx - 150x - 40 \tag{5-90}$$

For Dx negative

$$1.55\,D^2x = -15.25\,Dx - 150x + 40 \tag{5-91}$$

Letting $v = Dx$ and substituting from Eqs. (5-90) and (5-91) into Eq. (5-85), we find:

For v positive

$$\frac{dv}{dx} = -\frac{15.25v + 150x + 40}{1.55v} \tag{5-92}$$

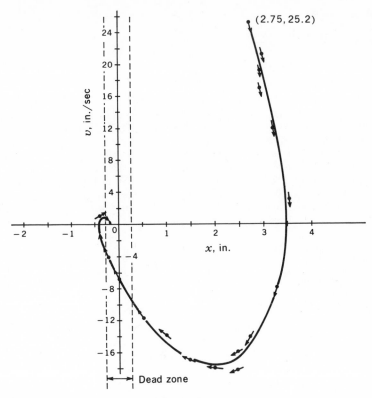

Fig. 5-10

For v negative

$$\frac{dv}{dx} = -\frac{15.25v + 150x - 40}{1.55v} \tag{5-93}$$

From the problem statement, the initial conditions are $x_0 = 2.75$ in. and $v_0 = 2.1$ fps $= 25.2$ in./sec. At the initial point,

$$\frac{dv}{dx} = -\frac{15.25 \times 25.2 + 150 \times 2.75 + 40}{1.55 \times 25.2} = -21.4 \text{ in./sec/in.}$$

As shown in Fig. 5-10, a short arrow indicating the slope -21.4 in./sec/in. is drawn through the point (2.75,25.2).

Now, rather than indiscriminately calculate values of dv/dx, we can use a little judgment and save a great deal of work. Let us locate the next arrow or arrows at $x = 3.0$ in. If we projected the trajectory as a straight line from the first point, we would find at $x = 3.0$ in. that $v \approx 20$ in./sec. However, with the spring force increasing and in the presence of damping we can expect even a greater change in velocity, and we shall therefore calculate dv/dx for values of $v \approx 19$ in./sec. For example, solving Eq. (5-92) for $x = 3.0$ and $v = 21$, 19, and 17 in./sec, we find

v, in./sec	dv/dx, in./sec/in.
21	-24.9
19	-26.4
17	-28

As can be seen, the values of dv/dx are not greatly different and one calculation would have been sufficient. Drawing the arrows, as before, and projecting the trajectory, it appears that it will intersect $x = 3.25$ in. at $v \approx 12$ in./sec. Calculating dv/dx for these conditions gives $dv/dx = -40.1$ in./sec/in. The arrow is drawn and the process is repeated until a point of zero velocity within the dead zone is reached. The arrows shown in Fig. 5-10 are the total calculations made for this example.

Summary From Fig. 5-10, the maximum deflection of the spring will be about 3.47 in. and there will be no oscillation. The mass will overshoot the equilibrium point, to about $x = -0.40$ in., but it will come to rest without the displacement again becoming positive.

chapter 6

MULTI-DEGREES-OF-FREEDOM
SYSTEMS

In reality there are no systems with a single degree of freedom because (1) there are no masses with infinite rigidity and (2) there are no massless springs. In many situations the error introduced by ignoring the additional degrees of freedom is truly negligible and the discussion in the previous two chapters can be used with confidence. In many additional cases the understanding that was gained of the relationship between the physical properties of masses, springs, and dampers to system behavior should help in qualitative analyses of existing systems where trouble has been encountered.

Nevertheless, there are a number of important situations in which an attempt to use a single-degree-of-freedom approximation can lead only to unsatisfactory—even disastrous—results. Consequently, some proficiency with systems of more than one degree of freedom is important to the designer.

As will be evident, mathematical solutions rapidly become more complex as the systems become more complex and a computer becomes a practical necessity for all except some special cases of relatively simple systems.

Multi-degrees-of-freedom systems are classified as either lumped-parameter or continuous systems. Again there are no true lumped-parameter systems—just as there are no true single-degree-of-freedom systems, but for all practical purposes we can separate many important systems into distinct elements of inertia (mass), energy storage (springs), and energy dissipation (damping). As discussed in Sec. 3-5, the number of degrees of freedom corresponds to the number of coordinates required to locate all elements of the system at any time. Therefore, since any

163

body can be located precisely in terms of three lineal and three angular coordinates, as in Fig. 3-10, it can be seen that a system with a finite number of rigid inertial elements must have a finite number of degrees of freedom.

On the other hand, a continuous system is one that cannot be broken down into masses, springs, and dampers (or their equivalents) and must be considered in its entirety as containing an infinite number of particles, thereby requiring an infinite number of coordinates and possessing an infinite number of degrees of freedom.

Both classifications will be considered in this chapter, but major emphasis will be given to the lumped-parameter systems because of their greater importance in mechanical design.

6-1. Systems with Two Degrees of Freedom Several of the most frequently encountered systems with two degrees of freedom are shown in Fig. 6-1. The coordinates indicated are not the only possibilities; but in each case two angles, two displacements, or an angle and a displacement are required to locate the inertia elements.

The system in Fig. 6-1a, consisting of two rotors with polar moments of inertia I_1 and I_2 on a shaft that is mounted in bearings, is representative of many situations found in machines of all kinds. The rotors may be gears, pulleys, flywheels, turbine or compressor rotors, electric-motor

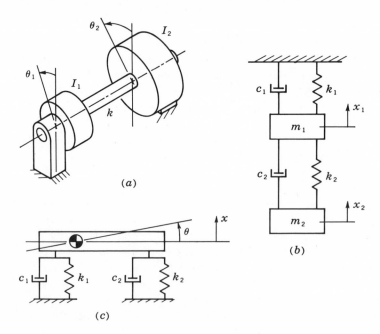

Fig. 6-1 *Systems with two degrees of freedom.*

armatures, etc.   The shaft has both lateral and torsional elasticity, and the bearing supports are also elastic.   As a result the designer must consider three types of vibrations or dynamic behavior:

1 Lateral vibrations
2 Shaft whirl
3 Torsional vibrations

Lateral vibrations and shaft whirl were considered for single-degree-of-freedom systems in Secs. 4-19 and 4-20, and torsional vibrations will now be considered.

If the inertia of the rotors is much greater than that of the shaft and the internal damping in the shaft is negligible, the system in Fig. 6-1a becomes two rotors connected by a torsional spring with a spring constant k.   In this case the units of k are usually pound-inches per radian.

Each rotor will be subjected to torques due to (1) friction in the adjacent bearing, (2) the twist of the shaft, and (3) an external forcing function, if present.

Assuming viscous damping in the bearings and letting c be the damping coefficient, we find that, in general, the magnitude of the damping torque will be

$$T_\mu = c\,D\theta \tag{6-1}$$

where $D\theta = d\theta/dt$.

The magnitude of the spring torque will be

$$T_k = k(\Delta\theta) \tag{6-2}$$

where $\Delta\theta$ = angle of twist of the shaft.

If we measure the angular displacements θ_1 and θ_2 of the rotors from the at-rest position, i.e., with neither external nor internal torques acting on the system, and if we assume that (1) $\theta_2 > \theta_1$, (2) both $D\theta_1$ and $D\theta_2$ are positive, and (3) the rotors are acted upon by forcing functions $T_1(t)$ and $T_2(t)$, respectively, the free-body diagrams of the rotors in Fig. 6-1 will be as shown in Fig. 6-2.

Applying Newton's second law, $\Sigma T = I\alpha$, to the rotors, we find:

For rotor 1
$$I_1\,D^2\theta_1 + c_1\,D\theta_1 + k(\theta_1 - \theta_2) = T_1(t) \tag{6-3}$$

For rotor 2
$$I_2\,D^2\theta_2 + c_2\,D\theta_2 + k(\theta_2 - \theta_1) = T_2(t) \tag{6-4}$$

Since both Eqs. (6-3) and (6-4) are functions of θ_1 and θ_2, we are faced with the task of solving simultaneous differential equations.   As discussed in Chap. 4, solving a single second-order differential equation with an arbitrary forcing function is not a simple matter; in this case it becomes almost impossible without the use of an analog or digital computer.   The analog computer becomes particularly valuable in solving problems of

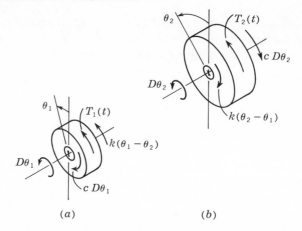

(a)   (b)

Fig. 6-2   *Free-body diagrams of rotors in Fig. 6-1a.*

this (or greater) complexity because with it only a little more effort is required than for the simplest possible case.

As for the single-degree-of-freedom systems, the most important forcing function in engineering applications is one that varies sinusoidally with time. In many cases, such as when one rotor is the turbine and the other the compressor of a gas-turbine power plant, both rotors will be subjected to periodic disturbances as the rotor blades pass the stator blades. In other cases, such as when one rotor is the flywheel of a diesel engine and the other the propeller of a ship or the rotor of a generator, the forcing function can often be assumed to act on only one of the rotors at a time.

In most applications the long-term, or steady-state, sinusoidal response is of major importance. If only one forcing function is involved, the method of complex algebra (Sec. 4-12) can be used to get a solution without undue difficulty.[1]

Assuming $T_1(t) = T_0 \sin \omega t$ and $T_2(t) = 0$, we can write

$$T_1(t) = T_0 e^{j\omega t} \tag{6-5}$$
$$T_2(t) = 0 \tag{6-6}$$
$$\theta_1 = \Theta_1 e^{j(\omega t + \phi_1)} \tag{6-7}$$
and
$$\theta_2 = \Theta_2 e^{j(\omega t + \phi_2)} \tag{6-8}$$

where Θ_1 = amplitude of θ_1
  Θ_2 = amplitude of θ_2
  ϕ_1 = phase angle of Θ_1 relative to T_0
  ϕ_2 = phase angle of Θ_2 relative to T_0

---

[1] When dealing with linear systems with more than one forcing function, the principle of superposition can be used to advantage. That is, the response to a sum of inputs will be the sum of responses to the individual inputs.

Differentiating Eqs. (6-7) and (6-8) the proper number of times, substituting from Eqs. (6-5) through (6-8) and the differentiated terms into Eqs. (6-3) and (6-4), collecting terms, and dividing through by $e^{j\omega t}$ gives, respectively,

$$(k - I_1\omega^2 + jc_1\omega)\Theta_1 e^{j\phi_1} - k\Theta_2 e^{j\phi_2} = T_0 \tag{6-9}$$

and

$$-k\Theta_1 e^{j\phi_1} + (k - I_2\omega^2 + jc_2\omega)\Theta_2 e^{j\phi_2} = 0 \tag{6-10}$$

Applying Cramer's rule to Eqs. (6-9) and (6-10) to solve for Θ_1, we find

$$\Theta_1 = \frac{\begin{vmatrix} T_0 & -ke^{j\phi_2} \\ 0 & (k - I_2\omega^2 + jc_2\omega)e^{j\phi_2} \end{vmatrix}}{\begin{vmatrix} (k - I_1\omega^2 + jc_1\omega)e^{j\phi_1} & -ke^{j\phi_2} \\ -ke^{j\phi_1} & (k - I_2\omega^2 + jc_2\omega)e^{j\phi_2} \end{vmatrix}} \tag{6-11}$$

from which

$$\Theta_1 = \frac{(k - I_2\omega^2 + jc_2\omega)T_0 e^{-j\phi_1}}{(k - I_1\omega^2 + jc_1\omega)(k - I_2\omega^2 + jc_2\omega) - k^2}$$

$$= \frac{(k - I_2\omega^2 + jc_2\omega)T_0 e^{-j\phi_1}}{I_1 I_2\omega^4 - k(I_1 + I_2)\omega^2 - c_1 c_2\omega^2 - j[(I_1 c_2 + I_2 c_1)\omega^3 - k(c_1 + c_2)\omega]} \tag{6-12}$$

As discussed in Sec. 4-12, the sum of a real and an imaginary term can be written as a magnitude and an angle, e.g.,

$$A + jB = \sqrt{A^2 + B^2}\, e^{j\alpha} \tag{6-13}$$

where

$$\alpha = \tan^{-1}\frac{B}{A} \tag{6-14}$$

Using the relationships in Eqs. (6-13) and (6-14), we can rewrite Eq. (6-12) as

$$\Theta_1 =$$

$$\frac{\sqrt{(k - I_2\omega^2)^2 + (c_2\omega)^2}\, e^{j\alpha_1} T_0 e^{-j\phi_1}}{\sqrt{[I_1 I_2\omega^4 - (kI_1 + kI_2 + c_1 c_2)\omega^2]^2 + [(I_1 c_2 + I_2 c_1)\omega^3 - k(c_1 + c_2)\omega]^2}\, e^{j\alpha_2}} \tag{6-15}$$

where

$$\alpha_1 = \tan^{-1}\frac{c_2\omega}{k - I_2\omega^2} \tag{6-16}$$

and

$$\alpha_2 = \tan^{-1}\frac{-(I_1 c_2 + I_2 c_1)\omega^2 + k(c_1 + c_2)}{I_1 I_2\omega^3 - (kI_1 + kI_2 + c_1 c_2)\omega} \tag{6-17}$$

Since Θ_1 has magnitude only, the exponential terms must cancel. Therefore,

$$\frac{e^{j\alpha_1}e^{-j\phi_1}}{e^{j\alpha_2}} = 1 \tag{6-18}$$

from which

$$\alpha_1 - \phi_1 - \alpha_2 = 0$$

or

$$\phi_1 = \alpha_1 - \alpha_2 \tag{6-19}$$

From functions of multiple angles,

$$\tan \phi_1 = \tan (\alpha_1 - \alpha_2) = \frac{\tan \alpha_1 - \tan \alpha_2}{1 + \tan \alpha_1 \tan \alpha_2} \tag{6-20}$$

Substituting from Eqs. (6-16) and (6-17) into Eq. (6-20) and simplifying gives

$$\tan \phi_1 = \frac{I_2{}^2 c_1 \omega^4 - (2k I_2 c_1 - c_1 c_2{}^2)\omega^2 + (c_1 + c_2)k^2}{I_1 I_2{}^2 \omega^5 - (2k I_1 I_2 + k I_2{}^2 - I_1 c_2{}^2)\omega^3 + (k^2 I_1 + k^2 I_2 - k c_2{}^2)\omega} \tag{6-21}$$

From Eq. (6-15) we have for the amplitude of the oscillation of rotor 1

$$\Theta_1 = \frac{\sqrt{(k - I_2 \omega^2)^2 + (c_2 \omega)^2}\, T_0}{\sqrt{[I_1 I_2 \omega^4 - (k I_1 + k I_2 + c_1 c_2)\omega^2]^2 + [(I_1 c_2 + I_2 c_1)\omega^3 - k(c_1 + c_2)\omega]^2}} \tag{6-22}$$

In a similar manner, we find

$$\Theta_2 = \frac{k T_0}{\sqrt{[I_1 I_2 \omega^4 - (k I_1 + k I_2 + c_1 c_2)\omega^2]^2 + [(I_1 c_2 + I_2 c_1)\omega^3 - k(c_1 + c_2)\omega]^2}} \tag{6-23}$$

and

$$\phi_2 = \tan^{-1} \frac{(I_1 c_2 + I_2 c_1)\omega^2 - k(c_1 + c_2)}{I_1 I_2 \omega^3 - (k I_1 + k I_2 + c_1 c_2)\omega} \tag{6-24}$$

In most cases, the phase angles are not particularly important but the variation of Θ_1 and Θ_2 with frequency and damping will be. Curves, such as in Fig. 4-20, in terms of dimensionless ratios would be useful to the designer. Unfortunately, the number of variables is now large enough that it is not possible to present information covering all cases in a two-dimensional plot. Consequently, each set of values becomes a separate case and the calculations can become laborious.

Since in most engineering applications the magnitude of the forcing function and the values of the damping coefficients can only be approximated, the efforts of the designer are usually limited to avoiding critical frequencies and possible destructive amplitudes of vibration. As discussed in Sec. 4-14, the critical frequency of a single-degree-of-freedom system was found to be close to the undamped natural frequency for reasonably low values of the damping ratio, and, thus, knowledge of the undamped natural frequency is often sufficient for predicting whether or not resonance will be a problem. To determine if the same reasoning can be applied to multiple-degrees-of-freedom systems, let us consider a particular system in detail.

For undamped free vibrations of the system in Figs. 6-1*a* and 6-2, the equations of motion become:

For rotor 1

$$I_1 D^2 \theta_1 + k(\theta_1 - \theta_2) = 0 \tag{6-25}$$

For rotor 2

$$I_2 D^2\theta_2 + k(\theta_2 - \theta_1) = 0 \tag{6-26}$$

Natural (undamped-free) vibrations of single-degree-of-freedom systems were shown in Sec. 4-4 to be simple harmonic. Intuitively, one might expect this to carry over in some manner to systems with more than one degree of freedom. Let us assume that θ_1 and θ_2 are simple harmonic motions and determine the conditions required for this to be true. Thus, we write

$$\theta_1 = \Theta_1 e^{j\omega t} \tag{6-27}$$

and
$$\theta_2 = \Theta_2 e^{j(\omega t + \beta)} \tag{6-28}$$

where β is the phase angle of Θ_2 relative to Θ_1. Differentiating Eqs. (6-27) and (6-28) the proper number of times, substituting into Eqs. (6-25) and (6-26), and collecting like terms results in

$$(k - I_1\omega^2)\Theta_1 - k\Theta_2 e^{j\beta} = 0 \tag{6-29}$$

and
$$-k\Theta_1 + (k - I_2\omega^2)\Theta_2 e^{j\beta} = 0 \tag{6-30}$$

Applying Cramer's rule to Eqs. (6-29) and (6-30) to solve for Θ_1, we find

$$\Theta_1 = \frac{\begin{vmatrix} 0 & -ke^{j\beta} \\ 0 & (k - I_2\omega^2)e^{j\beta} \end{vmatrix}}{\begin{vmatrix} k - I_1\omega^2 & -ke^{j\beta} \\ -k & (k - I_2\omega^2)e^{j\beta} \end{vmatrix}}$$

$$= \frac{0}{\omega^2[I_1 I_2\omega^2 - k(I_1 + I_2)]e^{j\beta}} \tag{6-31}$$

When the denominator of Eq. (6-31) does not equal zero, $\Theta_1 = 0$. This corresponds to the trivial case of the steady-state sinusoidal response to a forcing function with zero amplitude.

When the denominator of Eq. (6-31) equals zero, we find

$$\Theta_1 = \frac{0}{0} \tag{6-32}$$

and Θ_1 can have an infinite number of values. The denominator of Eq. (6-31) will be zero when

$$\omega^2 = 0 \tag{6-33}$$

or
$$\omega^2 = \frac{k(I_1 + I_2)}{I_1 I_2} \tag{6-34}$$

These frequencies are the only frequencies at which there can be free vibrations of the system with nonzero amplitudes of oscillation. Thus, these frequencies are the natural frequencies of the system and from Eqs.

(6-33) and (6-34) we have, respectively,

$$\omega_{n1} = 0 \tag{6-35}$$

and

$$\omega_{n2} = \sqrt{\frac{k(I_1 + I_2)}{I_1 I_2}} \tag{6-36}$$

The term *principal mode* is used to describe the configuration of a system when vibrating at one of its natural frequencies. In general, the phase angle between the amplitudes of the several parts is the feature that distinguishes one principal mode from another.

To determine the values of β for the two principal modes, let us take a closer look at Eq. (6-29). Since by definition Θ_1 and Θ_2 are magnitudes and, therefore, always positive numbers, the ratio must be positive. From Eq. (6-29),

$$\frac{\Theta_1}{\Theta_2} = \frac{k}{k - I_1 \omega^2} e^{j\beta} > 0 \tag{6-37}$$

At ω_{n1}, substituting from Eq. (6-35) into Eq. (6-37) leads to

$$\frac{\Theta_1}{\Theta_2} = e^{j\beta} \tag{6-38}$$

which is positive when

$$\frac{\Theta_1}{\Theta_2} = 1 \tag{6-39}$$

and

$$\beta = 0 \tag{6-40}$$

From Eqs. (6-39) and (6-40) we see that at $\omega_{n1} = 0$ the two rotors rotate together as a solid member.

At ω_{n2}, substituting from Eq. (6-36) into Eq. (6-37) leads to

$$\frac{\Theta_1}{\Theta_2} = -\frac{I_2}{I_1} e^{j\beta} \tag{6-41}$$

which is positive when

$$\frac{\Theta_1}{\Theta_2} = \frac{I_2}{I_1} \tag{6-42}$$

and

$$\beta = \pm 180° \tag{6-43}$$

For $\beta = \pm 180°$, θ_1 and θ_2 must always be in opposite directions and the rotors will be out of phase.

If the shaft has a uniform diameter between the rotors, the twist will be constant along the shaft and the information in Eqs. (6-42) and (6-43) can be presented graphically, as in Fig. 6-3. The point on the shaft where the angle of rotation is always zero is called a *node*. The principal modes of free vibration of multi-degrees-of-freedom systems are most

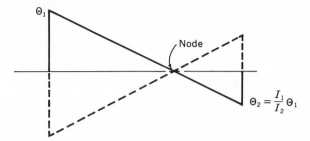

Fig. 6-3 *Node location for principal-mode vibrations of a two-rotor torsional system.*

readily described by the location of the nodes.   If either amplitude is made equal to 1.0, the curve in Fig. 6-3 becomes the *normal-mode* curve.

Since a vibration at a frequency of zero cycles per unit time behaves as a constant-speed system, in any finite length of time, it is customary to ignore the zero frequency and to consider only the second frequency as *the* natural frequency.   This type of system is often termed *degenerate*

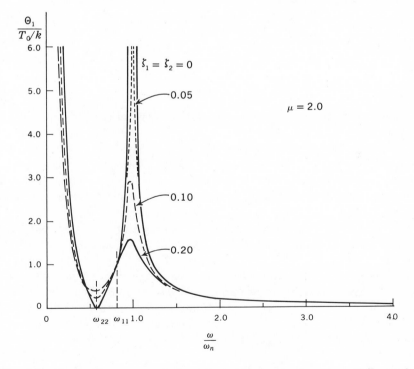

Fig. 6-4 *Steady-state response of a two-rotor torsional system to a sinusoidally varying torque. Rotor on which torque acts.*

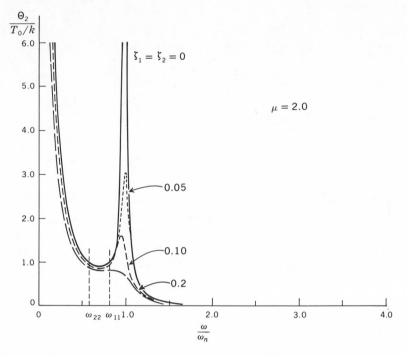

Fig. 6-5 *Steady-state response of a two-rotor torsional system to a sinusoidally varying torque. Rotor at other end of shaft from the rotor on which the torque acts.*

because the number of nonzero natural frequencies is less than the degrees of freedom.

To continue our investigation of the relationship of the undamped natural frequency to the steady-state sinusoidal response of systems with more than one degree of freedom, let us consider Figs. 6-4 and 6-5, which show in nondimensional form the results for a particular system.[1]   The curves were plotted using results obtained by numerical solutions of Eqs. (6-22) and (6-23) for the case with zero damping and by analog-computer solutions of the simultaneous differential equations, Eqs. (6-3) and (6-4), for the other cases.   The reasons for using the analytical solution for zero damping were: (1) It is difficult to use an analog computer in the case of forced vibration with zero damping because the transient introduced during starting the solution never dies out, and (2) the analytical solution is fairly simple for zero damping and it also affords a good check on the results obtained by use of the analog computer.

[1] R. M. Phelan, An Introduction to the Dynamics of Shaft Systems, *Engineering Proc. P-41, High Speed Flexible Couplings*, pp. 61–80, The Pennsylvania State University, 1963.

To provide a maximum degree of generality by use of nondimensional ratios, the following nomenclature was used:

$\mu = I_2/I_1$

ω_n = undamped natural frequency, Eq. (6-36)

$\omega_{11} = \sqrt{k/I_1}$ = natural frequency when rotor 2 is held fixed

$\omega_{22} = \sqrt{k/I_2}$ = natural frequency when rotor 1 is held fixed

$\zeta_1 = c_1/2\sqrt{I_1 k}$ = damping ratio when rotor 2 is held fixed

$\zeta_2 = c_2/2\sqrt{I_2 k}$ = damping ratio when rotor 1 is held fixed

T_0/k = angle of rotation if one rotor is held fixed and a static torque T_0 is applied to the other rotor

The high ratios of amplitudes at low values of ω/ω_n may be a little disturbing at first glance.   Actually, this just means that the rotors turn together in one direction as long as the torque is applied, and the twist of the shaft is negligible in relation to the angle through which the rotors turn.   For example, for $\omega/\omega_n = 0$, the torque never reverses and both Θ_1 and $\Theta_2 \to \infty$ because the shaft rotates in one direction forever.   Consequently, the only real critical region is located about $\omega/\omega_n = 1$, and damping can be seen to have only a minor effect on the location of the maximum amplitude ratios.   Therefore, as for single-degree-of-freedom systems, the undamped natural frequency is the most important characteristic of the system and operation near it should be avoided.

As a matter of practical interest, since bearings are designed to provide minimum resistance to rotation, the damping ratios in real situations will be almost zero.   For example, a value of $\zeta_1 = 0.00001$ would be reasonable for a well-designed journal bearing in a typical application.

If operating conditions indicate that a much lower natural frequency would be helpful and/or increased damping would be desirable, the solution is often to insert a flexible coupling[1] between the rotors or to add a vibration absorber (Secs. 6-2 through 6-4) to the system.

The system in Fig. 6-6a, consisting of two simple pendulums connected by a massless linear spring, can behave in a particularly interesting manner when $l_1 = l_2 = l$, $a_1 = a_2 = a$, and $m_1 = m_2 = m$, as shown in Fig. 6-6b.   Under these conditions, and if we assume that the axis of the spring is always horizontal, or so close to it that the effect of the angle can be neglected, the free-body diagrams showing the forces acting on the pendulums will be as shown in Fig. 6-7.   Applying Newton's second law

[1] *Engineering Proc.* P-41, *High Speed Flexible Couplings*, The Pennsylvania State University, 1963.

R. M. Phelan, "Fundamentals of Mechanical Design," 2d ed., chap. 11, McGraw-Hill Book Company, New York, 1962.

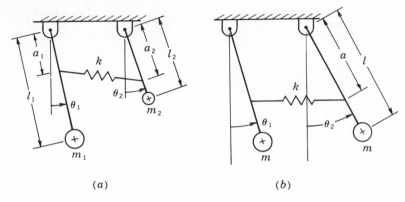

(a)   (b)

Fig. 6-6 *Coupled pendulums.*

to rotation about O_1, we find for pendulum 1

$$I\,D^2\theta_1 = ml^2\,D^2\theta_1 = -Wl\sin\theta_1 - ka^2(\sin\theta_1 - \sin\theta_2)\cos\theta_1 \quad (6\text{-}44)$$

As the reader will recognize, Eq. (6-44) is nonlinear, but it can be linearized for small angles of oscillation by using the linear approximations for $\sin\theta$ and $\cos\theta$ about the equilibrium point, $\theta = 0$. From Examples 5-1 and 5-3, for small angles,

$$\sin\theta \approx \theta \quad (6\text{-}45)$$
and $$\cos\theta \approx 1 \quad (6\text{-}46)$$

Substituting from Eqs. (6-45) and (6-46) into Eq. (6-44) gives

$$ml^2\,D^2\theta_1 = -Wl\theta_1 - ka^2(\theta_1 - \theta_2) \quad (6\text{-}47)$$

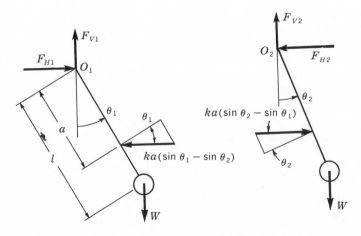

Fig. 6-7 *Free-body diagrams of coupled identical pendulums.*

Substituting mg for W and rearranging Eq. (6-47) results in

$$ml^2 D^2\theta_1 + mgl\theta_1 + ka^2\theta_1 - ka^2\theta_2 = 0 \qquad (6\text{-}48)$$

In a similar manner we find for pendulum 2

$$ml^2 D^2\theta_2 + mgl\theta_2 + ka^2\theta_2 - ka^2\theta_1 = 0 \qquad (6\text{-}49)$$

As above, letting $\theta_1 = \Theta_1 e^{j\omega t}$ and $\theta_2 = \Theta_2 e^{j(\omega t+\beta)}$ leads, for Eqs. (6-48) and (6-49), respectively, to

$$(mgl + ka^2 - \omega^2 ml^2)\Theta_1 - ka^2\Theta_2 e^{j\beta} = 0 \qquad (6\text{-}50)$$

and $\qquad -ka^2\Theta_1 + (mgl + ka^2 - \omega^2 ml^2)\Theta_2 e^{j\beta} = 0 \qquad (6\text{-}51)$

Applying Cramer's rule to Eqs. (6-50) and (6-51) leads to

$$\omega_{n1} = \sqrt{\frac{g}{l}} \qquad (6\text{-}52)$$

and $\qquad\qquad \omega_{n2} = \sqrt{\frac{g}{l} + \frac{2ka^2}{ml^2}} \qquad (6\text{-}53)$

Substituting from Eq. (6-52) into Eq. (6-50) and solving for the ratio of amplitudes gives

$$\left.\frac{\Theta_1}{\Theta_2}\right|_{\omega_{n1}} = e^{j\beta} \qquad (6\text{-}54)$$

from which

$$\left.\frac{\Theta_1}{\Theta_2}\right|_{\omega_{n1}} = 1 \qquad (6\text{-}55)$$

and $\qquad\qquad \beta_{\omega_{n1}} = 0° \qquad (6\text{-}56)$

Similarly, we find at ω_{n2}

$$\left.\frac{\Theta_1}{\Theta_2}\right|_{\omega_{n2}} = -e^{j\beta} \qquad (6\text{-}57)$$

from which

$$\left.\frac{\Theta_1}{\Theta_2}\right|_{\omega_{n2}} = 1 \qquad (6\text{-}58)$$

and $\qquad\qquad \beta_{\omega_{n2}} = \pm 180° \qquad (6\text{-}59)$

From Eqs. (6-55) and (6-56) we see that at ω_{n1} the pendulums oscillate in phase with identical amplitudes and, consequently, the spring is neither stretched nor compressed and we have the equivalent of a simple pendulum—for which $\omega_n = \sqrt{g/l}$.

From Eqs. (6-58) and (6-59) we see that at ω_{n2} the pendulums oscillate out of phase with identical amplitudes and, consequently, the node is at the midlength of the spring. As shown in Fig. 6-8, this is the equivalent of having a single pendulum and half the spring. The new spring rate is twice that of the whole spring.

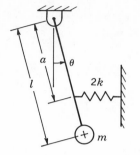

Fig. 6-8 *Single-pendulum equivalent for higher-frequency principal-mode vibrations of a system consisting of coupled identical pendulums.*

In general, a disturbance will result in a system oscillating in all of its natural modes at the same time. One of the most significant characteristics of *linear systems* is that superposition applies and the resulting free vibrations of undamped systems can always be shown to be the sum of natural-mode vibrations. For example, let us consider the system in Fig. 6-6b when its initial conditions are $\theta_1 = 0$ and $\theta_2 = \Theta$. These initial conditions correspond to the sum of natural-mode oscillations of the two modes as follows:

For mode 1

$$\omega_{n1} = \sqrt{\frac{g}{l}} \text{ and } \beta = 0°$$

$$\theta_1 = \frac{\Theta}{2} e^{j\omega_{n1}t} \quad \text{and} \quad \theta_2 = \frac{\Theta}{2} e^{j\omega_{n1}t} \tag{6-60}$$

or $\qquad \theta_1 = \frac{\Theta}{2} \cos \omega_{n1}t \quad \text{and} \quad \theta_2 = \frac{\Theta}{2} \cos \omega_{n1}t \tag{6-61}$

For mode 2

$$\omega_{n2} = \sqrt{\frac{g}{l} + \frac{2ka^2}{ml^2}} \quad \text{and} \quad \beta = 180°$$

$$\theta_1 = -\frac{\Theta}{2} e^{j\omega_{n2}t} \quad \text{and} \quad \theta_2 = \frac{\Theta}{2} e^{j\omega_{n2}t} \tag{6-62}$$

or $\qquad \theta_1 = -\frac{\Theta}{2} \cos \omega_{n2}t \quad \text{and} \quad \theta_2 = \frac{\Theta}{2} \cos \omega_{n2}t \tag{6-63}$

From Eqs. (6-61) and (6-63) we have by superposition

$$\theta_1 = \frac{\Theta}{2} \cos \omega_{n1}t - \frac{\Theta}{2} \cos \omega_{n2}t \tag{6-64}$$

and $\qquad \theta_2 = \frac{\Theta}{2} \cos \omega_{n1}t + \frac{\Theta}{2} \cos \omega_{n2}t \tag{6-65}$

Figure 6-9 shows a graphical representation of the superposition of the two modes at $t = 0$.

Rearranging Eqs. (6-64) and (6-65) gives

$$\theta_1 = \frac{\Theta}{2}(\cos \omega_{n1}t - \cos \omega_{n2}t) \qquad (6\text{-}66)$$

and

$$\theta_2 = \frac{\Theta}{2}(\cos \omega_{n1}t + \cos \omega_{n2}t) \qquad (6\text{-}67)$$

Referring to tables of trigonometric relationships, we find we can write

$$\cos \omega_{n1}t - \cos \omega_{n2}t = -2\left(\sin \frac{\omega_{n1} + \omega_{n2}}{2}t\right)\sin \frac{\omega_{n1} - \omega_{n2}}{2}t \qquad (6\text{-}68)$$

and

$$\cos \omega_{n1}t + \cos \omega_{n2}t = 2\left(\cos \frac{\omega_{n1} + \omega_{n2}}{2}t\right)\cos \frac{\omega_{n1} - \omega_{n2}}{2}t \qquad (6\text{-}69)$$

Since the order in which we have been designating the natural frequencies is such that $\omega_{n1} < \omega_{n2}$, etc., $\omega_{n1} - \omega_{n2}$ will be negative. To eliminate the need for considering negative frequencies, let us make use of the fact that $\sin(-\theta) = -\sin \theta$ and $\cos(-\theta) = \cos \theta$ and rewrite Eqs. (6-68) and (6-69) as, respectively,

$$\cos \omega_{n1}t - \cos \omega_{n2}t = 2\left(\sin \frac{\omega_{n1} + \omega_{n2}}{2}t\right)\sin \frac{\omega_{n2} - \omega_{n1}}{2}t \qquad (6\text{-}70)$$

and

$$\cos \omega_{n1}t + \cos \omega_{n2}t = 2\left(\cos \frac{\omega_{n1} + \omega_{n2}}{2}t\right)\cos \frac{\omega_{n2} - \omega_{n1}}{2}t \qquad (6\text{-}71)$$

Substituting from Eqs. (6-70) and (6-71) into Eqs. (6-66) and (6-67), respectively, we find

$$\theta_1 = \Theta\left(\sin \frac{\omega_{n1} + \omega_{n2}}{2}t\right)\sin \frac{\omega_{n2} - \omega_{n1}}{2}t \qquad (6\text{-}72)$$

and

$$\theta_2 = \Theta\left(\cos \frac{\omega_{n1} + \omega_{n2}}{2}t\right)\cos \frac{\omega_{n2} - \omega_{n1}}{2}t \qquad (6\text{-}73)$$

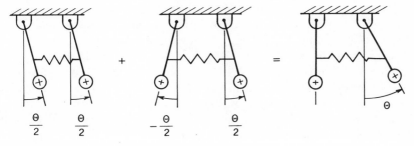

Fig. 6-9 *Superposition of principal modes of vibration of a system consisting of coupled identical pendulums.*

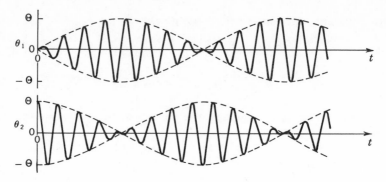

Fig. 6-10   *Beat phenomenon for system in Fig. 6-9 when $\omega_{n1} \approx \omega_{n2}$.*

In Eqs. (6-72) and (6-73) it can be seen that both θ_1 and θ_2 are the products of two harmonically varying terms, one with a frequency equal to the average of the two natural frequencies and the other with a frequency equal to one-half the difference between the natural frequencies. When the natural frequencies are almost the same, the average frequency will be about the same as either of the natural frequencies and the difference frequency will be very low. The result is a relatively high frequency oscillation with a magnitude that varies slowly between zero and Θ, as shown in Fig. 6-10. This phenomenon is called *beating*, and the beat frequency will be

$$\omega_{\text{beat}} = 2 \frac{\omega_{n2} - \omega_{n1}}{2} = \omega_{n2} - \omega_{n1} \qquad (6\text{-}74)$$

or
$$f_{\text{beat}} = f_{n2} - f_{n1} \qquad (6\text{-}75)$$

For the system in Fig. 6-6b, from Eqs. (6-52) and (6-53), it can be noted that the beating phenomenon will be pronounced for low values of ka^2/ml^2.

From Eqs. (6-72) and (6-73) or Fig. 6-10 it can be seen that, when θ_1 is a maximum, θ_2 is zero, and vice versa. Thus, this system not only has the continuous exchange of energy between the forms of kinetic and potential energy associated with free vibrations, but also has a continuous transfer of the exchange of energy between the pendulums.

6-2. Vibration Absorbers   It is not too unusual to find after a machine is built that the vibration of some particular part becomes excessive at one or more operating speeds. In some of these situations it is not feasible to redesign the machine or the offending parts to try to move the resonance frequency outside the operating range, and in most cases it is also not practical to introduce damping in the place where there is relative motion. In other cases, particularly crankshafts of internal-combustion engines, the dimensional and weight limitations are so severe and the frequency of the forcing function is so high that it is not possible to design a member that will be stiff enough to keep the lowest natural frequency well above the highest forcing frequency.

As discussed in Chap. 2, the unwanted dynamic forces resulting from the motion of rigid bodies can be minimized and often eliminated by adding or taking away mass at the proper places. The purpose was to introduce dynamic forces with magnitudes and directions tending to cancel the unwanted force, and the procedure was called balancing.

When unwanted vibrations are present, we are again dealing with a dynamic system, but displacement as well as force is involved because the systems can no longer be assumed to be made up of rigid bodies. As before, opposite forces or torques will be effective, but we now have an additional possibility. Vibration always implies motion of points relative to each other, and wherever there is relative motion there is a possibility of dissipating energy by introducing damping. As discussed in Secs. 4-12 through 4-17, a small degree of damping can be highly effective in reducing the amplitude of vibration at resonance. Devices that can be added to a system to reduce vibration amplitude by introducing equal and opposite dynamic forces and/or by introducing damping to dissipate energy are called *vibration absorbers*. The principles underlying operation and design of the most important types will be discussed below.

The mass-spring system in Fig. 6-11*a* represents a machine or a part of a machine that has been found to vibrate with excessive amplitude under conditions that are otherwise normal. Assuming that neither the mass nor the spring can be changed and that it is impractical to add damping in parallel with the spring, it becomes apparent that improvement in operation can be achieved only by adding some element with the right dynamic

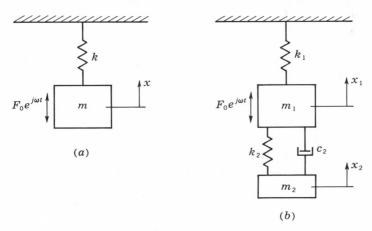

Fig. 6-11 (a) *Single-degree-of-freedom system with sinusoidal force disturbance;* (b) *system in* (a) *with addition of vibration absorber consisting of m_2, k_2, and c_2.*

properties to the system.   The simplest device one can think of to add is another mass-spring-damper system, as shown in Fig. 6-11*b*, where the subscripts 1 refer to the original system and subscripts 2 refer to the vibration absorber.   The main questions are (1) to what degree can the amplitude of the main mass be decreased and (2) what are the optimum values of *m*, *c*, and *k* for the absorber?   An additional question of practical importance is what will be the maximum deflection of the absorber spring?

Applying Newton's second law to the forces acting on m_1, we find

$$m_1 D^2 x_1 + c_2 D(x_1 - x_2) + k_1 x_1 + k_2(x_1 - x_2) = F_0 e^{j\omega t} \quad (6\text{-}76)$$

and for the forces acting on m_2,

$$m_2 D^2 x_2 + c_2 D(x_2 - x_1) + k_2(x_2 - x_1) = 0 \quad (6\text{-}77)$$

Following the usual procedure, we let

$$x_1 = X_1 e^{j(\omega t + \phi_1)} \quad (6\text{-}78)$$

and
$$x_2 = X_2 e^{j(\omega t + \phi_2)} \quad (6\text{-}79)$$

Substituting values of x_1 and x_2 and their derivatives from Eqs. (6-78) and (6-79) into Eqs. (6-76) and (6-77) and simplifying gives, respectively,

$$(-m_1\omega^2 + k_1 + k_2 + jc_2\omega)X_1 e^{j\phi_1} - (k_2 + jc_2\omega)X_2 e^{j\phi_2} = F_0 \quad (6\text{-}80)$$

and
$$-(k_2 + jc_2\omega)X_1 e^{j\phi_1} + (-m_2\omega^2 + k_2 + jc_2\omega)X_2 e^{j\phi_2} = 0 \quad (6\text{-}81)$$

Applying Cramer's rule to Eqs. (6-80) and (6-81), we find, after considerable algebraic manipulation,

$$X_1 = \frac{(k_2 - m_2\omega^2 + jc_2\omega)e^{-j\phi_1}F_0}{m_1 m_2 \omega^4 - (m_1 k_2 + m_2 k_1 + m_2 k_2)\omega^2 + k_1 k_2 \\ - j[(m_1 c_2 + m_2 c_2)\omega^3 - k_1 c_2 \omega]} \quad (6\text{-}82)$$

and

$$X_2 = \frac{(k_2 + jc_2\omega)e^{-j\phi_2}F_0}{m_1 m_2 \omega^4 - (m_1 k_2 + m_2 k_1 + m_2 k_2)\omega^2 + k_1 k_2 \\ - j[(m_1 c_2 + m_2 c_2)\omega^3 - k_1 c_2 \omega]} \quad (6\text{-}83)$$

Equation (6-82) can be simplified by combining the real and imaginary terms into magnitudes and angles, as in Sec. 4-12, to give

$$X_1 = \frac{\sqrt{(k_2 - m_2\omega^2)^2 + (c_2\omega)^2}\; e^{j\alpha_1}e^{-j\phi_1}F_0}{\sqrt{[m_1 m_2 \omega^4 - (m_1 k_2 + m_2 k_1 + m_2 k_2)\omega^2 + k_1 k_2]^2 \\ + [(m_1 c_2 + m_2 c_2)\omega^3 - k_1 c_2 \omega]^2}\; e^{j\alpha_2}} \quad (6\text{-}84)$$

where
$$\alpha_1 = \tan^{-1}\frac{c_2\omega}{k_2 - m_2\omega^2} \quad (6\text{-}85)$$

and
$$\alpha_2 = \tan^{-1}\frac{-[(m_1 c_2 + m_2 c_2)\omega^3 - k_1 c_2 \omega]}{m_1 m_2 \omega^4 - (m_1 k_2 + m_2 k_1 + m_2 k_2)\omega^2 + k_1 k_2} \quad (6\text{-}86)$$

Since F_0 and X_1 have magnitude only,

$$\alpha_1 - \phi_1 - \alpha_2 = 0 \tag{6-87}$$

Therefore,

$$\phi_1 = \alpha_1 - \alpha_2 \tag{6-88}$$

Equation (6-88) can be solved for ϕ, as in Sec. 6-1, but there is little need for the information and we shall not bother with it at this time.

The deflection of the absorber spring will be

$$\delta_{s_2} = x_2 - x_1 \tag{6-89}$$

Substituting from Eqs. (6-82) and (6-83) into Eqs. (6-78) and (6-79), respectively, results in

$$x_1 = \frac{(k_2 - m_2\omega^2 + jc_2\omega)F_0e^{j\omega t}}{m_1m_2\omega^4 - (m_1k_2 + m_2k_1 + m_2k_2)\omega^2 + k_1k_2 \\ \qquad - j[(m_1c_2 + m_2c_2)\omega^3 - k_1c_2\omega]} \tag{6-90}$$

and

$$x_2 = \frac{(k_2 + jc_2\omega)F_0e^{j\omega t}}{m_1m_2\omega^4 - (m_1k_2 + m_2k_1 + m_2k_2)\omega^2 + k_1k_2 \\ \qquad - j[(m_1c_2 + m_2c_2)\omega^3 - k_1c_2\omega]} \tag{6-91}$$

Then substituting from Eqs. (6-90) and (6-91) into Eq. (6-89) and simplifying gives

$$\delta_{s_2} = \frac{m_2\omega^2F_0e^{j\omega t}}{m_1m_2\omega^4 - (m_1k_2 + m_2k_1 + m_2k_2)\omega^2 + k_1k_2 \\ \qquad - j[(m_1c_2 + m_2c_2)\omega^3 - k_1c_2\omega]} \tag{6-92}$$

Therefore, the amplitude of the spring deflection Δ_s will be

$$\Delta_s = \frac{m_2\omega^2F_0}{\sqrt{[m_1m_2\omega^4 - (m_1k_2 + m_2k_1 + m_2k_2)\omega^2 + k_1k_2]^2 \\ \qquad + [(m_1c_2 + m_2c_2)\omega^3 - k_1c_2\omega]^2}} \tag{6-93}$$

Equations (6-84) and (6-93) can be used as they stand, but since we are dealing with the addition of a separate single-degree-of-freedom system to an existing one, we can develop a better understanding of what happens if each system is considered separately, in so far as is possible. For example, we can write

$$\frac{k_1}{m_1} = \omega_1^2 \tag{6-94}$$

where ω_1 is the natural frequency of the original system,

$$\frac{k_2}{m_2} = \omega_2^2 \tag{6-95}$$

where ω_2 is the natural frequency of the absorber by itself, that is, m_1 is held stationary, and

$$\frac{c_2}{k_2} = \frac{2\zeta}{\omega_2} \tag{6-96}$$

where ζ is the damping ratio for the absorber by itself.
An additional ratio that will be useful is that of the masses,

$$\mu = \frac{m_2}{m_1} \tag{6-97}$$

If, in Eqs. (6-84) and (6-93) we factor k_2 out of the numerator and k_1k_2 out of the denominator, substitute from Eqs. (6-94) through (6-97), and solve for $X_1/(F_0/k_1)$ and $\Delta_s/(F_0/k_1)$, respectively, we find

$$\frac{X_1}{F_0/k_1} = \frac{\sqrt{\left[1 - \left(\frac{\omega}{\omega_2}\right)^2\right]^2 + \left(2\zeta\frac{\omega}{\omega_2}\right)^2}}{\sqrt{\left[\frac{\omega^4}{\omega_1^2\omega_2^2} - (1+\mu)\left(\frac{\omega}{\omega_1}\right)^2 - \left(\frac{\omega}{\omega_2}\right)^2 + 1\right]^2 + 4\zeta^2\left[\frac{(1+\mu)\omega^3}{\omega_1^2\omega_2} - \frac{\omega}{\omega_2}\right]^2}} \tag{6-98}$$

and

$$\frac{\Delta_s}{F_0/k_1} = \frac{\left(\frac{\omega}{\omega_2}\right)^2}{\sqrt{\left[\frac{\omega^4}{\omega_1^2\omega_2^2} - (1+\mu)\left(\frac{\omega}{\omega_1}\right)^2 - \left(\frac{\omega}{\omega_2}\right)^2 + 1\right]^2 + 4\zeta^2\left[\frac{(1+\mu)\omega^3}{\omega_1^2\omega_2} - \frac{\omega}{\omega_2}\right]^2}} \tag{6-99}$$

In general, the natural frequency of the main system ω_1 will be known and the designer must specify the optimum values for m_2, c_2, and k_2, or, as used in Eqs. (6-98) and (6-99), the optimum values of ω_2, μ, and ζ. The main difficulty here, as in most optimization problems, is to decide with respect to what is the system to be optimized? In many engineering problems the optimum answer is the one that is most economical, everything considered; in others it may be related to size, weight, convenience, appearance, etc.

In general, the factors are different for each situation and thus no single set of values can apply in every case. Nevertheless, we can say that, for practical reasons, (1) the absorber should be small relative to the main system and (2) the spring deflection cannot be too large. Requirement 1 means that μ should be small, for example, $\mu \leq 0.1$, and requirement 2 is

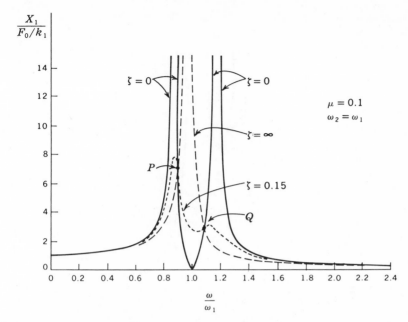

Fig. 6-12 *Steady-state response of the main system to a sinusoidal force disturbance when the absorber is tuned to the natural frequency of the main system and the mass of the absorber is one-tenth that of the main system.*

closely related to the practical considerations of stress, stability, etc., encountered in designing springs.[1]

Examination of Eqs. (6-98) and (6-99) shows that they will be simplified somewhat if $\omega_2 = \omega_1$. Such a system is called a *tuned system* because the absorber is "tuned" to the natural frequency of the main system. Several curves for a particular tuned vibration absorber are presented in Figs. 6-12 and 6-13.   The mass ratio is 0.1 and the damping ratios are 0, 0.15, and ∞.   A number of observations are worth noting:

1   When $\zeta = 0$, there are two frequencies at which $X_1/(F_0/k_1) = \infty$, although $X_1/(F_0/k_1) = 0$ at $\omega/\omega_1 = 1.0$.
2   When $0 < \zeta < \infty$, $X_1/(F_0/k_1)$ remains finite at all times.
3   All curves pass through the points P and Q.
4   When $\zeta = \infty$, the system has only one critical frequency.

[1] A. M. Wahl, "Mechanical Springs," 2d ed., McGraw-Hill Book Company, New York, 1963.

N. P. Chironis (ed.), "Spring Design and Application," McGraw-Hill Book Company, New York, 1961.

Phelan, "Fundamentals of Mechanical Design," chap. 9.

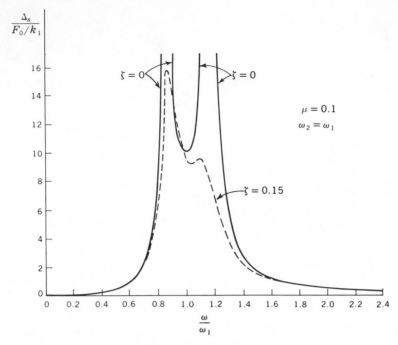

Fig. 6-13 *Maximum deflection of absorber spring for a system with the response curves in Fig. 6-12.*

From observation 1, it appears that the undamped vibration absorber will be absolutely effective when the system is operating at the natural frequency of the main system. Actually, a more general statement can be made by noting that when damping is zero, the numerator of Eq. (6-98) will be zero when $\omega/\omega_2 = 1$. Thus, *the undamped absorber will be absolutely effective whenever, and only whenever, the natural frequency of the absorber corresponds to the frequency of the forcing function.*

Observing that the main mass can be held motionless for finite values of F_0 leads to the conclusion that the effective inertia of the undamped absorber becomes infinite at its point of attachment when the frequency of the forcing function equals the natural frequency of the absorber. It should be noted that the improvement in operation at the single natural frequency of the main system has been accomplished at the expense of introducing an additional natural frequency such that there is now a resonant frequency on each side of, and not very far from, the original resonant frequency of the main system. Thus, *the undamped vibration absorber is useful only when operation is at or almost at a constant speed.* Examples are machines driven by synchronous motors, or other types of motors with slips not much greater than 10 percent, and devices, such as some types of hair clippers and electric shavers, that operate as vibrating systems with a 120-cps forcing function due to the variation of the magnetic field excited by 60-cycle alternating current.

Solving Eq. (6-99) for the case of an undamped vibration absorber tuned to the frequency of the forcing function, that is, $\omega = \omega_2$, we find

$$\frac{\Delta_s}{F_0/k_1}\bigg|_{\omega_2} = \frac{1}{\mu(\omega_2/\omega_1)^2} = \frac{1}{k_2/k_1} \qquad (6\text{-}100)$$

from which

$$\Delta_s = \frac{F_0}{k_2} \qquad (6\text{-}101)$$

or

$$\Delta_s k_2 = F_0 \qquad (6\text{-}102)$$

Therefore, from Eq. (6-102), we can conclude that *at the natural frequency of the absorber, the exciting force goes into deflecting the absorber spring* rather than into moving the main mass. If the phase angle for Δ_s is determined for this condition, it will be found to equal $-180°$. These conclusions can be deduced on a logical basis by noting that since the main mass remains stationary at $\omega = \omega_2$, there must be a reaction force that is equal and opposite to the forcing function at all times and the spring deflection is the only possible source for the reaction force.

It should be noted that if the resultant spring force and the exciting force are not collinear, there will be an unbalanced couple even though the forces are balanced. From observation 2 it can be seen that the presence of damping in the absorber permits its use over the entire frequency range, provided the maximum values of X_1 and Δ_s can be tolerated.

Although, in the interest of simplicity, curves for only three values of damping ratio are presented in Figs. 6-12 and 6-13, they cover all possible values and we can conclude that every curve will pass through the points P and Q. Thus, for $\omega_2 = \omega_1$, it can be seen that the optimum damping would be the value giving a curve that has its maximum at P. The value of optimum damping can be found, but our time will be more profitably spent in first seeing what can be done to decrease the magnitude at P. Since P and Q are determined by the intersection of the curves for $\zeta = 0$ and $\zeta = \infty$, we can see that X_P will decrease and X_Q will increase if the curve for $\zeta = 0$ is shifted to the left relative to the curve for $\zeta = \infty$. The only choices open for the designer are values of m_2 and k_2. Thus, the variables are the relationship between the masses, μ, and the relationship between the natural frequencies of the main system and the absorber, ω_2/ω_1.

Considering that μ, in general, must be small and the value will be specified, let us consider it as the independent variable and ω_2/ω_1 as the dependent variable. Since shifting the undamped curve to the left results in an increase in X_Q, as well as a decrease in X_P, it seems logical to expect the optimum conditions will be close to those resulting in $X_P = X_Q$. Therefore, let us write

$$\frac{\omega_2}{\omega_1} = f(\mu) \tag{6-103}$$

or, more simply,

$$\frac{\omega_2}{\omega_1} = f \tag{6-104}$$

where $f = f(\mu)$, and determine the value of f that will make $X_P = X_Q$. For $\zeta = \infty$ and $\omega_2 = \omega_1 f$, Eq. (6-98) becomes

$$\frac{X_1}{F_0/k_1}\bigg|_{\zeta=\infty} = \frac{1}{|(1+\mu)(\omega/\omega_1)^2 - 1|} \tag{6-105}$$

which is independent of f. The absolute-value signs complicate matters somewhat.

For $X_P = X_Q$, we can write, from Eq. (6-105),

$$\frac{1}{|(1+\mu)(\omega_P/\omega_1)^2 - 1|} = \frac{1}{|(1+\mu)(\omega_Q/\omega_1)^2 - 1|} \tag{6-106}$$

From Fig. 6-12 we can see that $\omega_P/\omega_1 < 1$ and $\omega_Q/\omega_1 > 1$. Thus, the right-hand part of Eq. (6-106) is always positive and the absolute-value signs can be dropped. The equation then becomes either

$$\frac{1}{(1+\mu)(\omega_P/\omega_1)^2 - 1} = \frac{1}{(1+\mu)(\omega_Q/\omega_1)^2 - 1} \tag{6-107}$$

or

$$\frac{1}{-(1+\mu)(\omega_P/\omega_1)^2 + 1} = \frac{1}{(1+\mu)(\omega_Q/\omega_1)^2 - 1} \tag{6-108}$$

If Eq. (6-107) is used, the answer is $\omega_P/\omega_1 = \omega_Q/\omega_1$. This result applies to the trivial case where $X_P/(F_0/k_1) = X_Q/(F_0/k_1) = 1$ at $\omega/\omega_1 = 0$. Therefore, Eq. (6-108) is the appropriate equation and, after rearranging terms, we find

$$\left(\frac{\omega_P}{\omega_1}\right)^2 + \left(\frac{\omega_Q}{\omega_1}\right)^2 = \frac{2}{1+\mu} \tag{6-109}$$

A somewhat similar procedure could be applied to the case for zero damping, but the algebraic manipulations will be decreased and the problems resulting from there being four nonzero values of ω/ω_1 that give equal values of $X_1/(F_0/k_1)$ will be eliminated if a slightly different approach is used. For example, at ω_P/ω_1 and ω_Q/ω_1, $X_1/(F_0/k_1)$ must be identical for both the case with infinite damping and that with zero damping.

Substituting $\zeta = 0$ and $\omega_2 = \omega_1 f$ into Eq. (6-98) gives

$$\frac{X_1}{F_0/k_1}\bigg|_{\zeta=0} = \frac{|1 - (1/f^2)(\omega/\omega_1)^2|}{|(1/f^2)(\omega/\omega_1)^4 - (1 + \mu + 1/f^2)(\omega/\omega_1)^2 + 1|} \quad (6\text{-}110)$$

Since, at ω_P/ω_1 and ω_Q/ω_1, $X_1/(F_0/k_1)$ will be the same for $\zeta = \infty$ and $\zeta = 0$, from Eqs. (6-105) and (6-110), we can write

$$\frac{1}{|(1 + \mu)(\omega/\omega_1)^2 - 1|} = \frac{|1 - (1/f^2)(\omega/\omega_1)^2|}{|(1/f^2)(\omega/\omega_1)^4 - (1 + \mu + 1/f^2)(\omega/\omega_1)^2 + 1|} \quad (6\text{-}111)$$

Ignoring the trivial case at zero, the lower of the two values of ω/ω_1 satisfying Eq. (6-111) will be ω_P/ω_1 and the higher will be ω_Q/ω_1. To solve Eq. (6-111) we must again remove the absolute-value signs. There will be only two possibilities, as before, and the simplest approach is to try each and see which one actually applies. Thus, Eq. (6-111) can be written as

$$\pm \frac{1}{(1 + \mu)(\omega/\omega_1)^2 - 1} = \frac{1 - (1/f^2)(\omega/\omega_1)^2}{(1/f^2)(\omega/\omega_1)^4 - (1 + \mu + 1/f^2)(\omega/\omega_1)^2 + 1} \quad (6\text{-}112)$$

which becomes, for the left-hand term positive,

$$\frac{1}{f^2}\left(\frac{\omega}{\omega_1}\right)^4 - \left(1 + \mu + \frac{1}{f^2}\right)\left(\frac{\omega}{\omega_1}\right)^2 + 1 = (1 + \mu)\left(\frac{\omega}{\omega_1}\right)^2 - 1$$
$$- \frac{1}{f^2}(1 + \mu)\left(\frac{\omega}{\omega_1}\right)^4 + \frac{1}{f^2}\left(\frac{\omega}{\omega_1}\right)^2 \quad (6\text{-}113)$$

and, for the left-hand term negative,

$$\frac{1}{f^2}\left(\frac{\omega}{\omega_1}\right)^4 - \left(1 + \mu + \frac{1}{f^2}\right)\left(\frac{\omega}{\omega_1}\right)^2 + 1 = -(1 + \mu)\left(\frac{\omega}{\omega_1}\right)^2 + 1$$
$$+ \frac{1}{f^2}(1 + \mu)\left(\frac{\omega}{\omega_1}\right)^4 - \frac{1}{f^2}\left(\frac{\omega}{\omega_1}\right)^2 \quad (6\text{-}114)$$

After rearranging and simplifying, Eqs. (6-113) and (6-114) become, respectively,

$$\frac{1}{f^2}(2 + \mu)\left(\frac{\omega}{\omega_1}\right)^4 + 2\left(1 + \mu + \frac{1}{f^2}\right)\left(\frac{\omega}{\omega_1}\right)^2 + 2 = 0 \quad (6\text{-}115)$$

and

$$\left(\frac{\omega}{\omega_1}\right)^4 = 0 \quad (6\text{-}116)$$

Since Eq. (6-115) is the only one with nonzero values of $(\omega/\omega_1)^2$, it applies to the problem at hand. By use of the binomial theorem, from

Eq. (6-115) we find

$$\left(\frac{\omega}{\omega_1}\right)^2 = \frac{f^2 + \mu f^2 + 1}{2 + \mu} \pm \sqrt{\left(\frac{f^2 + \mu f^2 + 1}{2 + \mu}\right)^2 - \frac{2f^2}{2 + \mu}} \quad (6\text{-}117)$$

From Eq. (6-117),

$$\left(\frac{\omega_P}{\omega_1}\right)^2 + \left(\frac{\omega_Q}{\omega_1}\right)^2 = 2\frac{f^2 + \mu f^2 + 1}{2 + \mu} \quad (6\text{-}118)$$

Equating the right-hand sides of Eqs. (6-109) and (6-118) and solving for f gives

$$f = \frac{1}{1 + \mu} \quad (6\text{-}119)$$

Therefore, to make $X_P = X_Q$, we must have, from Eqs. (6-104) and (6-119),

$$\omega_2 = \frac{1}{1 + \mu} \omega_1 \quad (6\text{-}120)$$

A vibration absorber tuned according to Eq. (6-120) is called a *most favorably tuned system.* Curves for solutions of Eqs. (6-98) and (6-99) for this case when the damping ratio is 0, 0.15, and ∞, and the mass ratio is 0.1, are given in Figs. 6-14 and 6-15, respectively.

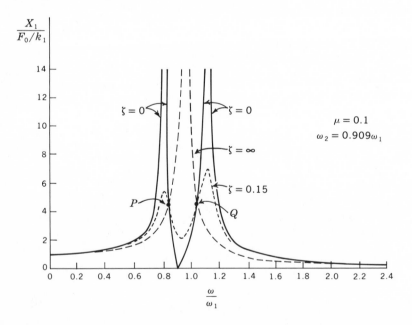

Fig. 6-14 *Steady-state response of the main system to a sinusoidal force disturbance when the absorber is most favorably tuned and the mass of the absorber is one-tenth that of the main system.*

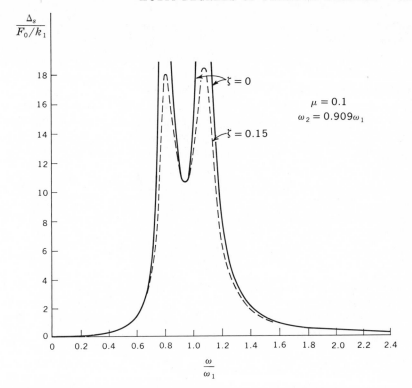

Fig. 6-15 *Maximum deflection of absorber spring for a system with the response curves in Fig. 6-14.*

Substituting from Eq. (6-119) into Eq. (6-117) leads to

$$\left(\frac{\omega_P}{\omega_1}\right)^2 = \frac{1}{1 + \mu}\left(1 - \sqrt{\frac{\mu}{2 + \mu}}\right) \qquad (6\text{-}121)$$

and

$$\left(\frac{\omega_Q}{\omega_1}\right)^2 = \frac{1}{1 + \mu}\left(1 + \sqrt{\frac{\mu}{2 + \mu}}\right) \qquad (6\text{-}122)$$

Substituting from either Eq. (6-121) or Eq. (6-122) into Eq. (6-105) results in

$$\frac{X_P}{F_0/k_1} = \frac{X_Q}{F_0/k_1} = \sqrt{\frac{2 + \mu}{\mu}} \qquad (6\text{-}123)$$

The optimum damping ratio for the most favorably tuned absorber would be the value resulting in the lowest possible peak value of $X_1/(F_0/k_1)$ over the entire operating range of frequencies. However, since the general case requires determining the frequency ratio, as well as the damping ratio, the problem becomes quite complicated and a general solution is not available.   But an approximate solution, adequate for all engineering applications, has been obtained by Brock.[1]

As can be seen in Fig. 6-14, the optimum curve cannot have a peak value lower than at P and Q.   Thus, a curve that would be a straight line through P and Q would be ideal.   Unfortunately, such a curve is not possible and the best one can do is to find the value of ζ that will make the curve pass horizontally through either P or Q, but not both.   According to Brock, the optimum damping ratios[2] for the most favorably tuned system are, for the curve horizontal at P,

$$\zeta^2 = \frac{\mu[3 - \sqrt{\mu/(2 + \mu)}]}{8(1 + \mu)} \tag{6-124}$$

and, for the curve horizontal at Q,

$$\zeta^2 = \frac{\mu[3 + \sqrt{\mu/(2 + \mu)}]}{8(1 + \mu)}. \tag{6-125}$$

Considering that μ is usually in the order of 0.1 or less, there will be little difference between the results given by Eqs. (6-124) and (6-125) and for almost all purposes the average of the two is quite satisfactory.   Thus, for the most favorably tuned system,

$$\zeta_{\text{opt}} = \sqrt{\frac{3\mu}{8(1 + \mu)}} \tag{6-126}$$

The preceding development can be applied directly to a torsional system by replacing each of the terms for rectilinear motion by its equivalent for angular motion.   However, the nature of rotation offers some new possibilities that are not generally found when the displacements are rectilinear, notably, (1) the displacement of the absorber rotor relative to the main rotor can be arbitrarily large without running out of room for travel and (2) the entire system will often be rotating at an appreciable speed.

The first possibility means that the practical problems associated with designing the spring can be eliminated by eliminating the spring itself. The second possibility means that centrifugal force can be used as a substitute for gravity and a pendulum can replace the mass and spring in the absorber.

[1] John E. Brock, A Note on the Damped Vibration Absorber, *Trans. ASME*, vol. 68, p. A284, 1946.

[2] The equations appear to be different from those presented by Brock.   This is due to his following the nomenclature of J. P. Den Hartog, "Mechanical Vibrations," 4th ed., p. 96, McGraw-Hill Book Company, New York, 1956, and defining the damping ratio in terms of the natural frequency of the main system rather than in terms of the natural frequency of the absorber.

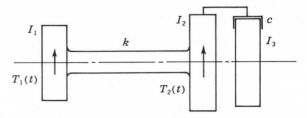

Fig. 6-16 *Lanchester damper with viscous damping.*

6-3. Lanchester Damper   The vibration absorber without a spring is called a Lanchester damper.   The damping may be either Coulomb or viscous.   In the latter case silicone fluid is particularly useful because of its high viscosity index, i.e., in comparison with common petroleum oils its viscosity is relatively independent of the temperature.

The simplest torsional case of engineering importance is a system consisting of two rotors on a shaft, as shown in Fig. 6-1a.   If a Lanchester damper is added, the system becomes that in Fig. 6-16.   There are now three masses, and the equations of motion for the system are

$$I_1 D^2\theta_1 + k(\theta_1 - \theta_2) = T_1(t) \qquad (6\text{-}127)$$
$$I_2 D^2\theta_2 + c(D\theta_2 - D\theta_3) + k(\theta_2 - \theta_1) = T_2(t) \qquad (6\text{-}128)$$
and $$I_3 D^2\theta_3 + c(D\theta_3 - D\theta_2) = 0 \qquad (6\text{-}129)$$

As shown in Sec. 6-1, a torsional system that is free to rotate is a degenerate system and actually has one less (nonzero) natural frequency than it has masses.   Thus, the system in Fig. 6-16 might be expected to behave as a two-degrees-of-freedom system.   As will be apparent, things are not this simple and this system is not analogous to that in Fig. 6-11b, which is a true two-degrees-of-freedom system.

In this case, the optimum damper may be the one that gives the minimum amplitude of θ_1 or θ_2 or, more likely, the one that results in a minimum amplitude of $\theta_1 - \theta_2$.   The general method will be similar to that given above, but the additional variables greatly increase the complexity of the problem.   Lewis[1] has considered this problem in some detail and presents equations applicable to the system in Fig. 6-16 when $T_1(t) = 0$ and $T_2 = T_0 \sin \omega t$.   Absence of a spring between I_2 and the absorber inertia I_3 results in further degeneration of the system so that it behaves as a single-degree-of-freedom system and there is only one point at which the curves for all values of damping intersect.   Designating the point of intersection as P, we find, after Lewis,

$$\frac{\omega_P}{\omega_1} = \sqrt{\frac{I_2(2I_1 + 2I_2 + I_3)}{(I_1 + I_2)(2I_2 + I_3)}} \qquad (6\text{-}130)$$

[1] F. M. Lewis, The Extended Theory of the Viscous Vibration Damper, *J. Appl. Mech., Trans. ASME*, vol. 77, pp. 377–382, 1955.

where $\omega_1 = \sqrt{k(I_1 + I_2)/I_1 I_2}$ = natural frequency of the main system,

$$\frac{\Delta\Theta_P}{T_0/k} = \frac{2I_2 + I_3}{I_3} \tag{6-131}$$

where $\Delta\Theta_P = |\theta_1 - \theta_2|_{\omega_P}$, amplitude of shaft twist at ω_P, and

$$c_{\text{opt}} = \frac{\omega_P}{\omega_1} \sqrt{\frac{kI_3{}^2(I_1 + I_2)}{I_1(I_2 + I_3)}} \tag{6-132}$$

Equation (6-132) has not been nondimensionalized, because damping ratio has no meaning for the absorber system without a spring.

If, for the system in Fig. 6-16, $I_1 \gg I_2$, we can consider I_1 to be infinite and the system becomes analogous to that in Fig. 6-11b with $k_2 = 0$. The reader is referred to Den Hartog[1] for information about this particular case.

The major disadvantage of the Lanchester damper is that the optimum will still be *much* less effective than the equivalent optimum absorber using both a spring and damping.

6-4. Centrifugal Pendulum Absorber The undamped vibration absorber was shown in Eq. (6-98) to be absolutely effective when tuned to the frequency of the forcing function. Thus, the undamped absorber could be practical for a torsional system if the frequency of the forcing function is constant. Many torsional systems that are most likely to have vibration problems must run through a wide range of speeds, and in most cases the period of the forcing function will be directly related to the nominal speed of rotation. Consequently, in this case, the type of undamped vibration absorber discussed above with its additional resonant frequency would result in worse rather than improved operation. However, as will be shown, an undamped system that utilizes a pendulum acting in the centrifugal-force field associated with the nominal speed of rotation can be effective throughout the operating range of the system.

In Fig. 6-17a a pendulum has been added to a rotating member. It should be recalled that the effective inertia of an undamped absorber becomes infinite when the frequency of the forcing function equals the natural frequency of the absorber. Thus, we are interested in determining the conditions under which the natural frequency of the pendulum can equal the frequency of the forcing function and thereby make the pendulum effective as a vibration absorber.

Figure 6-17b is an enlarged view for the situation when the carrier is rotating at a nominal (average) speed Ω and the pendulum has been displaced through a small angle ψ. If we assume that the carrier speed is relatively high and, therefore, the effects of gravity and the angular velocity of the pendulum relative to the carrier can be neglected, the

---

[1] *Op. cit.*, pp. 102–105.

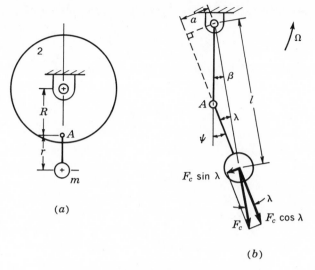

Fig. 6-17   *Centrifugal-pendulum absorber.*

inertia force on the pendulum is

$$F_c = ml\Omega^2 \tag{6-133}$$

and the torque on the carrier is

$$\begin{aligned} T &= (F_c \cos \lambda)a \\ &= (F_c \cos \lambda)R \sin \psi \end{aligned} \tag{6-134}$$

The pendulum absorber is a nonlinear system, but as in Sec. 5-2, we can linearize the system for small angles of oscillation by letting $\sin \theta = \theta$ and $\cos \theta = 1$. Therefore, from Fig. 6-17b,

$$\begin{aligned} l &= R \cos \beta + r \cos \lambda \\ &\approx R + r \end{aligned} \tag{6-135}$$

and Eq. (6-133) becomes

$$F_c = m(R + r)\Omega^2 \tag{6-136}$$

Similarly, for small angles,

$$F_c \cos \lambda \approx F_c \tag{6-137}$$

and

$$a \approx R\psi \tag{6-138}$$

Substituting from Eqs. (6-137) and (6-138) into Eq. (6-134) gives

$$T = mR(R + r)\Omega^2\psi \tag{6-139}$$

Considering the pendulum only, we find, by applying Newton's second law,

$$F_c \sin \lambda = -mr \, D^2\psi \qquad (6\text{-}140)$$

After substituting from Eq. (6-136) for F_c, letting $\sin \lambda = \lambda$, and rearranging terms, Eq. (6-140) becomes

$$D^2\psi + \frac{1}{r}(R + r)\Omega^2\lambda = 0 \qquad (6\text{-}141)$$

From the geometry of Fig. 6-17b,

$$\lambda = \frac{R}{R + r}\psi \qquad (6\text{-}142)$$

Substituting from Eq. (6-142) into Eq. (6-141) results in

$$D^2\psi + \frac{R}{r}\Omega^2\psi = 0 \qquad (6\text{-}143)$$

Therefore, from Sec. 4-4, we can write

$$\omega_n = \Omega \sqrt{\frac{R}{r}} \qquad (6\text{-}144)$$

The major application of this type of absorber has been to internal-combustion engines in which the major component of the forcing function results from combustion of the fuel-air mixture and the resultant periodic variation in torque consists of numerous harmonics,[1] each of which is directly related to the speed. Therefore, in the general case it is convenient to write

$$T = T_0 \sin N\Omega t \qquad (6\text{-}145)$$

where N is the number of cycles of the forcing function per revolution. Since for the absorber to be effective ω_n must equal the frequency of the forcing function, we want to have

$$\omega_n = N\Omega \qquad (6\text{-}146)$$

Substituting from Eq. (6-146) into Eq. (6-144) and rearranging terms gives

$$\frac{R}{r} = N^2 \qquad (6\text{-}147)$$

Equation (6-147) provides the basic information required to design a system that is tuned properly. The remaining design problem is concerned with providing a torque that counterbalances the forcing function.

---

[1] See Example 6-1 for a brief discussion of harmonics of engine torque curves.

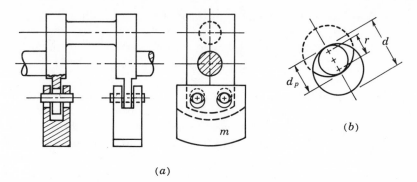

(a)

(b)

Fig. 6-18   *Bifilar absorber.*

For sinusoidal oscillation of the pendulum relative to the carrier,

$$\psi = \psi_0 \sin \omega t \tag{6-148}$$

Substituting from Eq. (6-148) into Eq. (6-139) gives

$$T = mR(R + r)\Omega^2\psi_0 \sin \omega t \tag{6-149}$$

Since the amplitude from Eq. (6-149) must equal that of the forcing function in Eq. (6-145), we shall have

$$T_0 = mR(R + r)\Omega^2\psi_0 \tag{6-150}$$

Equations (6-147) and (6-150) contain all of the relationships needed for design.   In general, values of N, T_0, and Ω would be specified.   There will be practical space limitations on the value of R, and ψ_0 must remain small or the possibly severe nonlinear aspects[1] cannot be ignored.

Study of Eqs. (6-147) and (6-150) leads to the conclusion that in most practical situations where $N > 1$ and T is significant, it will not be possible to construct pendulums that look like that in Fig. 6-17, because sufficient mass cannot be located at the required small value of r.   A number of methods for getting around this difficulty have been developed.   A good example is the *bifilar* or Sarazin-Chilton absorber in Fig. 6-18a, where the constraint provided by the rolling contact between the pins and holes is such that in effect the pendulum mass is supported by a parallel linkage with a constant radius.   Thus, all points in m have the same radius and we have a simple pendulum.   From Fig. 6-18b we find

$$r = d - d_p \tag{6-151}$$

[1] D. E. Newland, Nonlinear Aspects of the Performance of Centrifugal Pendulum Vibration Absorbers, *ASME Paper* 63-WA-275, 1963.

where d = diameter of the holes in the carrier and pendulum

d_p = diameter of the pin

The literature[1] should be consulted for information about other types of vibration absorbers and for discussions of additional practical considerations.

6-5. Coupled Vibrations—Systems with Two Degrees of Freedom

The concept of principal modes of vibration has been used in the preceding sections of this chapter, where it was shown that there would be a principal mode for each natural frequency. If we assume that the rigid body in Fig. 6-19a is constrained to have plane motion, it can translate in the vertical and horizontal directions and can rotate about an axis perpendicular to the plane. Thus, we have three degrees of freedom and three coordinates are required to specify the location of the body at any instant. As shown in Fig. 6-19b, the coordinates most often used are the x and y displacements of the center of gravity and the angle of rotation in the xy plane—all measured from the equilibrium position.

Although, in general, vibration in any of the principal modes will result in displacements in the direction of every coordinate, it is convenient to consider the basic modes of vibration to be those corresponding directly to independent motions in the direction of the coordinates, such as in the x, y, and θ directions in Fig. 6-19b. For lack of a better term, we shall call these the *coordinate modes*.

When a coordinate mode is also a principal mode, it can exist independently and the coordinate mode is considered to be *decoupled*. Coordinate modes are coupled when two or more of the coordinates are required to

[1] W. K. Wilson, "Practical Solution of Torsional Vibration Problems," 3d ed., John Wiley & Sons, Inc., New York, 1956.

E. J. Nestorides (compiler), "A Handbook on Torsional Vibration," Cambridge University Press, London, 1958.

C. M. Harris and C. E. Crede (eds.), "Shock and Vibration Handbook," vols. I–III, McGraw-Hill Book Company, New York, 1961.

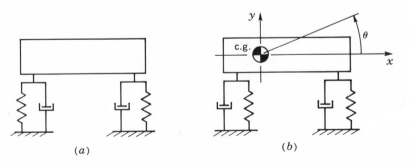

(a)   (b)

Fig. 6-19 *System with three degrees of freedom.*

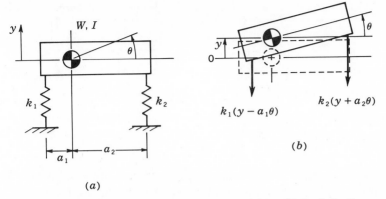

(a)

Fig. 6-20   (a) *System constrained to two degrees of freedom;* (b) *free-body diagram for system in* (a).

describe a principal mode.   In these terms, the system in Fig. 6-19b can have three coupled modes (the general case), one decoupled and two coupled modes, or three decoupled modes.   This section will be concerned with determining the conditions under which the modes for a simplified two-degrees-of-freedom system will be coupled or decoupled, and the following section will be concerned with the application to the real engineering problem involving six degrees of freedom.

Let us assume that a frictionless guide is added to the system in Fig. 6-19 so that there can be no motion in the x direction and that damping is negligible.   These assumptions result in the system in Fig. 6-20a, where the spring rates are k_1 and k_2, the weight is W, and the mass moment of inertia about the center of gravity is I.   If we give the body a positive displacement y and a small positive rotation θ, the free-body diagram at the instant of release becomes that shown in Fig. 6-20b. Applying Newton's second law to this system, we can write

$$\frac{W}{g} D^2 y = -k_1(y - a_1\theta) - k_2(y + a_2\theta) \qquad (6\text{-}152)$$

and
$$I D^2\theta = k_1(y - a_1\theta)a_1 - k_2(y + a_2\theta)a_2 \qquad (6\text{-}153)$$

After rearranging and collecting like terms, Eqs. (6-152) and (6-153) become, respectively,

$$\frac{W}{g} D^2 y + (k_1 + k_2)y - (k_1a_1 - k_2a_2)\theta = 0 \qquad (6\text{-}154)$$

and
$$-(k_1a_1 - k_2a_2)y + I D^2\theta + (k_1a_1{}^2 + k_2a_2{}^2)\theta = 0 \qquad (6\text{-}155)$$

As can be seen, both equations are functions of both y and θ and the modes will be coupled.   The discussion in Sec. 6-1 can be followed to determine the location of the nodes, which will characterize the principal

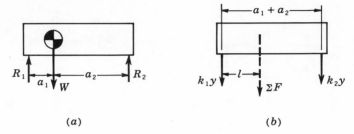

(a)   (b)

Fig. 6-21

modes of vibration.   However, at this time we are interested in the question of coupling of the coordinate modes.   Referring to Eqs. (6-154) and (6-155), we see that if $k_1a_1 = k_2a_2$, the modes will be decoupled because Eq. (6-154) becomes a function of y only and Eq. (6-155) becomes a function of θ only.

For a physical interpretation of decoupled modes of vibration, let us consider the conditions for static and dynamic equilibrium.   For static equilibrium, the system becomes that in Fig. 6-21a.   Taking moments about the right-hand end, we find

$$R_1 = \frac{a_2}{a_1 + a_2} W \qquad (6\text{-}156)$$

and taking moments about the left-hand end, we find

$$R_2 = \frac{a_1}{a_1 + a_2} W \qquad (6\text{-}157)$$

From definition of the spring rate,

$$\delta_{st_1} = \frac{R_1}{k_1} = \frac{a_2 W}{k_1(a_1 + a_2)} \qquad (6\text{-}158)$$

and $$\delta_{st_2} = \frac{R_2}{k_2} = \frac{a_1 W}{k_2(a_1 + a_2)} \qquad (6\text{-}159)$$

Dividing Eq. (6-158) by Eq. (6-159) gives

$$\frac{\delta_{st_1}}{\delta_{st_2}} = \frac{k_2 a_2}{k_1 a_1} \qquad (6\text{-}160)$$

but, from above, for decoupled modes, $k_1a_1 = k_2a_2$.   Therefore, for decoupled modes,

$$\frac{\delta_{st_1}}{\delta_{st_2}} = 1 \qquad (6\text{-}161)$$

Since, from Eq. (4-52), the natural frequency of a simple mass-spring system is directly proportional to $\sqrt{1/\delta_{st}}$, Eq. (6-161) can be interpreted

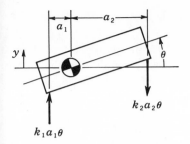

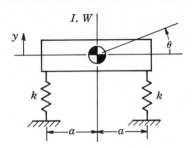

Fig. 6-22

Fig. 6-23 *System with decoupled y- and θ-coordinate modes of vibration.*

to mean that the modes will be decoupled when the natural frequency corresponding to a simple system made up of the spring and the mass supported by the spring at one end is identical to that at the other end.

When $k_1a_1 = k_2a_2$, the total external force on the body becomes, from Eq. (6-152),

$$\Sigma F = -(k_1 + k_2)y \qquad (6\text{-}162)$$

and the free-body diagram for positive displacement y is given in Fig. 6-21b.   The line of action of the resultant spring force can be found by determining the point about which the sum of moments of the spring forces is zero.   From Fig. 6-21b,

$$k_1yl - k_2y(a_1 + a_2 - l) = 0 \qquad (6\text{-}163)$$

Substituting from $k_1a_1 = k_2a_2$ into Eq. (6-163) and solving for l results in

$$l = a_1 \qquad (6\text{-}164)$$

Therefore, the resultant spring force acts through the center of gravity and displacement in the y direction does not introduce a couple.

In a similar manner, substituting from $k_1a_1 = k_2a_2$ into Eq. (6-153) results in

$$\Sigma T = -(k_1a_1{}^2 + k_2a_2{}^2)\theta \qquad (6\text{-}165)$$

and the free-body diagram for a positive displacement θ is shown in Fig. 6-22.   Summing up the forces in the y direction gives

$$\Sigma F_y = k_1a_1\theta - k_2a_2\theta \qquad (6\text{-}166)$$

which, for $k_1a_1 = k_2a_2$, becomes

$$\Sigma F_y = 0 \qquad (6\text{-}167)$$

Thus, in general, we find that, when a translational mode is decoupled, a force applied to the center of gravity results only in translation in the direction of the applied force and that, when a rotational mode is decoupled, a couple applied to the body results only in a rotation in the direction of the couple.

The simplest way to decouple the coordinate modes is to make $k_1 a_1 = k_2 a_2 = ka$. If this is done, the system becomes symmetrical about the y axis, as shown in Fig. 6-23, and the equations of motion, Eqs. (6-154) and (6-155), become, respectively,

$$\frac{W}{g} D^2 y + 2ky = 0 \tag{6-168}$$

and $\qquad\qquad I D^2\theta + 2ka^2\theta = 0 \tag{6-169}$

Equation (6-168) can be rewritten as

$$D^2 y + \omega_{ny}^2 y = 0 \tag{6-170}$$

where $\qquad\qquad \omega_{ny} = \sqrt{\dfrac{2kg}{W}} \tag{6-171}$

and Eq. (6-169) can be rewritten as

$$D^2\theta + \omega_{n\theta}^2\theta = 0 \tag{6-172}$$

where $\qquad\qquad \omega_{n\theta} = \sqrt{\dfrac{2ka^2}{I}} \tag{6-173}$

By definition,

$$I = m\rho^2 = \frac{W}{g}\rho^2 \tag{6-174}$$

where ρ = radius of gyration. Substituting from Eq. (6-174) into Eq. (6-173) results in

$$\omega_{n\theta} = \sqrt{\frac{2kg}{W}\frac{a^2}{\rho^2}} \tag{6-175}$$

and from Eqs. (6-175) and (6-171) we find

$$\omega_{n\theta} = \frac{a}{\rho}\omega_{ny} \tag{6-176}$$

From Eq. (6-176) it can be seen that if $a = \rho$, $\omega_{n\theta} = \omega_{ny}$.

Additional physical understanding can be gained by replacing the distributed-mass body by a dynamically equivalent body with lumped masses, as discussed in Sec. 1-6. In terms of the equivalent body, the system in Fig. 6-23 becomes that in Fig. 6-24a, which for $a = \rho$ becomes that in Fig. 6-24b.

As can be seen, the system in Fig. 6-24b is in effect two independent single-degree-of-freedom systems with the same natural frequency. The decoupled vibration in the y direction results when the two systems are vibrating in phase with the same amplitude, and the decoupled vibration in the θ direction results when the two systems are vibrating out of phase with the same amplitude.

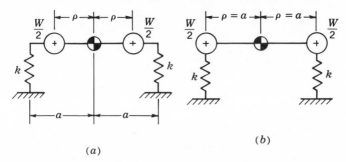

(a)   (b)

Fig. 6-24   (a) *System with decoupled coordinate modes of vibration;* (b) *system with identical natural frequencies of decoupled coordinate modes of vibration.*

It should be noted that if $\rho \approx a$, $\omega_{ny} \approx \omega_{n\theta}$ and, as discussed in Sec. 6-1, the proper choice of initial conditions will result in the phenomenon of beating. It should also be noted that a disturbance applied to one end of the system in Fig. 6-24b will have no effect on the other end. This is the ideal situation for automobile suspension systems. However, for practical and esthetic reasons the radius of gyration is usually less than half the wheel base and $\omega_{n\theta}$ is higher than ω_{ny}.

In the above discussion, the coordinates were chosen as the displacement of the center of gravity and the angle of rotation. This was purely a matter of convenience because it simplifies describing the appearance of the resulting vibrations. For example, using the displacements y_1 and y_2 as the coordinates in Fig. 6-25 will lead to the same values for natural frequencies that would be found using y and θ as the coordinates; but the concept of decoupling of coordinate modes will no longer have much significance.

6-6. Coupled Vibrations—Systems with Six Degrees of Freedom
One of the most important "real" problems in dynamics of machinery is the isolation of a three-dimensional body from its surroundings. In most cases the body will be supported at four points and the system can be

Fig. 6-25 *Alternate coordinates for system constrained to two degrees of freedom.*

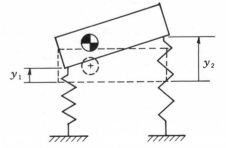

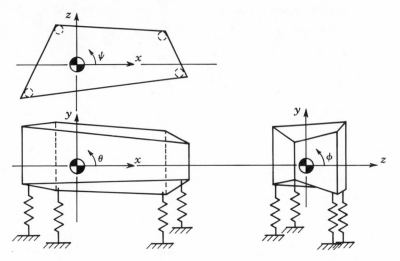

Fig. 6-26 *General suspension system with six degrees of freedom.*

represented as shown in Fig. 6-26, where the six coordinates x, y, z, θ, ϕ, and ψ are required to locate the body at all times. The system will have six degrees of freedom, and the determination of natural frequencies and/or the response to a forcing function will require the simultaneous solution of a set of six differential equations. Just writing the equations

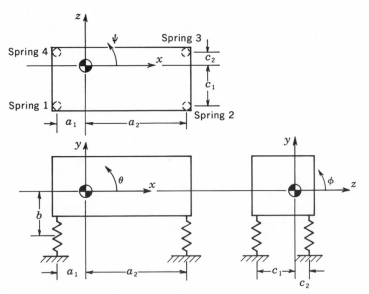

Fig. 6-27 *Typical suspension system with six degrees of freedom in which the isolators are located in a plane at corners of a rectangle.*

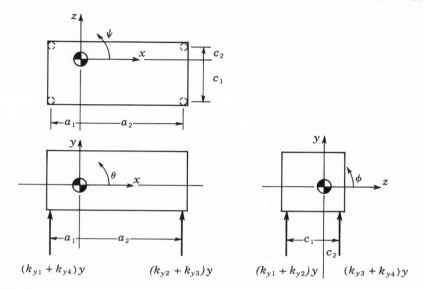

$(k_{y1} + k_{y4})y$ $\qquad$ $(k_{y2} + k_{y3})y$ $\qquad$ $(k_{y1} + k_{y2})y$ $\qquad$ $(k_{y3} + k_{y4})y$

Fig. 6-28 *Isolator forces acting on the body in Fig.* 6-27 *when displaced in the negative y direction.*

for the general case is a formidable task, and it is impractical to obtain solutions without the use of computers. Further treatment of the general case is beyond the scope of this book, and the following discussion will be limited to the cases where the geometry and mass distribution of the body and the elastic properties of the isolators are such that the coordinate modes are decoupled to the extent that all six natural frequencies can be determined by considering not more than two degrees of freedom at one time. In general, we shall consider only systems that are linear (or can be considered linear for small amplitudes) and have negligible damping. We shall also limit the discussion to systems in which four isolators are located at corners of a plane rectangle, as shown in Fig. 6-27, and we shall be interested primarily in determining the six natural frequencies of the system. Additional minor limitations will be introduced when appropriate.

The first step will be to determine the conditions under which the y-coordinate mode is decoupled from all others. As discussed in Sec. 6-5, this requires that the resultant of the spring forces due to a deflection in the y direction act through the center of gravity. Since the moments due to spring deflections in the y direction act in space, the summation of moments must be written as a vector equation. However, in this case we can simplify the calculation by considering the projections of the moments on the xy and zy planes.

Figure 6-28 shows the isolator forces acting when the body is displaced in the negative y direction and all other coordinate motions are constrained to be zero. The resultant spring force is

$$F_{Ry} = \Sigma F_y = -k_{y1}y - k_{y2}y - k_{y3}y - k_{y4}y = -y\Sigma k_y \quad (6\text{-}177)$$

where k_y is the spring rate for longitudinal or y deflections and the number subscripts designate the particular spring being considered.

In preceding discussions the locations of forces have been specified by dimensions, without regard to sign, and the signs of the moments have been determined by inspection. In parts of the following discussion the results will be more generally useful if the distances from the center of gravity to the line of action of the forces include sign as well as magnitude. Therefore, in the interest of consistency, distances *in the remainder of this section and in Sec.* 6-7 will be considered to have direction as well as magnitude. For example, in Fig. 6-28, a_1 and c_1 are negative numbers whereas a_2 and c_2 are positive numbers.

For the resultant spring force to act through the center of gravity, the summation of moments about the center of gravity must be zero in both the xy and zy planes. Thus,

$$\Sigma T_{xy} = 0 \quad -(k_{y1} + k_{y4})ya_1 - (k_{y2} + k_{y3})ya_2 = 0 \quad (6\text{-}178)$$

and $\quad \Sigma T_{zy} = 0 \quad -(k_{y1} + k_{y2})yc_1 - (k_{y3} + k_{y4})yc_2 = 0 \quad (6\text{-}179)$

Figure 6-28 becomes the free-body diagram(s) for the body at the instant of release after being displaced in the negative y direction from the equilibrium position. Applying Newton's second law to this case gives

$$\Sigma F_y = m\,D^2y \quad (6\text{-}180)$$

When the y-coordinate mode is decoupled, we can substitute from Eq. (6-177) into Eq. (6-180) and rearrange terms to give

$$D^2y + \frac{\Sigma k_y}{m}\,y = 0 \quad (6\text{-}181)$$

from which we recognize

$$\omega_{ny} = \sqrt{\frac{\Sigma k_y}{m}} \quad (6\text{-}182)$$

In a design situation, the weight and the dimensions will probably be known before the isolators are designed or selected. The natural frequency that the suspension system is to provide will be related to the degree of isolation desired or the maximum amplitude of motion that can be tolerated, and the choice will be based on the discussion in Secs. 4-12 through 4-17. Thus, the sum of the spring rates can be calculated from Eq. (6-182), and the designer must then satisfy Eqs. (6-178) and (6-179). Since this leaves us with four unknowns and only three equations, we must arbitrarily specify another condition.   If we expand Eq. (6-178) and rearrange terms, we find

$$-k_{y1}a_1 - k_{y4}a_1 = k_{y2}a_2 + k_{y3}a_2 \tag{6-183}$$

One set of conditions that will satisfy Eq. (6-183) is

$$k_{y1}a_1 = -k_{y2}a_2 \tag{6-184}$$

and $\qquad\qquad k_{y4}a_1 = -k_{y3}a_2 \tag{6-185}$

Referring to Fig. 6-28, we see that Eq. (6-184) means that the resultant force from springs 1 and 2 acts in the xy plane at $x = 0$.   Similarly, Eq. (6-185) means that the resultant force from springs 3 and 4 also acts in the xy plane at $x = 0$.   Since both results are reasonable, i.e., they agree with the requirements for decoupling in a plane, as discussed in Sec. 6-5, we shall let this be the arbitrary condition.

From Eq. (6-184),

$$k_{y2} = -\frac{a_1}{a_2}k_{y1} \tag{6-186}$$

which, since we are now limiting our discussion to the case where a_1 is negative and a_2 is positive, can be written as

$$k_{y2} = \left|\frac{a_1}{a_2}\right| k_{y1} \tag{6-187}$$

and, from Eq. (6-185),

$$k_{y3} = -\frac{a_1}{a_2}k_{y4} = \left|\frac{a_1}{a_2}\right| k_{y4} \tag{6-188}$$

Substituting from Eqs. (6-187) and (6-188) into Eq. (6-179) and simplifying results in

$$k_{y4} = -\frac{c_1}{c_2}k_{y1} = \left|\frac{c_1}{c_2}\right| k_{y1} \tag{6-189}$$

Now, substituting from Eq. (6-189) into Eq. (6-188) gives

$$k_{y3} = \left|\frac{a_1}{a_2}\right|\left|\frac{c_1}{c_2}\right| k_{y1} \tag{6-190}$$

From Eqs. (6-187), (6-189), and (6-190),

$$\Sigma k_y = k_{y1}\left(1 + \left|\frac{a_1}{a_2}\right| + \left|\frac{c_1}{c_2}\right| + \left|\frac{a_1}{a_2}\right|\left|\frac{c_1}{c_2}\right|\right) \tag{6-191}$$

Combining Eqs. (6-182) and (6-191) and solving for k_{y1}, we find

$$k_{y1} = \frac{m\omega_{ny}{}^2}{1 + |a_1/a_2| + |c_1/c_2| + |a_1/a_2|\,|c_1/c_2|} \tag{6-192}$$

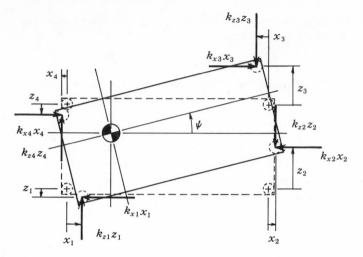

Fig. 6-29 *Isolator forces acting on the body in Fig. 6-27 when rotated a positive angle ψ in the xz plane.*

Therefore, to ensure decoupling of the y-coordinate mode of vibration for a system such as that shown in Fig. 6-28, we need only to design or select isolators having the spring rates corresponding to Eqs. (6-192), (6-187), (6-189), and (6-190).

Looking at the top (plan) view of the system in Fig. 6-28, we see that the coordinate modes of vibration correspond to the translations x and z and the rotation ψ. If we consider the translation x and rotation ψ and the translation z and the rotation ψ as separate cases, each becomes directly analogous to the two-degrees-of-freedom system in Sec. 6-5. Therefore, the ψ-coordinate mode will be decoupled from the x- and z-coordinate modes if the resultant of the isolator forces due to rotation about the center of gravity is a pure couple. Thus, the conditions can be expressed as

$$\Sigma F_{x,xz} = 0 \qquad (6\text{-}193)$$
$$\Sigma F_{z,xz} = 0 \qquad (6\text{-}194)$$

and
$$\Sigma T_{xz} = I_{zz}\, D^2\psi \qquad (6\text{-}195)$$

The isolator forces in the xz plane due to rotation without translation are shown in Fig. 6-29.

To satisfy Eq. (6-193),

$$-k_{x1}x_1 - k_{x2}x_2 - k_{x3}x_3 - k_{x4}x_4 = 0 \qquad (6\text{-}196)$$

From the geometry of Fig. 6-28, for *small displacements* where $\sin\psi \approx \psi$ and $\cos\psi \approx 1$,

$$x_1 = x_2 = -c_1\psi \qquad (6\text{-}197)$$

and
$$x_3 = x_4 = -c_2\psi \qquad (6\text{-}198)$$

Substituting from Eqs. (6-197) and (6-198) into Eq. (6-196) and simplifying gives

$$k_{x1}c_1 + k_{x2}c_1 + k_{x3}c_2 + k_{x4}c_2 = 0 \qquad (6\text{-}199)$$

and the problem is that we now have four unknowns and only one equation. It would appear that Eq. (6-195) contains some useful information, but since it involves both x and z forces, trying to use it at this point will only complicate matters.

From a practical viewpoint, symmetry will practically always result in $k_x = k_z$ and it will be convenient to classify isolators by the ratio of the lateral to the axial spring rates. Therefore, *in the remainder of this section and in Sec. 6-7 we shall let*[1]

$$\eta = \frac{k_x}{k_y} = \frac{k_z}{k_y} \qquad (6\text{-}200)$$

Substituting from Eq. (6-200) into Eq. (6-199) and simplifying gives

$$k_{y1}c_1 + k_{y2}c_1 + k_{y3}c_2 + k_{y4}c_2 = 0 \qquad (6\text{-}201)$$

Applying the above reasoning to the forces in the z direction will lead to

$$k_{y1}a_1 + k_{y2}a_2 + k_{y3}a_2 + k_{y4}a_1 = 0 \qquad (6\text{-}202)$$

The relationships in Eqs. (6-201) and (6-202) between the spring rates and dimensions are identical with those in Eqs. (6-179) and (6-178), respectively. Thus, the relationships in Eqs. (6-187), (6-189), and (6-190) apply directly to the lateral as well as to the axial spring rates—provided $k_x = k_z = \eta k_y$.

The relationship between the natural frequency and the moment of inertia and the spring rates must be found by solving Eq. (6-195). When $k_x \neq k_z$ for each spring, the summation of the torques becomes rather lengthy. However, since we are now limiting our discussion to the case where the isolators are symmetrical so that $k_x = k_z = \eta k_y$, the lateral spring rate in any direction can be expressed as ηk_y and the summation of the torques can be simplified by redrawing Fig. 6-29 as Fig. 6-30, showing

---

[1] Almost all texts and handbooks related to mechanical design discuss the relationship of material properties and dimensions to the axial spring rate (k_y) for helical springs, but there are not many references for correct information about the lateral spring rate $(k_x$ and $k_z)$. The discussion in Appendix C will be adequate for most purposes. For additional information, see:

C. E. Crede and J. P. Walsh, The Design of Vibration-isolating Bases for Machinery, *Trans. ASME*, vol. 69, pp. A7–14, 1947.

C. E. Crede, "Vibration and Shock Isolation," pp. 244–247, John Wiley & Sons, Inc., New York, 1951.

A. M. Wahl, "Mechanical Springs," 2d ed., McGraw-Hill Book Company, pp. 70–72, New York, 1964.

polar coordinates only. In terms of the forces and geometry in Fig. 6-30, Eq. (6-195) becomes

$$-\eta[k_{y1}(a_1{}^2 + c_1{}^2) + k_{y2}(a_2{}^2 + c_1{}^2) + k_{y3}(a_2{}^2 + c_2{}^2)$$
$$+ k_{y4}(a_1{}^2 + c_2{}^2)]\psi = I_{xz}\, D^2\psi \quad (6\text{-}203)$$

Rearranging Eq. (6-203) gives

$$I_{xz}\, D^2\psi + \eta[k_{y1}(a_1{}^2 + c_1{}^2) + k_{y2}(a_2{}^2 + c_1{}^2) + k_{y3}(a_2{}^2 + c_2{}^2)$$
$$+ k_{y4}(a_1{}^2 + c_2{}^2)]\psi = 0 \quad (6\text{-}204)$$

in which we recognize, from Sec. 4-4,

$$\omega_{nxz}$$
$$= \sqrt{\frac{\eta[k_{y1}(a_1{}^2 + c_1{}^2) + k_{y2}(a_2{}^2 + c_1{}^2) + k_{y3}(a_2{}^2 + c_2{}^2) + k_{y4}(a_1{}^2 + c_2{}^2)]}{I_{xz}}}$$
$$(6\text{-}205)$$

Considering the xy plane in Fig. 6-27, the three coordinate modes are y, x, and θ. We have just seen how the y-coordinate mode will be decoupled if the resultant spring force in the plane acts along the y axis. When this logic is applied to the x-coordinate mode, it becomes apparent that, since all of the spring forces resulting from a displacement of the body in the x direction act well below the center of gravity, the resultant will also act below the center of gravity and will therefore give a moment tending to rotate the body in the xy plane. Consequently, for the system in Fig. 6-27, the x- and θ-coordinate modes will be coupled. Similarly, we find that the z- and ϕ-coordinate modes will also be coupled in the

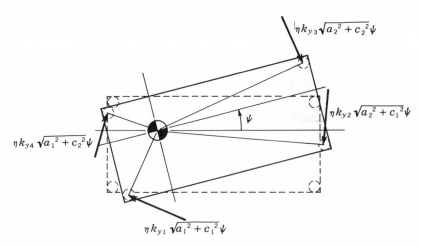

Fig. 6-30 *Resultant isolator forces acting on the body in Fig. 6-27 when the lateral spring rate is the same in all directions and the body has been rotated through a positive angle ψ in the xz plane.*

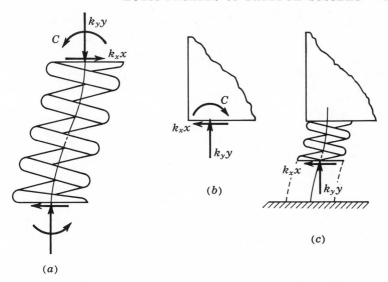

Fig. 6-31   *Helical compression spring under combined axial and lateral loads.*

zy plane.   These four coupled modes of vibration can best be designated by the planes in which they appear.   We shall designate them as ω_{nxy_1}, ω_{nxy_2}, ω_{nzy_1}, and ω_{nzy_2}, where the sub-subscripts 1 and 2 refer to the lower and higher values, respectively.

The calculation of the natural frequencies can be done exactly as in earlier sections of this chapter, but first it is necessary to take a closer look at the manner in which the isolator reaction to a lateral (x or z) displacement acts on the body.   Assuming that the isolator is a helical compression spring, or acts like one, the free body of the isolator when subjected to both axial and lateral loads will be as shown in Fig. 6-31a. The reactions on the body at the point of attachment are shown in Fig. 6-31b.   The couple C acts to rotate the body in the xy plane and introduces undesirable complications.   As shown in Fig. 6-31c, the couple can be eliminated by taking the free body at the midpoint of the spring where the bending moment is zero.   Thus, as shown in Fig. 6-27, b will be the distance (with sign) from the center of gravity to the midpoint of the elastic element of the isolator.

Figure 6-32a shows the system in Fig. 6-27 after being given a positive displacement x and a positive angular displacement θ, and Fig. 6-32b shows the free-body diagram for small displacements at the instant of release.[1]   For small displacements, where, from Sec. 5-2, $\sin \theta \approx \theta$ and

---

[1] It should be noted that an additional restraint has been introduced in that b and d are assumed to be the same for all isolators.

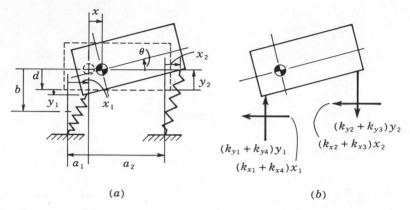

(a)   (b)

Fig. 6-32 (a) System with coupled x- and θ-coordinate modes of vibration; (b) isolator forces acting when the body in (a) is displaced in the positive x direction and rotated through a positive angle θ.

$\cos \theta \approx 1$, the geometry of Fig. 6-32a leads to

$$x_1 = x_2 = x - d\theta \qquad (6\text{-}206)$$
$$y_1 = a_1\theta \qquad (6\text{-}207)$$
and $$y_2 = a_2\theta \qquad (6\text{-}208)$$

Applying Newton's second law to the forces acting in the x direction and to the moments acting in the xy plane leads to

$$-(k_{x1} + k_{x4})x_1 - (k_{x2} + k_{x3})x_2 = m\,D^2x \qquad (6\text{-}209)$$
and

$$(k_{x1} + k_{x4})x_1b + (k_{x2} + k_{x3})x_2b - (k_{y1} + k_{y4})y_1a_1 - (k_{y2} + k_{y3})y_2a_2$$
$$= I_{xy}\,D^2\theta \quad (6\text{-}210)$$

where I_{xy} is the moment of inertia in the xy plane about the center of gravity.

Substituting from Eqs. (6-206) through (6-208) into Eqs. (6-209) and (6-210) and simplifying gives, respectively,

$$m\,D^2x + (\Sigma k_x)x - (\Sigma k_x)d\theta = 0 \qquad (6\text{-}211)$$
and

$$-(\Sigma k_x)bx + I_{xy}\,D^2\theta$$
$$+ [(\Sigma k_x)bd + (k_{y1} + k_{y4})a_1{}^2 + (k_{y2} + k_{y3})a_2{}^2]\theta = 0 \quad (6\text{-}212)$$

Since, for reasons of practicality and convenience, we are limiting subsequent developments to the case where $k_x = k_z = \eta k_y$, Eqs. (6-211) and (6-212) can be rewritten in simpler form as, respectively,

$$m\,D^2x + \eta(\Sigma k_y)x - \eta(\Sigma k_y)d\theta = 0 \qquad (6\text{-}213)$$

and

$$-\eta(\Sigma k_y)bx + I_{xy}D^2\theta$$
$$+ [\eta(\Sigma k_y)bd + (k_{y1} + k_{y4})a_1^2 + (k_{y2} + k_{y3})a_2^2]\theta = 0 \quad (6\text{-}214)$$

Comparing Eq. (6-213) with Eq. (6-214) leads to an interesting observation in relation to coupling of the coordinate modes. For example, if $d = 0$, the force in the x direction, Eq. (6-213), becomes independent or decoupled from the angle of rotation θ; and if $b = 0$, the torque on the body in the xy plane, Eq. (6-214), becomes independent or decoupled from the displacement x. Therefore, since physically b cannot equal d, it is apparent that complete decoupling of the coordinate modes is not possible when using helical springs for isolators.

To derive expressions for the two natural frequencies for the coupled modes of vibration in the xy plane, we shall follow the discussion in Sec. 6-1 by writing

$$x = Xe^{j\omega t} \quad (6\text{-}215)$$

and
$$\theta = \Theta e^{j(\omega t + \beta)} \quad (6\text{-}216)$$

Differentiating Eqs. (6-215) and (6-216) the correct numbers of times, substituting in Eqs. (6-213) and (6-214), and rearranging results in, respectively,

$$(\eta\Sigma k_y - m\omega^2)X - \eta(\Sigma k_y)d\Theta e^{j\beta} = 0 \quad (6\text{-}217)$$

and

$$-\eta(\Sigma k_y)bX + [\eta(\Sigma k_y)bd + (k_{y1} + k_{y4})a_1^2$$
$$+ (k_{y2} + k_{y3})a_2^2 - I_{xy}\omega^2]\Theta e^{j\beta} = 0 \quad (6\text{-}218)$$

Applying Cramer's rule to Eqs. (6-217) and (6-218) leads to

$$mI_{xy}\omega^4 - \{\eta(\Sigma k_y)(I_{xy} + bdm) + m[(k_{y1} + k_{y4})a_1^2 + (k_{y2} + k_{y3})a_2^2]\}\omega^2$$
$$+ \eta(\Sigma k_y)[(k_{y1} + k_{y4})a_1^2 + (k_{y2} + k_{y3})a_2^2] = 0 \quad (6\text{-}219)$$

By use of the binomial theorem, we find from Eq. (6-219)

$$\omega_{nxy}^2 = \frac{A_{xy} \pm \sqrt{B_{xy}}}{2mI_{xy}} \quad (6\text{-}220)$$

where

$$A_{xy} = \eta(\Sigma k_y)(I_{xy} + bdm) + m[(k_{y1} + k_{y4})a_1^2 + (k_{y2} + k_{y3})a_2^2]$$
and

$$B_{xy} = \{\eta(\Sigma k_y)(I_{xy} + bdm) + m[(k_{y1} + k_{y4})a_1^2 + (k_{y2} + k_{y3})a_2^2]\}^2$$
$$- 4mI_{xy}\eta(\Sigma k_y)[(k_{y1} + k_{y4})a_1^2 + (k_{y2} + k_{y3})a_2^2]$$

In a similar manner, we can show that

$$\omega_{nzy}^2 = \frac{A_{zy} \pm \sqrt{B_{zy}}}{2mI_{zy}} \quad (6\text{-}221)$$

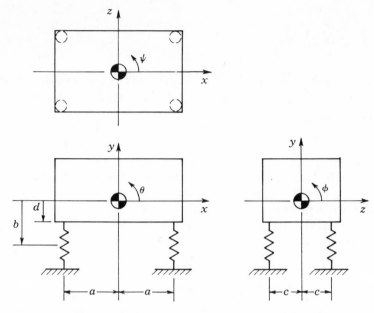

Fig. 6-33   *System with two planes of symmetry.*

where

$$A_{zy} = \eta(\Sigma k_y)(I_{zy} + bdm) + m[(k_{y1} + k_{y2})c_1^2 + (k_{y3} + k_{y4})c_2^2]$$

and

$$B_{zy} = \{\eta(\Sigma k_y)(I_{zy} + bdm) + m[(k_{y1} + k_{y2})c_1^2 + (k_{y3} + k_{y4})c_2^2]\}^2$$
$$- 4mI_{zy}\eta(\Sigma k_y)[(k_{y1} + k_{y2})c_1^2 + (k_{y3} + k_{y4})c_2^2]$$

6-7. Coupled Vibrations—Two and Three Planes of Symmetry

The discussion in the preceding section was concerned with the general case in which coordinate modes were decoupled to the extent that all six natural frequencies can be calculated without considering more than two degrees of freedom at one time.   Further simplifications can be made in the equations if additional restraints related to symmetry are introduced. In the interest of providing a simpler basis for understanding, and since the results can be used in many design situations, this section will be given to the special cases in which the isolator forces are symmetrical with the center of gravity in at least two out of the three possible planes.

If in Fig. 6-28, $|a_1| = |a_2| = a$, $|c_1| = |c_2| = c$, and

$$k_{y1} = k_{y2} = k_{y3} = k_{y4} = k_y$$

the system becomes that in Fig. 6-33.   The forces resulting from a deflection in the y direction alone are symmetrical with respect to the center of gravity in both the xy and zy planes.   This symmetry in two planes automatically makes the resultant vertical force pass through the center of gravity and, therefore, automatically provides for decoupling of the y-coordinate mode of vibration.

In Fig. 6-33, the forces resulting from an x deflection or a z deflection are symmetrical with respect to the center of gravity only in the top view. The result of this is that the ψ-coordinate mode is also automatically decoupled, whereas the x- and θ- and the z- and ϕ-coordinate modes are coupled.

For two planes of symmetry, Eq. (6-182) becomes

$$\omega_{ny} = \sqrt{\frac{4k_y}{m}} \tag{6-222}$$

Eq. (6-205) becomes

$$\omega_{nxz} = \sqrt{\frac{4\eta k_y(a^2 + c^2)}{I_{xz}}} \tag{6-223}$$

Eq. (6-220) becomes

$$\omega_{nxy}^2 = \frac{2[\eta k_y I_{xy} + m(\eta k_y bd + k_y a^2)]}{mI_{xy}}$$
$$\pm \frac{2\sqrt{[\eta k_y I_{xy} + m(\eta k_y bd + k_y a^2)]^2 - 4\eta I_{xy}mk_y^2 a^2}}{mI_{xy}} \tag{6-224}$$

and Eq. (6-221) becomes

$$\omega_{nzy}^2 = \frac{2[\eta k_y I_{zy} + m(\eta k_y bd + k_y c^2)]}{mI_{zy}}$$
$$\pm \frac{2\sqrt{[\eta k_y I_{zy} + m(\eta k_y bd + k_y c^2)]^2 - 4\eta I_{zy}mk_y^2 c^2}}{mI_{zy}} \tag{6-225}$$

Since the moment of inertia is related to the mass by the square of the radius of gyration ($I = m\rho^2$), we can rewrite Eqs. (6-223), (6-224), and (6-225) as, respectively,

$$\omega_{nxz} = \sqrt{\frac{4k_y}{m}} \sqrt{\frac{\eta(a^2 + c^2)}{\rho_{xz}^2}} \tag{6-226}$$

$$\omega_{nxy}^2 = \frac{1}{2}\left(\frac{4k_y}{m}\right)\left\{\eta\left(1 + \frac{bd}{\rho_{xy}^2}\right) + \frac{a^2}{\rho_{xy}^2}\right.$$
$$\left. \pm \sqrt{\left[\eta\left(1 + \frac{bd}{\rho_{xy}^2}\right) + \frac{a^2}{\rho_{xy}^2}\right]^2 - \frac{4\eta a^2}{\rho_{xy}^2}}\right\} \tag{6-227}$$

and

$$\omega_{nzy}^2 = \frac{1}{2}\left(\frac{4k_y}{m}\right)\left\{\eta\left(1 + \frac{bd}{\rho_{zy}^2}\right) + \frac{c^2}{\rho_{zy}^2}\right.$$
$$\left. \pm \sqrt{\left[\eta\left(1 + \frac{bd}{\rho_{zy}^2}\right) + \frac{c^2}{\rho_{zy}^2}\right]^2 - \frac{4\eta c^2}{\rho_{zy}^2}}\right\} \tag{6-228}$$

Recognizing that $\sqrt{4k_y/m} = \omega_{ny}$, Eqs. (6-226) through (6-228) can be rewritten as, respectively,

$$\frac{\omega_{nxz}}{\omega_{ny}} = \sqrt{\frac{\eta(a^2 + c^2)}{\rho_{xz}{}^2}} \qquad (6\text{-}229)$$

$$\frac{\omega_{nxy}^2}{\omega_{ny}{}^2} = \frac{1}{2}\left\{\eta\left(1 + \frac{bd}{\rho_{xy}{}^2}\right) + \frac{a^2}{\rho_{xy}{}^2} \right.$$
$$\left. \pm \sqrt{\left[\eta\left(1 + \frac{bd}{\rho_{xy}{}^2}\right) + \frac{a^2}{\rho_{xy}{}^2}\right]^2 - \frac{4\eta a^2}{\rho_{xy}{}^2}}\right\} \qquad (6\text{-}230)$$

and

$$\frac{\omega_{nzy}^2}{\omega_{ny}{}^2} = \frac{1}{2}\left\{\eta\left(1 + \frac{bd}{\rho_{zy}{}^2}\right) + \frac{c^2}{\rho_{zy}{}^2} \right.$$
$$\left. \pm \sqrt{\left[\eta\left(1 + \frac{bd}{\rho_{zy}{}^2}\right) + \frac{c^2}{\rho_{zy}{}^2}\right]^2 - \frac{4\eta c^2}{\rho_{zy}{}^2}}\right\} \qquad (6\text{-}231)$$

To decouple the x- and the θ-coordinate modes in the xy plane requires that the forces in the x direction be symmetrical about the center of gravity in the xy and the xz planes. This would indicate that the only additional requirement for the system in Fig. 6-33 is that $b = 0$. However, as discussed in relation to Eqs. (6-213) and (6-214), both b and d must be zero for complete decoupling. However, if $b = -d$, the coupling will be minimal and in most cases if either b or d equals zero, the effect of coupling can still be ignored. It should be noted that the above discussion applies also to the z- and ϕ-coordinate modes in the zy plane.

When $|a_1| = |a_2|$, $|c_1| = |c_2|$, the springs are identical, and $b = -d$, we can consider, for most purposes, that we have *three planes of symmetry* and that all six modes of vibration are decoupled. For this case we shall expect $bd \ll \rho_{xy}{}^2$ and $bd \ll \rho_{zy}{}^2$, and Eqs. (6-230) and (6-231) become, respectively,

$$\frac{\omega_{nxy}^2}{\omega_{ny}{}^2} = \frac{1}{2}\left[\eta + \frac{a^2}{\rho_{xy}{}^2} \pm \left(\eta - \frac{a^2}{\rho_{xy}{}^2}\right)\right] \qquad (6\text{-}232)$$

and

$$\frac{\omega_{nzy}^2}{\omega_{ny}{}^2} = \frac{1}{2}\left[\eta + \frac{c^2}{\rho_{zy}{}^2} \pm \left(\eta - \frac{c^2}{\rho_{zy}{}^2}\right)\right] \qquad (6\text{-}233)$$

Solving for the roots of Eq. (6-232) gives

$$\frac{\omega_{nxy_1}}{\omega_{ny}} = \sqrt{\eta} \qquad (6\text{-}234)$$

and

$$\frac{\omega_{nxy_2}}{\omega_{ny}} = \frac{a}{\rho_{xy}} \qquad (6\text{-}235)$$

Similarly, from Eq. (6-233),

$$\frac{\omega_{nzy_1}}{\omega_{ny}} = \sqrt{\eta} \qquad (6\text{-}236)$$

and

$$\frac{\omega_{nzy_2}}{\omega_{ny}} = \frac{c}{\rho_{zy}} \qquad (6\text{-}237)$$

With "three planes of symmetry" we now have the possibility of selecting parameters so that all six natural frequencies are identical. From Eqs. (6-229) and (6-234) through (6-237), the requirements for identical natural frequencies are $\eta = 1$, $a = \rho_{xy}$, and $c = \rho_{zy}$.

The response of a system with six degrees of freedom will be the sum of the responses of each of the modes to the forcing function. In many cases the line of action of the forcing function is such that all of the modes are excited at one or more frequencies. The complete solution is so complex that the efforts of the designer are often limited to avoiding resonance in all of the modes. It is usually desirable to decouple as many coordinate modes as possible. As shown, two planes of symmetry permit breaking the system down into two systems, each with a single degree of freedom, and two systems, each with two degrees of freedom, whereas three (or almost three) planes of symmetry permit breaking the system down into six systems, each with a single degree of freedom. It was also shown that with "three" planes of symmetry the proper choice of parameters would provide the same value of natural frequency for all modes. Although not so obvious, when bd is not negligibly small in comparison with the square of the radii of gyration, the quantities under the radicals in Eqs. (6-230) and (6-231) become larger relative to the terms not under the radicals and the natural frequencies are spread further apart than when bd is negligibly small. Thus, designing for "three" planes of symmetry can be useful when it is desired to keep the six natural frequencies close together.

6-8. Torsional Vibrations of Systems with Many Degrees of Freedom   The general method of applying Newton's second law to each mass or rotor to arrive at the set of differential equations describing the motions in the system, as discussed in earlier sections in this chapter, can be extended to systems with any number of degrees of freedom. When there are more than two degrees of freedom (or three that degenerate into two) the work involved in obtaining an analytical solution of the set of simultaneous differential equations becomes formidable for even the simplest type of forcing function.

If the system is nonlinear or if the response to an arbitrary forcing function is required, the use of an analog computer is indicated. However, if the system is linear and the designer's needs can be satisfied by knowing only the critical frequencies and the response to sinusoidal forcing functions, *and if the constraints on the system are such that the position of each mass or inertial body can be defined completely by use of a single coordinate, and if the forces acting on each body can be specified completely in terms of the coordinates of the body itself and the immediately adjacent bodies,* as in Fig. 6-34a and b but *not c,* the following discussion can be applied with reasonable ease and great effectiveness.

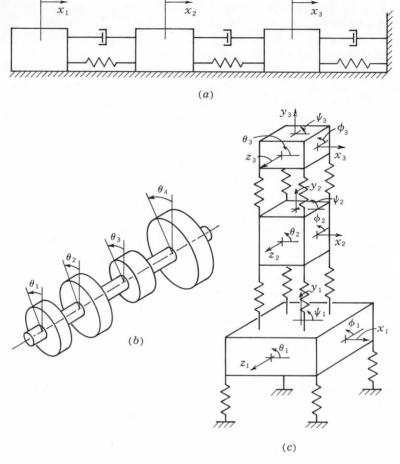

Fig. 6-34   *Systems with many degrees of freedom.*

The critical frequencies and sinusoidal response of systems such as in Fig. 6-34a and b can be found by making use of the following observations:

1   As discussed in Sec. 4-14, a critical frequency is defined as a frequency at which the response to a given amplitude of forcing function is a maximum.

2   For linear systems the amplitude of the response is directly (linearly) proportional to the amplitude of the forcing function.

Observation 1 is used in determining the critical frequencies, and observation 2 is used in determining the sinusoidal response to a forcing function applied at a particular point in the system. Although the method can be applied equally well to systems with translatory motion

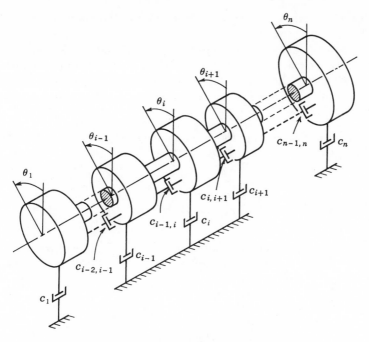

Fig. 6-35 *General multirotor torsional system.*

and to systems with oscillatory motion, only the latter will be considered here, because there are many more real situations in which the oscillatory model is a good representation of the true system.   If the need arises, the following discussion can be made to apply directly to translatory systems by substituting the equivalent translation terms for the rotation terms.

The system shown in Fig. 6-35 can be considered to be the general case of a multirotor torsional system.   The mass of the shaft is considered negligible, the dashpots or dampers between the rotors and the frame represent energy dissipation in bearings, windage losses, etc., and the dashpots between the rotors represent energy dissipation by internal or solid damping, relative motion in flexible couplings, etc.

Critical frequencies   To determine the critical frequencies, we shall make use of observation 1, above, by restating it as: *For a specified amplitude of response, the amplitude of forcing function required to maintain the vibration will be a minimum at a critical frequency.*   The procedure will be to assume an amplitude of oscillation of one rotor at a given frequency and find the forcing function at some point required to maintain the oscillation.   The calculation must be carried out a sufficient number of times to permit drawing a curve of torque amplitude vs. frequency from which the critical frequencies, corresponding to the minimum amplitudes of torque, can be found.

The simplest approach is to specify the amplitude of oscillation of the rotor at one end of the shaft and find the torque amplitude required on the rotor at the other end. For the system in Fig. 6-35, let us assume that the torque will be applied to rotor n. Applying Newton's second law to this case results in

$$I_1 D^2\theta_1 + (c_1 + c_{1,2}) D\theta_1 + k_{1,2}\theta_1 - c_{1,2} D\theta_2 - k_{1,2}\theta_2 = 0 \quad (6\text{-}238)$$
$$I_i D^2\theta_i + (c_i + c_{i-1,i} + c_{i,i+1}) D\theta_i + (k_{i-1,i} + k_{i,i+1})\theta_i$$
$$- c_{i-1,i} D\theta_{i-1} - k_{i-1,i}\theta_{i-1} - c_{i,i+1} D\theta_{i+1} - k_{i,i+1}\theta_{i+1} = 0 \quad (6\text{-}239)$$
and

$$I_n D^2\theta_n + (c_n + c_{n-1,n}) D\theta_n + k_{n-1,n}\theta_n - c_{n-1,n} D\theta_{n-1}$$
$$- k_{n-1,n}\theta_{n-1} = T \quad (6\text{-}240)$$

Since we shall be assuming the amplitude of oscillation of rotor 1, it will be convenient to specify phase angles relative to θ_1. Therefore, we shall write

$$\theta_1 = \Theta_1 e^{j\omega t} \quad (6\text{-}241)$$
$$\theta_i \bigg|_{i=2 \text{ to } n} = \Theta_i e^{j(\omega t + \phi_i)} \quad (6\text{-}242)$$

and
$$T = T_0 e^{j(\omega t + \phi_T)} \quad (6\text{-}243)$$

where Θ_i = amplitude of oscillation of rotor i

ϕ_i = phase angle of Θ_i relative to Θ_1

T_0 = amplitude of external torque

ϕ_T = phase angle of T_0 relative to Θ_1

Substituting from Eqs. (6-241) through (6-243) and their derivatives into Eqs. (6-238) through (6-240) and simplifying results in, respectively,

$$[k_{1,2} - \omega^2 I_1 + j\omega(c_1 + c_{1,2})]\Theta_1 - (k_{1,2} + j\omega c_{1,2})\Theta_2 e^{j\phi_2} = 0 \quad (6\text{-}244)$$
$$- (k_{i-1,i} + j\omega c_{i-1,i})\Theta_{i-1} e^{j\phi_{i-1}}$$
$$+ [(k_{i-1,i} + k_{i,i+1} - \omega^2 I_i) + j\omega(c_i + c_{i-1,i} + c_{i,i+1})]\Theta_i e^{j\phi_i}$$
$$- (k_{i,i+1} + j\omega c_{i,i+1})\Theta_{i+1} e^{j\phi_{i+1}} = 0 \quad (6\text{-}245)$$

and

$$- (k_{n-1,n} + j\omega c_{n-1,n})\Theta_{n-1} e^{j\phi_{n-1}}$$
$$+ [k_{n-1,n} - \omega^2 I_n + j\omega(c_n + c_{n-1,n})]\Theta_n e^{j\phi_n} = T_0 e^{j\phi_T} \quad (6\text{-}246)$$

From Eq. (6-244) we can write

$$\Theta_2 = \Theta_1 \frac{\sqrt{(k_{1,2} - \omega^2 I_1)^2 + [\omega(c_1 + c_{1,2})]^2}\; e^{j\alpha_1}}{\sqrt{k_{1,2}^2 + (\omega c_{1,2})^2}\; e^{j\alpha_2} e^{j\phi_2}} \quad (6\text{-}247)$$

where
$$\alpha_1 = \tan^{-1} \frac{\omega(c_1 + c_{1,2})}{k_{1,2} - \omega^2 I_1} \quad (6\text{-}248)$$

and
$$\alpha_2 = \tan^{-1} \frac{\omega c_{1,2}}{k_{1,2}} \quad (6\text{-}249)$$

Since Θ_1 and Θ_2 have magnitude only, from Eq. (6-247) we can write

$$\alpha_1 = \alpha_2 + \phi_2$$

which can be rearranged as

$$\phi_2 = \alpha_1 - \alpha_2 \tag{6-250}$$

Letting $\Theta_1 = 1$ rad, as discussed above, we find

$$\Theta_2 = \frac{\sqrt{(k_{1,2} - \omega^2 I_1)^2 + [\omega(c_1 + c_{1,2})]^2}}{\sqrt{k_{1,2}{}^2 + (\omega c_{1,2})^2}} \tag{6-251}$$

Thus, Eqs. (6-251) and (6-250) describe completely the motion of rotor 2 that is required to make rotor 1 oscillate with an amplitude of 1 rad. Knowing Θ_2 and ϕ_2, we next find Θ_3 and ϕ_3. The process is repeated step by step until the last rotor is reached and T_0 and ϕ_T are determined.

Handling the phase angles becomes somewhat more complicated than before and further discussion is desirable at this point. Considering rotor $i + 1$, we shall find, from Eq. (6-245),

$$\Theta_{i+1} = \frac{\sqrt{(k_{i-1,i} + k_{i,i+1} - \omega^2 I_i)^2 + [\omega(c_i + c_{i-1,i} + c_{i,i+1})]^2}\; \Theta_i e^{j(\phi_i + \alpha_1)}}{\sqrt{k_{i,i+1}^2 + (\omega c_{i,i+1})^2}\; e^{j(\phi_{i+1} + \alpha_3)}}$$
$$- \frac{\sqrt{k_{i-1,i}^2 + (\omega c_{i-1,i})^2}\; \Theta_{i-1} e^{j(\phi_{i-1} + \alpha_2)}}{\sqrt{k_{i,i+1}^2 + (\omega c_{i,i+1})^2}\; e^{j(\phi_{i+1} + \alpha_3)}} \tag{6-252}$$

Every term except ϕ_{i+1} in Eq. (6-252) will be known from previous steps or can now be calculated directly. The problem here is that the numerator contains two terms with different phase angles and we must combine them into a single term with an amplitude and a phase angle. To do this, let us rewrite Eq. (6-252) as

$$\Theta_{i+1} = \frac{C_i e^{j(\phi_i + \alpha_1)} - C_{i-1} e^{j(\phi_{i-1} + \alpha_2)}}{C_{i+1} e^{j(\phi_{i+1} + \alpha_3)}} = \frac{C e^{j\beta}}{C_{i+1} e^{j(\phi_{i+1} + \alpha_3)}} \tag{6-253}$$

where $C_i = \sqrt{(k_{i-1,i} + k_{i,i+1} - \omega^2 I_i)^2 + [\omega(c_i + c_{i-1,i} + c_{i,i+1})]^2}\; \Theta_i$
$\quad\quad C_{i-1} = \sqrt{k_{i-1,i}^2 + (\omega c_{i-1,i})^2}\; \Theta_{i-1}$
$\quad\quad C_{i+1} = \sqrt{k_{i,i+1}^2 + (\omega c_{i,i+1})^2}$

Since Θ_{i+1} has magnitude only, we can write, from Eq. (6-253),

$$\Theta_{i+1} = \frac{C}{C_{i+1}} \tag{6-254}$$

and

$$\phi_{i+1} + \alpha_3 = \beta \tag{6-255}$$

To determine C and β we observe that, as shown in Fig. 6-36, the numerator of Eq. (6-253) is the sum of two vectors, $C_i e^{j(\phi_i + \alpha_1)}$ and $-C_{i-1} e^{j(\phi_{i-1} + \alpha_2)}$. Although the vector $Ce^{j\beta}$ can be determined graphically,

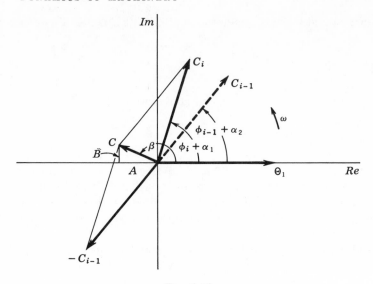

Fig. 6-36

as in Fig. 6-36, it can also be found analytically by noting that the real and imaginary parts of $Ce^{j\beta}$ are the sums of the real and imaginary parts, respectively, of the vectors $C_i e^{j(\phi_i + \alpha_1)}$ and $-C_{i-1} e^{j(\phi_{i-1} + \alpha_2)}$. Thus, we write

$$C_i e^{j(\phi_i + \alpha_1)} = C_i \cos(\phi_i + \alpha_1) + jC_i \sin(\phi_i + \alpha_1) \qquad (6\text{-}256)$$

and

$$-C_{i-1} e^{j(\phi_{i-1} + \alpha_2)} = -C_{i-1} \cos(\phi_{i-1} + \alpha_2) - jC_{i-1} \sin(\phi_{i-1} + \alpha_2) \qquad (6\text{-}257)$$

Adding Eqs. (6-256) and (6-257) gives

$$Ce^{j\beta} = A + jB \qquad (6\text{-}258)$$

where $\qquad A = C_i \cos(\phi_i + \alpha_1) - C_{i-1} \cos(\phi_{i-1} + \alpha_2)$

and $\qquad B = C_i \sin(\phi_i + \alpha_1) - C_{i-1} \sin(\phi_{i-1} + \alpha_2)$

Equation (6-258) can then be written as

$$Ce^{j\beta} = \sqrt{A^2 + B^2}\, e^{j\gamma} \qquad (6\text{-}259)$$

where $\gamma = \tan^{-1}(B/A)$. Therefore, from Eq. (6-259),

$$C = \sqrt{A^2 + B^2} \qquad (6\text{-}260)$$

and $\qquad\qquad \beta = \gamma = \tan^{-1}\dfrac{B}{A} \qquad (6\text{-}261)$

and θ_{i+1} and ϕ_{i+1} can now be determined by substituting from Eqs. (6-260) and (6-261) into Eqs. (6-254) and (6-255).

The values of Θ_n and ϕ_n will have been calculated, as above, when we reach rotor $n - 1$. Thus, the amplitude T_0 and phase angle ϕ_T of the external torque required to maintain the vibration are the remaining unknowns. From Eq. (6-246), we can write

$$T_0 = \sqrt{(k_{n-1,n} - \omega^2 I_n)^2 + [\omega(c_n + c_{n-1,n})]^2}\; \Theta_n e^{j(\phi_n + \alpha_1 - \phi_T)}$$
$$- \sqrt{k_{n-1,i}^2 + (\omega c_{n-1,n})^2}\; \Theta_{n-1} e^{j(\phi_{n-1} + \alpha_2 - \phi_T)} \quad (6\text{-}262)$$

which in turn can be written as

$$T_0 = (C_n e^{j(\phi_n + \alpha_1)} - C_{n-1} e^{j(\phi_{n-1} + \alpha_2)}) e^{-j\phi_T} \quad (6\text{-}263)$$

where $\quad C_n = \sqrt{(k_{n-1,n} - \omega^2 I_n)^2 + [\omega(c_n + c_{n-1,n})]^2}\; \Theta_n$

and $\quad C_{n-1} = \sqrt{k_{n-1,n}^2 + (\omega c_{n-1,n})^2}\; \Theta_{n-1}$

As discussed above, the vector sum in the right-hand side of Eq. (6-263) must be performed in order to separate the magnitude from the phase angle. Thus, we want to write

$$T_0 = C e^{j\beta} e^{-j\phi_T} \quad (6\text{-}264)$$

from which
$$T_0 = C \quad (6\text{-}265)$$

and $\qquad \phi_T = \beta \quad (6\text{-}266)$

In many cases, the damping is so small that assuming it to be zero results in insignificant errors in values of critical frequencies and in values of sinusoidal response at all frequencies except those very close, say, within 5 percent, of a critical frequency. Assuming the damping to be zero greatly simplifies the calculations because there are no $j\omega$ terms and the phase angles are either $0°$ or $\pm 180°$, or a multiple thereof, and $e^{j\phi}$ can be written as $+1$ when $\phi = 0°$, $\pm 360°$, etc., and as -1 when $\phi = \pm 180°$, $\pm 540°$, etc.

For zero damping, Eqs. (6-244) through (6-246) become, respectively,

$$(k_{1,2} - \omega^2 I_1)\Theta_i - k_{1,2}\Theta_2 e^{j\phi_2} = 0 \quad (6\text{-}267)$$
$$-k_{i-1,i}\Theta_{i-1} e^{j\phi_{i+1}} + (k_{i-1,i} + k_{i,i+1} - \omega^2 I_i)\Theta_i e^{j\phi_i}$$
$$- k_{i,i+1}\Theta_{i+1} e^{j\phi_{i+1}} = 0 \quad (6\text{-}268)$$

and
$$-k_{n-1,n}\Theta_{n-1} e^{j\phi_{n-1}} + (k_{n-1,n} - \omega^2 I_n)\Theta_n e^{j\phi_n} = T_0 e^{j\phi_T} \quad (6\text{-}269)$$

From Eq. (6-267), we find

$$\Theta_2 = \frac{(k_{1,2} - \omega^2 I_1)\Theta_1}{k_{1,2} e^{j\phi_2}} \quad (6\text{-}270)$$

Since Θ_2 has magnitude only, $e^{j\phi_2}$ must equal $+1$, and therefore $\phi_2 = 0°$, when $k_{1,2} - \omega^2 I_1$ is positive. Conversely, when the numerator is negative, $e^{j\phi_2}$ must equal -1, and therefore $\phi_2 = \pm 180°$ or simply $180°$.
From Eq. (6-268),

$$\Theta_{i+1} = \frac{(k_{i-1,i} + k_{i,i+1} - \omega^2 I_i)\Theta_i e^{j\phi_i} - k_{i-1,i}\Theta_{i-1} e^{j\phi_{i-1}}}{k_{i,i+1} e^{j\phi_{i+1}}} \quad (6\text{-}271)$$

Θ_i, ϕ_i, Θ_{i-1}, and ϕ_{i-1} will be known at this point in the calculations, and the equation can be solved for Θ_{i+1}. The numerator can be positive or negative, but Θ_{i+1} can have magnitude only. Thus, ϕ_{i+1} must be 0° if the numerator is positive and 180° if the numerator is negative.

From Eq. (6-269),

$$T_0 = \frac{(k_{n-1,n} - \omega^2 I_n)\Theta_n e^{j\phi_n} - k_{n-1,n}\Theta_{n-1}e^{j\phi_{n-1}}}{e^{j\phi_T}} \tag{6-272}$$

Again, Θ_n, ϕ_n, Θ_{n-1}, and ϕ_{n-1} will be known and Eq. (6-272) can be solved for T_0 and ϕ_T.

As just shown, the fact that phase angles are either zero or 180° when there is no damping greatly simplifies the calculation of the torque required to sustain the vibration at a given amplitude of oscillation of one rotor. An additional minor simplification can be made by letting the amplitude be positive or negative, thereby indicating it is in phase ($\phi = 0°$) or out of phase ($\phi = 180°$) with respect to the reference rotor.

The procedure can be reduced to filling in blanks in a table and is particularly well suited for solution by use of a digital computer. The tabular method is known as Holzer's method.[1]

Frequency response   To determine the response to a forcing function having a specified frequency and acting on a particular rotor, the procedure outlined above must be modified somewhat. In this case, we still assume the amplitude of oscillation at one end of the system and then work our way along the shaft, finding the amplitudes of oscillation necessary at each rotor until we reach the one at which the forcing function acts. At this point we must assume an amplitude for the forcing function and add it in as another torque. The phase angle between the rotor oscillation and the forcing function cannot be determined at this time, and the problem becomes quite complicated and lengthy if damping is not negligible. After adding in the forcing function, the solution must be continued until the external torque required on the last rotor has been calculated. When both the assumed torque amplitude and the assumed phase angle of the forcing function are correctly related to the value initially assumed for the first rotor, the external torque required on the last rotor will be zero.

Since both the amplitude and phase angle of the forcing function must be found, a large number of iterations may be required. If damping is

[1] J. P. Den Hartog, "Mechanical Vibrations," pp. 187–197, McGraw-Hill Book Company, New York, 1956.

W. T. Thomson, "Mechanical Vibrations," 2d ed., pp. 145–157, Prentice-Hall, Inc., Englewood Cliffs, N.J., 1953.

C. M. Harris and C. E. Crede (eds.), "Shock and Vibration Handbook," vol. III, pp. 38-6–38-25, McGraw-Hill Book Company, New York, 1961.

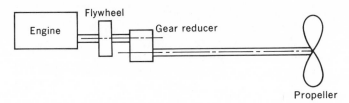

Fig. 6-37

small, the phase angle will be about 0 or 180°, except when close to a natural frequency, and the more nearly correct value will be that requiring the lower magnitude of external torque on the last rotor.   In general, the approach to follow will be to use the phase angle of 0 or 180°, whichever has been found to be closer to the correct value, and carry out enough solutions with different magnitudes of forcing function that the value resulting in a minimum magnitude of external torque can be read from a plot of external-torque amplitude vs. forcing-function amplitude.   If the external-torque amplitude is too far from zero for engineering purposes, the phase angle must be adjusted and a new series of solutions made. The process can be repeated until the external-torque amplitude is as small as desired.

Since, from observation 2 above, the amplitude of oscillation is directly proportional to the amplitude of the forcing function, we can now find the amplitude of oscillation of every rotor for any desired amplitude of forcing function—for the given frequency and point of application used in the iterative solution just discussed.

Example 6-1   The proposed drive for an ocean-going fishing vessel is shown schematically in Fig. 6-37.   We are asked to determine (*a*) the critical frequencies for torsional vibrations of the system and (*b*) the maximum amplitude of variable shear stress, neglecting stress concentration, in the shaft resulting from the second-order harmonic of the engine torque[1] when the diesel engine is running at its rated speed of 850 rpm and the amplitude of the second-order harmonic is 15,000 lb-in.

Solution.   General.   Since we shall be interested in determining the sinusoidal response to a forcing function that originates in the engine, we shall find it most convenient to designate the propeller as the first rotor and the engine inertia as the last rotor.   To minimize confusion, by continuing to work from left to right, the system is redrawn in Fig. 6-38. I_5 is the effective rotating inertia of the internal parts of the 4-cylinder 4-cycle diesel engine.   The diameters of the steel shafts, as well as other dimensions, have been determined by calculations based on space considerations and on assuming maximum inertia and load torques as steady or static values.

[1] For discussions of methods of calculation and tables of values of harmonic coefficients, see F. P. Porter, Harmonic Coefficients of Engine Torque Curves, *Trans. ASME*, 1943, or "Design Data and Methods," pp. 173–188, American Society of Mechanical Engineers, New York, 1953.
Den Hartog, *op. cit.*, pp. 197–206.
Harris and Crede, *op. cit.*, vol. III, pp. 38-15–38-20.

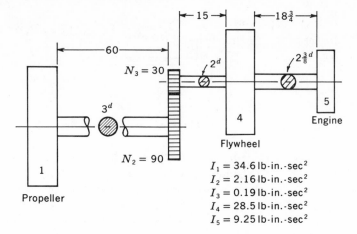

Fig. 6-38

We shall assume that damping is negligible and that the gear teeth are so much stiffer than the shafts that elasticity or flexibility at this point can be neglected. Thus, Eqs. (6-270) through (6-272) will be applicable.

Letting $\Theta_1 = 1$ rad, Eq. (6-270) becomes

$$\Theta_2 = \frac{k_{1,2} - \omega^2 I_1}{k_{1,2} e^{j\phi_2}} \tag{6-273}$$

where $\phi_2 = 0°$ if the numerator is positive and $180°$ if the numerator is negative.

For $i = 2$, Eq. (6-271) becomes

$$\Theta_3 = \frac{(k_{1,2} + k_{2,3} - \omega^2 I_2)\Theta_2 e^{j\phi_2} - k_{1,2}\Theta_1 e^{j\phi_1}}{k_{2,3} e^{j\phi_3}} \tag{6-274}$$

However, Eq. (6-274) is directly useful only when there is an elastic member between rotor 2 and the rotor on each side of it. In this case, rotor 3 is a gear and we have assumed that the teeth are rigid. Consequently, rotors 2 and 3 must be considered simultaneously. When only critical frequencies are to be determined, it is convenient to simplify the calculations by deriving an equivalent system without gears. This approach will be developed in the closure below, but to emphasize the behavior of the real system we shall continue with the gears in place.

Applying Newton's second law to gear 2, we find

$$-k_{1,2}\theta_1 + I_2 D^2\theta_2 + k_{1,2}\theta_2 - F_{32}R_2 = 0 \tag{6-275}$$

where F_{32} is the transmitted force of gear 3 on gear 2 (considered positive when $F_{32}R_2$ is a positive torque) and R_2 is the pitch radius of gear 2.

Similarly, for gear 3 we find

$$-F_{23}R_3 + I_3 D^2\theta_3 + k_{3,4}\theta_3 - k_{3,4}\theta_4 = 0 \tag{6-276}$$

From kinematics of gear trains, we know

$$R_3 = \frac{1}{r} R_2 \tag{6-277}$$

and

$$\theta_3 = r\theta_2 e^{j180°} = -r\theta_2 \tag{6-278}$$

where $r = N_2/N_3$ = output speed/input speed, considering gear 2 as the input or reference member.

Applying Newton's third law to the forces acting between the gear teeth, we find that

$$|F_{23}| = |F_{32}| \tag{6-279}$$

We also find that, for spur gears, when F_{32} results in a positive torque on gear 2, F_{23} results in a positive torque on gear 3. Therefore, we can keep the signs straight by simply writing

$$F_{23} = F_{32} \tag{6-280}$$

Substituting from Eqs. (6-277) through (6-280) into Eq. (6-276) results in

$$-\frac{1}{r} F_{32} R_2 - r I_3 D^2 \theta_2 - r k_{3,4} \theta_2 - k_{3,4} \theta_4 = 0 \tag{6-281}$$

Multiplying Eq. (6-281) by $-r$ gives

$$F_{32} R_2 + r^2 I_3 D^2 \theta_2 + r^2 k_{3,4} \theta_2 + r k_{3,4} \theta_4 = 0 \tag{6-282}$$

Adding Eqs. (6-275) and (6-282) results in

$$-k_{1,2}\theta_1 + (I_2 + r^2 I_3) D^2 \theta_2 + (k_{1,2} + r^2 k_{3,4})\theta_2 + r k_{3,4} \theta_4 = 0 \tag{6-283}$$

Letting $\theta_1 = \Theta_1 e^{i\omega t}$, $\theta_2 = \Theta_2 e^{i(\omega t + \phi_2)}$, and $\theta_4 = \Theta_4 e^{i(\omega t + \phi_4)}$, and substituting these values and their derivatives into Eq. (6-283) results in

$$\Theta_4 = \frac{k_{1,2}\Theta_1 - [k_{1,2} + r^2 k_{3,4} - (I_2 + r^2 I_3)\omega^2]\Theta_2 e^{i\phi_2}}{r k_{3,4} e^{i\phi_4}} \tag{6-284}$$

where $\phi_4 = 0°$ if the numerator is positive and $180°$ if the numerator is negative.

For $i = 4$, Eq. (6-271) becomes

$$\Theta_5 = \frac{(k_{3,4} + k_{4,5} - \omega^2 I_4)\Theta_4 e^{i\phi_4} - k_{3,4}\Theta_3 e^{i\phi_3}}{k_{4,5} e^{i\phi_5}} \tag{6-285}$$

But, from Eq. (6-278),

$$\Theta_3 e^{i\phi_3} = r\Theta_2 e^{i(\phi_2 + 180°)} = -r\Theta_2 e^{i\phi_2} \tag{6-286}$$

Thus, Eq. (6-285) can be written as

$$\Theta_5 = \frac{(k_{3,4} + k_{4,5} - \omega^2 I_4)\Theta_4 e^{i\phi_4} + r k_{3,4}\Theta_2 e^{i\phi_2}}{k_{4,5} e^{i\phi_5}} \tag{6-287}$$

where $\phi_5 = 0°$ if the numerator is positive and $180°$ if the numerator is negative.

In this problem $n = 5$, and Eq. (6-272) becomes

$$T_0 = \frac{(k_{4,5} - \omega^2 I_5)\Theta_5 e^{i\phi_5} - k_{4,5}\Theta_4 e^{i\phi_4}}{e^{i\phi_T}} \tag{6-288}$$

where $\phi_T = 0°$ if the numerator is positive and $180°$ if the numerator is negative.

From elementary strength of materials, the torsional spring rate of a shaft is

$$k = \frac{\Delta T}{\Delta \theta} = \frac{JG}{L} \qquad \text{lb-in./rad} \tag{6-289}$$

where J = polar moment of inertia of the section = $\pi d^4/32$ for a solid circular shaft
G = torsional modulus of elasticity = 11.5×10^6 psi for steel
L = length of shaft, in.
From Fig. 6-38 and Eq. (6-289),

$$k_{1,2} = \frac{\pi \times 3^4 \times 11.5 \times 10^6}{32 \times 60} = 1.525 \times 10^6 \text{ lb-in./rad}$$

$$k_{3,4} = \frac{\pi \times 2^4 \times 11.5 \times 10^6}{32 \times 15} = 1.205 \times 10^6 \text{ lb-in./rad}$$

$$k_{4,5} = \frac{\pi \times 2.375^4 \times 11.5 \times 10^6}{32 \times 18.75} = 1.917 \times 10^6 \text{ lb-in./rad}$$

Substituting these values of k and $r = N_2/N_3 = 90/30 = 3$ into Eqs. (6-273), (6-286), (6-284), (6-287), and (6-288) and simplifying gives, respectively, for $\Theta_1 = 1$ rad,

$$\Theta_2 = \frac{1.525 - \omega^2 \times 34.6 \times 10^{-6}}{1.525e^{j\phi_2}} \tag{6-290}$$

$$\Theta_3 = 3\Theta_2 \quad \text{and} \quad \phi_3 = \phi_2 \pm 180° \tag{6-291}$$

$$\Theta_4 = \frac{1.525 - (12.35 - \omega^2 \times 3.87 \times 10^{-6})\Theta_2 e^{j\phi_2}}{3.61e^{j\phi_4}} \tag{6-292}$$

$$\Theta_5 = \frac{(3.12 - \omega^2 \times 28.5 \times 10^{-6})\Theta_4 e^{j\phi_4} + 3.61\Theta_2 e^{j\phi_2}}{1.918e^{j\phi_5}} \tag{6-293}$$

and $\quad T_0 = \dfrac{[(1.917 - \omega^2 \times 9.25 \times 10^{-6})\Theta_5 e^{j\phi_5} - 1.917\Theta_4 e^{j\phi_4}]10^6}{e^{j\phi_T}} \tag{6-294}$

(a) *Critical-frequency Determination.* The system in Fig. 6-37 requires four coordinates to locate the masses. However, there is no spring between the system and the hull of the ship and, as shown in Sec. 6-1, this results in a degenerate system with one less nonzero critical frequency than one might expect solely on the basis of counting masses or coordinates. Therefore, we shall expect to find three nonzero critical frequencies, and Eq. (6-294) must be solved for enough frequencies to permit locating the minimum values of T_0. Since there is no damping, the minimum values of T_0 will be zero at the critical frequencies. In other words, the case becomes that of an undamped free vibration and the critical frequencies are natural frequencies.

To illustrate the method, let us consider the case when the frequency is that of the second-order harmonic. At 850 rpm, the first-order frequency is

$$\omega_{1st} = \frac{2\pi n}{60} = \frac{2\pi \times 850}{60} = 89.0 \text{ rad/sec}$$

and for the second-order harmonic,

$$\omega_{2nd} = 2\omega_{1st} = 178.0 \text{ rad/sec}$$

Therefore, $\omega^2 = 178.0^2 = 0.0317 \times 10^6$ (rad/sec)^2. Substituting $\omega^2 = 0.0317 \times 10^6$ into Eqs. (6-290) through (6-294) leads to

$$\Theta_2 = \frac{1.525 - 0.0317 \times 10^6 \times 34.6 \times 10^{-6}}{1.525e^{j\phi_2}} = \frac{0.281}{e^{j\phi_2}}$$

from which $\Theta_2 = 0.281$ rad and $\phi_2 = 0°$,

$$\Theta_3 = 3 \times 0.281 = 0.843 \text{ rad} \quad \text{and} \quad \phi_3 = 180°$$

$$\Theta_4 = \frac{1.525 - (12.35 - 0.0317 \times 10^6 \times 3.87 \times 10^{-6})0.281}{3.61e^{j\phi_4}} = \frac{-0.530}{e^{j\phi_4}}$$

from which $\Theta_4 = 0.530$ and $\phi_4 = 180°$,

$$\Theta_5 = \frac{(3.12 - 0.0317 \times 10^6 \times 28.5 \times 10^{-6})(-0.530) + 3.61 \times 0.281}{1.917e^{j\phi_5}} = \frac{-0.086}{e^{j\phi_5}}$$

from which $\Theta_5 = 0.086$ rad and $\phi_5 = 180°$, and

$$T_0 = \frac{[(1.917 - 0.0317 \times 10^6 \times 9.25 \times 10^{-6})(-0.086) - 1.917(-0.530)]10^6}{e^{j\phi_T}}$$

$$= \frac{0.877 \times 10^6}{e^{j\phi_T}}$$

from which $T_0 = 0.877 \times 10^6$ lb-in. and $\phi_T = 0°$.

The results of the sequence of calculations are summarized in Table 6-1.

After a number of series of calculations were made, the curves in Fig. 6-39 were plotted and the natural frequencies were noted as

$$\omega_{n_1} = 205 \text{ rad/sec}$$

or

$$f_{n_1} = \frac{205}{2\pi} \times 60 = 1,958 \text{ cpm}$$

and

$$\omega_{n_2} = 524 \text{ rad/sec}$$

or

$$f_{n_2} = \frac{524}{2\pi} \times 60 = 5,010 \text{ cpm}$$

and

$$\omega_{n_3} = 1,797 \text{ rad/sec}$$

or

$$f_{n_3} = \frac{1,797}{2\pi} \times 60 = 17,160 \text{ cpm}$$

Thus, the critical frequencies are 1,958, 5,010, and 17,160 cpm.

(b) *Response to Second-order Harmonic.* Table 6-1 lists the amplitudes and phase angles that will result if a sinusoidal torque with an amplitude of 0.877×10^6 lb-in.

Table 6-1

$\omega = 178.0$ rad/sec

Rotor	Θ, rad	ϕ, deg
1	1.0	0
2	0.281	0
3	0.843	180
4	0.530	180
5	0.086	180

$T_0 = 0.877 \times 10^6$ lb-in. and $\phi_T = 0°$

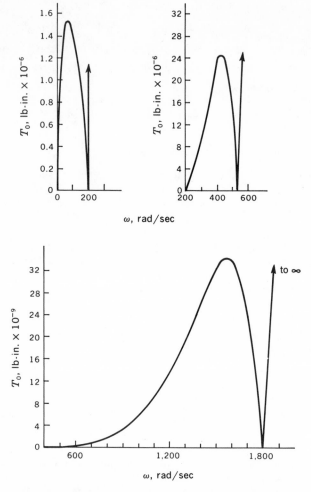

Fig. 6-39   *Amplitude of sinusoidal external torque required on rotor 5 to maintain an amplitude of oscillation of 1 rad for rotor 1 of the system in Fig. 6-38 as a function of frequency.*

and a frequency of 178.0 rad/sec (1,700 cpm) acts on rotor 5. Since the amplitudes will be proportional to the amplitude of the forcing function, the amplitudes of oscillation due to the second-harmonic torque amplitude of 15,000 lb-in. will be the values in Table 6-1 multiplied by $15,000/(0.877 \times 10^6) = 0.01710$. Thus,

$$\Theta_{1,2nd} = 0.01710 \times 1.0 = 0.01710 \text{ rad}$$
$$\Theta_{2,2nd} = 0.01710 \times 0.281 = 0.00481 \text{ rad}$$
$$\Theta_{3,2nd} = 0.01710 \times 0.843 = 0.01441 \text{ rad}$$
$$\Theta_{4,2nd} = 0.01710 \times 0.530 = 0.00907 \text{ rad}$$

and
$$\Theta_{5,2nd} = 0.01710 \times 0.086 = 0.001470 \text{ rad}$$

From elementary strength of materials the shear stress on the surface of a shaft, neglecting stress concentration, is

$$s_s = \frac{Tc}{J} = \frac{Td}{2J} \tag{6-295}$$

where c = radius of the shaft = $d/2$.

The torque acting on a section of shaft can be related to the twist of the shaft between rotors in either of the following ways:

$$T_{max} = k_t \, \Delta\theta_{max} = k_t |\theta_i - \theta_{i+1}|_{max} \tag{6-296}$$

or

$$T_{max} = \frac{JG}{L} \, \Delta\theta_{max} = \frac{JG}{L} |\theta_i - \theta_{i+1}|_{max} \tag{6-297}$$

Substituting from Eq. (6-297) into Eq. (6-295) and simplifying gives

$$s_{s,max} = \frac{Gd}{2L} |\theta_i - \theta_{i+1}|_{max} \tag{6-298}$$

The three sections of shaft have different diameters and lengths, and we must determine $s_{s,max}$ for each section. Thus, we must find $|\theta_i - \theta_{i+1}|_{max}$ for each section. In general,

$$\Delta\theta_{max} = |\theta_i - \theta_{i+1}|_{max} = |\Theta_i e^{j\phi_i} - \Theta_{i+1} e^{j\phi_{i+1}}| \tag{6-299}$$

Therefore, the twist of the shaft sections due to the second-order harmonic are as follows:

Between rotors 1 and 2

$$\Delta\theta_{max1,2} = |0.01710 e^{j0°} - 0.00481 e^{j0°}| = 0.01229 \text{ rad}$$

Between rotors 3 and 4

$$\Delta\theta_{max3,4} = |0.01441 e^{j180°} - 0.00907 e^{j180°}| = 0.00534 \text{ rad}$$

Between rotors 4 and 5

$$\Delta\theta_{max4,5} = |0.00907 e^{j180°} - 0.001470 e^{j180°}| = 0.00760 \text{ rad}$$

Substituting $G = 11.5 \times 10^6$ psi, the appropriate dimensions from Fig. 6-38, and the values of $\Delta\theta_{max}$ into Eq. (6-298), we find the maximum shear stresses, neglecting stress concentration, due to the second-order harmonic to be as follows:

Between rotors 1 and 2

$$s_{s,max} = \frac{11.5 \times 10^6 \times 3 \times 0.01229}{2 \times 60} = 3{,}530 \text{ psi}$$

Between rotors 3 and 4

$$s_{s,max} = \frac{11.5 \times 10^6 \times 2 \times 0.00534}{2 \times 15} = 4{,}090 \text{ psi}$$

Between rotors 4 and 5

$$s_{s,max} = \frac{11.5 \times 10^6 \times 2.375 \times 0.00760}{2 \times 18.75} = 5{,}540 \text{ psi}$$

Therefore, the maximum amplitude of the variable stress, neglecting stress concentration, in any section of the shaft that results from the second harmonic of the torque when the engine is running at 850 rpm, is 5,540 psi and it occurs in the shaft between the flywheel 4 and the engine 5.

Closure. This example is only an introduction to the problem of torsional vibrations of engine drive trains. The effect of many more harmonics would have to be considered. For example, the torque curve of a 4-cycle diesel engine includes harmonics for half and integral orders (for example, $\frac{1}{2}$, 1, $1\frac{1}{2}$, 2, etc.) that can be significant at orders up to 6—or even much higher under some circumstances. Porter gives values for coefficients up to the 18th harmonic.[1] In this example, with the engine running at 850 rpm, the half and integral orders would provide forcing functions at the following frequencies:

Order	Frequency, cpm
$\frac{1}{2}$	425
1	850
$1\frac{1}{2}$	1,275
2	1,700
$2\frac{1}{2}$	2,125
3	2,550
$3\frac{1}{2}$	2,975
4	3,400
$4\frac{1}{2}$	3,825
5	4,250
$5\frac{1}{2}$	4,675
6	5,100
$6\frac{1}{2}$	5,525
7	5,950

The behavior of any system in the vicinity of a critical frequency will be similar to that of a system with a single degree of freedom. Therefore, from Fig. 4-20, we can expect trouble if damping is negligible and the frequency of a forcing function lies within ±5 percent of a natural frequency.

Considering $f_{n_1} = 1,958$ cpm, we find for order 2

$$\frac{f_2}{f_{n_1}} = \frac{1,700}{1,958} = 0.868$$

and for order $2\frac{1}{2}$,

$$\frac{f_{2\frac{1}{2}}}{f_{n_1}} = \frac{2,125}{1,958} = 1.085$$

Considering $f_{n_2} = 5,010$ cpm, we find for order $5\frac{1}{2}$

$$\frac{f_{5\frac{1}{2}}}{f_{n_2}} = \frac{4,675}{5,010} = 0.933$$

and for order 6

$$\frac{f_6}{f_{n_2}} = \frac{5,100}{5,010} = 1.018$$

Although definite conclusions cannot be drawn without specific knowledge about the amplitudes of forcing functions (as for the second-order harmonic discussed above), it appears likely that order 6 will give trouble and further investigation would be required before we can consider the study completed.

[1] *Op. cit.*

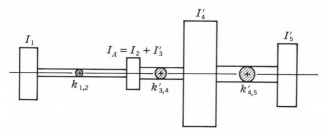

Fig. 6-40   *Single-shaft equivalent of geared system.*

Equivalent Systems.   When only critical frequencies are desired, the calculations can be simplified by developing an equivalent system in which the gears are eliminated and all rotors are on a single shaft.   For this example, the system in Fig. 6-38 becomes that in Fig. 6-40, where the gears are replaced by I_A, which is the sum of I_2 and I'_3, the equivalent inertia of gear 3 as sensed by the low-speed shaft; $k'_{3,4}$ and $k'_{4,5}$ are the spring rates of sections of the low-speed shaft that are equivalent to the high-speed shafts acting through the gear unit; and I'_4 and I'_5 are the equivalent inertias of I_4 and I_5 as sensed by the low-speed shaft.   The factors by which the actual inertias and spring rates are multiplied to obtain the equivalents can be derived in several ways.   However, our work has largely been done already.   For example, Eq. (6-276) was derived on the basis of the dimensions, inertias, angles, and spring rates related to the high-speed shaft, but these values were modified in developing Eq. (6-282), in which the gear radius R_2, $D^2\theta_2$, and θ_2 have replaced the actual values with 3 as subscripts. Thus, the equivalent inertia of I_3 referred to the low-speed shaft, i.e., to $D^2\theta_2$, is

$$I'_3 = r^2 I_3 \qquad (6\text{-}300)$$

Similarly, the equivalent spring rate of $k_{3,4}$ referred to the low-speed shaft, i.e., to θ_2, is,

$$k'_{3,4} = r^2 k_{3,4} \qquad (6\text{-}301)$$

Therefore, to replace a geared system by an equivalent system with a single shaft, we multiply the inertias and the spring rates by r^2, where r is the ratio of the speed of the output shaft to the speed of the input or reference shaft.   In this problem, $r = 3$ and

$$I'_3 = r^2 I_3 = 3^2 \times 0.19 = 1.71 \text{ lb-in.-sec}^2$$
$$I'_4 = r^2 I_4 = 3^2 \times 28.5 = 257 \text{ lb-in.-sec}^2$$
$$I'_5 = r^2 I_5 = 3^2 \times 9.25 = 83.3 \text{ lb-in.-sec}^2$$
$$k'_{3,4} = r^2 k_{3,4} = 3^2 \times 1.205 \times 10^6 = 10.84 \times 10^6 \text{ lb-in./rad}$$
and $\qquad k'_{4,5} = r^2 k_{4,5} = 3^2 \times 1.917 \times 10^6 = 17.25 \times 10^6 \text{ lb-in./rad}$

A plot of the torque amplitude vs. frequency for the system in Fig. 6-40 would lead to the same values for critical frequencies as found above, although the phase angles, amplitudes, etc., would apply directly to the equivalent system only.

In some cases approximate values of some of the critical frequencies, usually the lowest, will be sufficient, and in other cases time can be saved in making an exact analysis if approximate values are known or can be estimated.   Since it is easier to develop some feeling for behavior of systems with single shafts, the equivalent system can be quite useful under these conditions.   For example, if we redraw the equivalent system, as in Fig. 6-41, to show, qualitatively, the relative magnitudes of stiffnesses and inertias, we can often make a reasonable guess as to the mode of vibration at the critical frequencies. Then, by lumping appropriate inertias and stiffnesses together, the multi-degrees-of-freedom system can be reduced to a series of simpler systems, often with each having only a single degree of freedom. The greater the range in magnitudes of inertia and stiffness, the easier it is to break the system down into pieces that will result in useful answers.

In general, the nodes will be expected to be located fairly close to the largest inertia. Thus, the largest inertia should be included in every system. The actual breaking up of the system becomes a matter of estimating the relative importance of inertia and spring rates with respect to low and high values of natural frequency. For example, in Fig. 6-41, I_A is so small that its effect will not be expected to be significant at low frequencies. In addition, $k'_{3,4}$ is so much greater than $k_{1,2}$ that we shall expect relatively little twisting between I_A and I'_4. Thus, we shall lump I_A and I'_4 together for the first approximation. The stiffness of the shaft between I'_4 and I'_5 is over 11 times that between I_1 and I_A, and the inertia I'_5 is only about 2½ times that of I_1. Therefore, again, at low frequencies we shall expect relatively little twisting of the shaft between I'_4 and I'_5 and we shall also lump I'_5 with I_A and I'_4. The resulting approximate low-frequency system will be as shown in Fig. 6-42. By use of Eq. (6-36), we find

$$\omega_{n_1} = \sqrt{\frac{k(I_1 + I_B)}{I_1 I_B}} = \sqrt{\frac{1.527 \times 10^6 (34.6 + 344)}{34.6 \times 344}} = 220 \text{ rad/sec}$$

which agrees closely with the value of 205 rad/sec determined above for the complete system. If the spring rate for the entire shaft between I_1 and I'_4 had been used, the approximate value would have been found to be 207 rad/sec.

For higher frequencies, we can ignore I_1 because of the relatively great flexibility of the shaft between I_1 and I_A. Then, noting that $I_A \ll I'_5$, we can next approximate the system by considering it to be made up of I'_4, I'_5, and the shaft connecting them. By use of Eq. (6-36) we now find

$$\omega_{n_2} = \sqrt{\frac{k'_{4,5}(I'_4 + I'_5)}{I'_4 I'_5}} = \sqrt{\frac{17.25 \times 10^6 (257 + 83.3)}{257 \times 83.3}} = 524 \text{ rad/sec}$$

which agrees exactly with the value of 524 rad/sec found above for the complete system.

At the highest frequencies we can ignore both I_1 and I'_5 and the approximate system

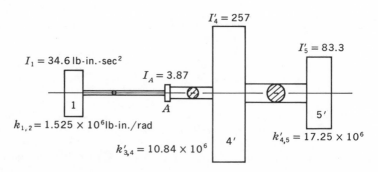

Fig. 6-41 *Single-shaft equivalent of system in Fig. 6-38.*

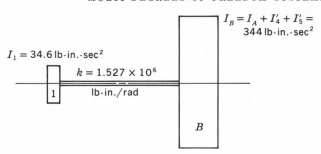

Fig. 6-42   *Approximate system for determining estimate of lowest natural frequency of system in Fig. 6-41.*

consists of I_A, I'_4, and the shaft between them.   By use of Eq. (6-36), we find

$$\omega_{n_3} = \sqrt{\frac{k'_{3,4}(I_A + I'_4)}{I_A I'_4}} = \sqrt{\frac{10.84 \times 10^6(3.87 + 257)}{3.87 \times 257}} = 1,687 \text{ rad/sec}$$

which agrees fairly well with the value of 1,797 rad/sec determined above for the complete system.

A word of caution.   The close agreement between the natural frequencies found from the analysis of the approximate systems with those found by analysis of the complete system in this example cannot be expected in every case, particularly where the inertias and spring rates of the several sections are comparable.   The breaking down of a complicated system into simple approximate systems should not be done indiscriminately but should be considered only as a means of getting a quick estimate of the lower critical frequencies.

6-9. Systems with Infinite Degrees of Freedom
The concept of a system with an infinite number of degrees of freedom has been encountered numerous times in earlier sections, and in Sec. 4-19 Rayleigh's method was presented as a means for calculating a close approximation to the lowest natural frequency of the particular case involving lateral vibrations of beams or shafts.   However, when the system is to be subjected to a forcing function with a frequency well above the fundamental natural frequency, or if the forcing function itself contains harmonics with significant amplitudes at frequencies greater than the fundamental natural frequency, the designer needs more information than can be found by applying Rayleigh's method.

In general, exact solutions for natural frequencies of systems with an infinite number of degrees of freedom can be obtained only for distributed-mass systems with relatively simple geometry, such as organ pipes, strings, membranes, and bars, rods, or beams with cross sections that are uniform or vary in special ways.   In the remaining cases, it is usually necessary to break the system down into a lumped-parameter approximation of the real system and apply methods that are extensions of

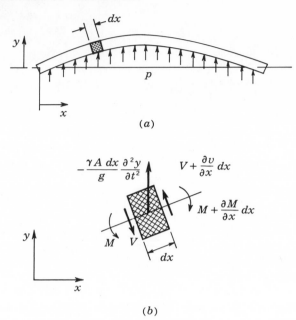

Fig. 6-43   (a) *Lateral vibrations of a uniform continuous beam;* (b) *free-body diagram showing forces and moments on an element of the beam.*

Rayleigh's method, using energy concepts, or of the approach in Sec. 6-8, using Newton's second law.

The three cases within the scope of this text are those related to undamped-free lateral, longitudinal, and torsional vibrations of bars or rods with uniform cross sections of homogeneous and isotropic materials that obey Hooke's law.   For other cases the reader should consult the literature.[1]

6-10. Lateral Vibrations of Uniform Beams   From strength of materials we know that, for small displacements, the loading and the deflection curve of a beam, such as in Fig. 6-43a, are related by

$$EI \frac{d^4y}{dx^4} = p \qquad (6\text{-}302)$$

[1] Harris and Crede, *op. cit.*, vol. I, chap. 7.
Den Hartog, *op. cit.*, chap. 4.
S. Timoshenko, "Vibration Problems in Engineering," 3d ed., chap. 5, D. Van Nostrand Company, Inc., Princeton, N.J., 1955.
Thomson, *op. cit.*, pp. 181–196 and chap. 7.
L. S. Jacobsen and R. S. Ayre, "Engineering Vibrations," chap. 10, McGraw-Hill Book Company, New York, 1958.

where E = modulus of elasticity of the material

I = moment of inertia of the beam section about the centroidal axis

p = unit loading, force per unit length

However, when the beam is vibrating, y is a function of time as well as position. Consequently, in this case we must use partial derivatives and Eq. (6-302) becomes

$$EI \frac{\partial^4 y}{\partial x^4} = p \qquad (6\text{-}303)$$

The forces and moments acting on an element of a vibrating beam are shown in Fig. 6-43b. The loading is the inertia force due to the acceleration, and it is

$$p = -\frac{\gamma A}{g} \frac{\partial^2 y}{\partial t^2} \qquad (6\text{-}304)$$

where γ = weight density of the material

A = area of the section

g = acceleration of gravity

Substituting from Eq. (6-304) into Eq. (6-303) and rearranging gives

$$\frac{\partial^4 y}{\partial x^4} + \frac{\gamma A}{gEI} \frac{\partial^2 y}{\partial t^2} = 0 \qquad (6\text{-}305)$$

Since the motion of all particles of a system with zero damping will be simple-harmonic when the system is vibrating in one of its principal modes, all points along the beam will be at their maximum deflections, in phase or out of phase, at the same time. Thus, we can write

$$y(x,t) = Y \cos \omega t \qquad (6\text{-}306)$$

where $Y = Y(x)$ = deflection curve of the beam in its extreme position.

If we now consider the instant when $t = 0$, we find from Eq. (6-306)

$$\frac{\partial^4 y}{\partial x^4}\bigg|_{t=0} = \frac{d^4 Y}{dx^4} \qquad (6\text{-}307)$$

and

$$\frac{\partial^2 y}{\partial t^2}\bigg|_{t=0} = -\omega^2 Y \qquad (6\text{-}308)$$

Since Eq. (6-305) must be satisfied for all values of t, we can substitute from Eq. (6-307) and (6-308) into Eq. (6-305). Doing this results in

$$\frac{d^4 Y}{dx^4} - \frac{\gamma A \omega^2}{gEI} Y = 0 \qquad (6\text{-}309)$$

which is an ordinary linear homogeneous fourth-order differential equation. As in Sec. 4-3, we assume the solution to be of the form

$$Y = e^{sx} \qquad (6\text{-}310)$$

from which

$$\frac{d^4Y}{dx^4} = s^4 e^{sx} \qquad (6\text{-}311)$$

Substituting from Eqs. (6-310) and (6-311) into Eq. (6-309) and simplifying results in

$$s^4 - \frac{\gamma A \omega^2}{gEI} = 0 \qquad (6\text{-}312)$$

If we let

$$\frac{\gamma A \omega^2}{gEI} = a^4 \qquad (6\text{-}313)$$

Eq. (6-312) becomes, after rearranging,

$$s^4 = a^4 \qquad (6\text{-}314)$$

and the obvious answer is $s = a$. However, the complementary solution to a fourth-order differential equation must have four constants of integration. In this case the problem is relatively simple because there are four distinct roots of Eq. (6-314), as follows:

$$s_1 = a \qquad s_2 = -a \qquad s_3 = ja \qquad s_4 = -ja \qquad (6\text{-}315)$$

Therefore, the solution to Eq. (6-309) is

$$Y = C_1 e^{ax} + C_2 e^{-ax} + C_3 e^{jax} + C_4 e^{-jax} \qquad (6\text{-}316)$$

in which the constants are determined by the boundary conditions.

Equation (6-316) can be used as it is, but it is usually more convenient to rewrite it in terms of trigonometric and hyperbolic functions. To do this we note that, from Sec. 4-4,

$$C_3 e^{jax} + C_4 e^{-jax} = C_7 \cos ax + C_8 \sin ax \qquad (6\text{-}317)$$

Also, by definition,

$$\cosh ax = \frac{e^{ax} + e^{-ax}}{2} \qquad (6\text{-}318)$$

and

$$\sinh ax = \frac{e^{ax} - e^{-ax}}{2} \qquad (6\text{-}319)$$

In terms of Eqs. (6-318) and (6-319),

$$C_1 e^{ax} + C_2 e^{-ax} = C_5 \cosh ax + C_6 \sinh ax \qquad (6\text{-}320)$$

where $C_5 = C_1 + C_2$ and $C_6 = C_1 - C_2$.

Substituting from Eqs. (6-317) and (6-320) into Eq. (6-316) gives

$$Y = C_5 \cosh ax + C_6 \sinh ax + C_7 \cos ax + C_8 \sin ax \qquad (6\text{-}321)$$

Table 6-2   Natural Frequencies for Lateral Vibrations of Uniform Bars*

Beam configuration	a_n			
	Fundamental mode, $n = 1$	Second mode, $n = 2$	Third mode, $n = 3$	Fourth mode, $n = 4$
Simply supported (hinged-hinged)	9.87	39.5	88.9	157.9
Cantilever (clamped-free)	3.52	22.0	61.7	120.9
Free-free	22.4	61.7	120.9	199.9
Clamped-clamped	22.4	61.7	120.9	199.9
Clamped-hinged	15.4	50.0	104.2	178.3
Hinged-free	0	15.4	50.0	104

* Based on values in table 7-3 in C. M. Harris and C. E. Crede (eds.) "Shock and Vibration Handbook," vol. I, pp. 7–14, and appendix V in J. P. Den Hartog, "Mechanical Vibrations," 4th ed., p. 432, McGraw-Hill Book Company, New York, 1956.

in which the constants are again determined by the boundary conditions. For example, if we have a cantilever, or clamped-free, beam, the boundary conditions at the clamped end are that the deflection (Y) and the slope (dY/dx) must be zero and that at the free end the moment (d^2Y/dx^2) and the shear (d^3Y/dx^3) must be zero. The amplitude at any particular point along the beam will be arbitrary because we have a free vibration, but the values of a, and thus ω, that will satisfy the boundary conditions will not be. Therefore, from Eq. (6-313), we find

$$\omega_n = a^2 \sqrt{\frac{gEI}{\gamma A}} \tag{6-322}$$

Based upon solutions for many cases, Eq. (6-322) can be put into more useful form as

$$\omega_n = a_n \sqrt{\frac{gEI}{\gamma A L^4}} \tag{6-323}$$

Values of a_n for the most frequently encountered beams are given in Table 6-2.

Example 6-2 Determine (a) the equation for the natural frequencies and (b) the shape of the deflection curve for the first three modes of free vibration of a uniform beam that is simply supported at the ends, as shown in Fig. 6-44.

(a) *Natural Frequencies.* The boundary conditions are that the deflection Y and moment d^2Y/dx^2 must be zero at $x = 0$ and $x = L$. For deflection we have [Eq. (6-321)]

$$Y = C_5 \cosh ax + C_6 \sinh ax + C_7 \cos ax + C_8 \sin ax \tag{6-324}$$

Fig. 6-44 *Simply supported beam.*

which, after differentiating twice with respect to x, gives

$$\frac{d^2Y}{dx^2} = a^2C_5 \cosh ax + C_6 \sinh ax - a^2C_7 \cos ax - a^2C_8 \sin ax \qquad (6\text{-}325)$$

At $x = 0$, $Y = d^2Y/dx^2 = 0$. Therefore, from Eq. (6-324),

$$0 = C_5 + 0 + C_7 + 0 \qquad (6\text{-}326)$$

and from Eq. (6-325),

$$0 = a^2C_5 + 0 - a^2C_7 - 0 \qquad (6\text{-}327)$$

which, since in general $a^2 \neq 0$, can be written as

$$0 = C_5 - C_7 \qquad (6\text{-}328)$$

Adding and then subtracting Eqs. (6-326) and (6-328) results in

$$C_5 = C_7 = 0 \qquad (6\text{-}329)$$

Substituting from Eq. (6-329) into Eqs. (6-324) and (6-325) and then solving for $Y = d^2Y/dx^2 = 0$ at $x = L$, we find, respectively,

$$0 = C_6 \sinh aL + C_8 \sin aL \qquad (6\text{-}330)$$

and

$$0 = a^2C_6 \sinh aL - a^2C_8 \sin aL \qquad (6\text{-}331)$$

which, again, since in general $a^2 \neq 0$, can be written as

$$0 = C_6 \sinh aL - C_8 \sin aL \qquad (6\text{-}332)$$

Adding and subtracting Eqs. (6-330) and (6-332) gives, respectively,

$$0 = C_6 \sinh aL \qquad (6\text{-}333)$$

and

$$0 = C_8 \sin aL \qquad (6\text{-}334)$$

Since $\sinh aL$ can equal zero only when $a = 0$, which in turn means $\omega = 0$, Eq. (6-333) will be satisfied in general only when $C_6 = 0$. $\sin aL$ will equal zero whenever $aL = 0$, $\pi, 2\pi, 3\pi, 4\pi, \ldots$; therefore, Eq. (6-334) will be satisfied when

$$a = \frac{n\pi}{L} \qquad (6\text{-}335)$$

where (excluding the trivial case of $a = 0$) $n = 1, 2, 3, \ldots$.

Substituting from Eq. (6-335) into Eq. (6-322) gives, for the natural frequencies,

$$\omega_n = \left(\frac{n\pi}{L}\right)^2 \sqrt{\frac{gEI}{\gamma A}} \qquad (6\text{-}336)$$

where $n = 1, 2, 3, \ldots$.

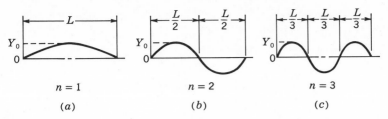

Fig. 6-45 *Mode shape for first three modes of lateral vibration of a simply supported uniform beam.*

(b) *Modes of Vibration.* Substituting $C_5 = C_6 = C_7 = 0$ and $a = n\pi/L$ into Eq. (6-324) gives

$$Y = C_8 \sin \frac{n\pi}{L} x \qquad (6\text{-}337)$$

which, if we let Y_0 be the maximum amplitude of any point on the beam, becomes

$$Y = Y_0 \sin \frac{n\pi}{L} x \qquad (6\text{-}338)$$

For $n = 1$ (fundamental mode)

$$Y = Y_0 \sin \frac{\pi}{L} x \qquad (6\text{-}339)$$

For $n = 2$ (2d mode)

$$Y = Y_0 \sin \frac{2\pi}{L} x \qquad (6\text{-}340)$$

For $n = 3$ (3d mode)

$$Y = Y_0 \sin \frac{3\pi}{L} x \qquad (6\text{-}341)$$

Figure 6-45 shows the beam vibrating in the first three modes.

6-11. Longitudinal Vibrations of Uniform Rods Figure 6-46*a* shows a long slender rod with an elemental length dx located a distance x from the left end of the rod. In Fig. 6-46*b* the rod is vibrating in the longitudinal direction. The displacement at the left face of the element

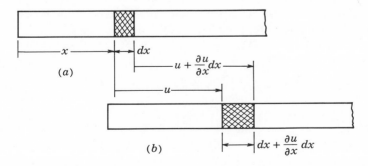

Fig. 6-46 *Displacements of an element of a uniform rod in longitudinal vibrations.*

is u, and the displacement at the right face of the element is $u + (\partial u / \partial x) \, dx$. By definition of unit strain,

$$\epsilon = \frac{\partial u}{\partial x} \tag{6-342}$$

and in accordance with Hooke's law,

$$\epsilon = \frac{P}{AE} = \frac{\partial u}{\partial x} \tag{6-343}$$

where P = force acting on the section
A = area of section
E = modulus of elasticity
The rate of change of the force with respect to x can be found by differentiating Eq. (6-343) with respect to x and rearranging to give

$$\frac{\partial P}{\partial x} = AE \frac{\partial^2 u}{\partial x^2} \tag{6-344}$$

The acceleration of the element is $\partial^2 u / \partial t^2$, and the mass of the element is $(\gamma A / g) \, dx$. The difference in force from one face of the element to the other is $(\partial P / \partial x) \, dx$. Applying Newton's second law to the element, we find

$$\Sigma F = mA$$
$$\frac{\partial P}{\partial x} \, dx = \frac{\gamma A}{g} \, dx \, \frac{\partial^2 u}{\partial t^2}$$

or

$$\frac{\partial P}{\partial x} = \frac{\gamma A}{g} \frac{\partial^2 u}{\partial t^2} \tag{6-345}$$

Substituting from Eq. (6-344) into Eq. (6-345) and rearranging results in

$$\frac{\partial^2 u}{\partial t^2} = \frac{Eg}{\gamma} \frac{\partial^2 u}{\partial x^2} \tag{6-346}$$

The problem now is to find a function $u = f(t,x)$ that satisfies Eq. (6-346). In this case, we are looking for the conditions under which principal-mode free vibrations can exist and we can expect u to vary simple-harmonically. Thus, one possibility is to express u as a product of functions of x and t as

$$u = f(x)(C_1 \cos \omega t + C_2 \sin \omega t) \tag{6-347}$$

and then find a function $f(x)$ that satisfies Eq. (6-347).

Since both the cosine and the sine are well behaved when differentiated twice, it is logical to try a similar expression for $f(x)$, such as

$$f(x) = C_3 \cos ax + C_4 \sin ax \tag{6-348}$$

Table 6-3   Natural Frequencies for Longitudinal Vibrations of Uniform Rods

Configuration	Natural frequencies
Clamped-free	$\omega_n = (n + \frac{1}{2})\pi \sqrt{\dfrac{Eg}{\gamma L^2}}$, where $n = 0, 1, 2, 3, \ldots$
Free-free or clamped-clamped	$\omega_n = n\pi \sqrt{\dfrac{Eg}{\gamma L^2}}$, where $n = 1, 2, 3, \ldots$

Substituting from Eq. (6-348) into Eq. (6-347) gives

$$u = (C_3 \cos ax + C_4 \sin ax)(C_1 \cos \omega t + C_2 \sin \omega t) \qquad (6\text{-}349)$$

Differentiating Eq. (6-349) twice with respect to t and twice with respect to x and substituting into Eq. (6-346) gives

$$\omega^2(C_3 \cos ax + C_4 \sin ax)(C_1 \cos \omega t + C_2 \sin \omega t)$$
$$= \frac{Eg}{\gamma} a^2(C_3 \cos ax + C_4 \sin ax)(C_1 \cos \omega t + C_2 \sin \omega t) \qquad (6\text{-}350)$$

from which

$$a = \frac{\omega}{\sqrt{Eg/\gamma}} = \frac{\omega}{c} \qquad (6\text{-}351)$$

where $c = \sqrt{Eg/\gamma}$. Substituting from Eq. (6-351) into Eq. (6-349) gives, for the general solution,

$$u = \left(C_3 \cos \frac{\omega}{c} x + C_4 \sin \frac{\omega}{c} x \right) (C_1 \cos \omega t + C_2 \sin \omega t) \qquad (6\text{-}352)$$

The constants will be determined by the boundary conditions. For example, a rod fixed at one end and free at the other will have zero displacement (u) at the fixed end and zero strain ($\partial u/\partial x$) at the free end. Since we are considering a free vibration, the amplitude will be arbitrary and we shall be looking for values of ω, which will be the natural frequencies.

Values of natural frequencies for the most frequently encountered cases are presented in Table 6-3.

A somewhat more general approach is to let

$$u = f_1(x + bt) + f_2(x - bt) \qquad (6\text{-}353)$$

which, for any arbitrary functions f_1 and f_2, can be shown[1] to satisfy Eq.

---

[1] For example, I. S. Sokolnikoff and R. M. Redheffer, "Mathematics of Physics and Modern Engineering," pp. 427-428, McGraw-Hill Book Company, New York, 1958.

(6-346). Considering that we are concerned about free vibrations, logical choices for f_1 and f_2 are

$$f_1 = A \sin B(x + bt) \tag{6-354}$$
and
$$f_2 = A \sin B(x - bt) \tag{6-355}$$

Substituting Eqs. (6-354) and (6-355) into Eq. (6-353) gives

$$u = A \sin B(x + bt) + A \sin B(x - bt) \tag{6-356}$$

Differentiating Eq. (6-356) twice with respect to x and twice with respect to t, then substituting into Eq. (6-346), leads to

$$b^2 = \frac{Eg}{\gamma} \tag{6-357}$$

But, we have already let $Eg/\gamma = c^2$. Therefore, we can say $b = c$, and Eq. (6-356) becomes

$$u = A \sin B(x + ct) + A \sin B(x - ct) \tag{6-358}$$

If we consider the instant when $t = 0$, each term on the right-hand side of Eq. (6-358) can be represented as shown in Fig. 6-47, where λ is the wavelength.

Noting that $A \sin Bx = 0$ when $Bx = \pi$ and $x = \lambda/2$, we can write

$$B = \frac{2\pi}{\lambda} \tag{6-359}$$

and Eq. (6-358) becomes

$$u = A \sin \frac{2\pi}{\lambda} (x + ct) + A \sin \frac{2\pi}{\lambda} (x - ct) \tag{6-360}$$

Now, if we consider a particular point located at x_1, we find that the first and the second terms on the right-hand side of Eq. (6-360) can be repre-

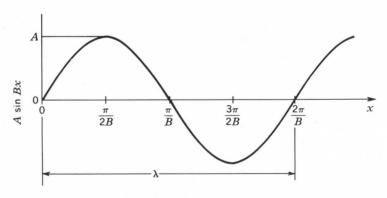

Fig. 6-47

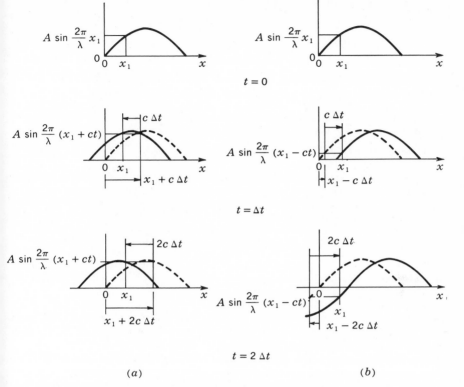

Fig. 6-48 *Traveling-wave representations of terms in Eq. (6-360).*

sented as time increases from 0 as shown in Fig. 6-48a and b, respectively. As can be seen, the first term is a wave moving to the left with a speed c and the second term is a wave moving to the right with a speed c. Therefore, c is the speed, or velocity, of wave propagation—which is commonly known as the velocity of sound.

In Fig. 6-48 it can be seen that the amplitude at a given position, e.g., at x_1, will vary through a complete cycle when $ct = \lambda$. Therefore, the period, or the time for one cycle, will be

$$T = \frac{\lambda}{c} \qquad (6\text{-}361)$$

and the frequency will be

$$f = \frac{1}{T} = \frac{c}{\lambda} \qquad (6\text{-}362)$$

From functions of sums of angles, Eq. (6-360) can be written as

$$u = 2A \left(\sin \frac{2\pi}{\lambda} x \right) \cos \frac{2\pi c}{\lambda} t \qquad (6\text{-}363)$$

which, upon substituting f for c/λ, becomes

$$u = 2A \left(\sin \frac{2\pi}{\lambda} x \right) \cos 2\pi ft \tag{6-364}$$

Equation (6-364) is the equation for a stationary, or standing, wave with nodal points at $x = \lambda/2$, λ, $3\lambda/2$, ... It should be noted that either Eq. (6-352) or Eq. (6-364) can be used with equal ease.

Example 6-3 For purposes of natural-frequency calculations, reciprocating devices can often be approximated by one of four models, as follows:

1 Systems in which the mass of the rod is negligible in relation to masses attached to the ends of the rod
2 Systems in which the masses attached at the ends are negligible relative to the mass of the rod
3 Systems in which the mass attached to one end is very large and the mass attached to the other end is very small
4 Systems in which the mass attached to one end is very large or very small and the mass attached to the other end is small, but not negligibly so

Model 1 becomes a lumped-parameter system with a spring and two masses, and the discussion in Sec. 6-1 can be applied to determining both the natural frequency and the response to a periodic forcing function.

Models 2 to 4 must be treated as continuous systems with an infinite number of natural frequencies. Model 2 is a rod with both ends free; model 3 is a rod with one end clamped and the other end free; and model 4 is a rod with one end clamped or free and the other end carrying a rigid mass.

We are asked to derive equations for calculating the natural frequencies of systems approximated by (a) model 3 and (b) model 4 when one end is clamped.

(a) The system is shown in Fig. 6-49, and the boundary conditions are that the displacement u is zero at $x = 0$ and the strain $\partial u/\partial x$ is zero at $x = L$.

Using Eq. (6-352), we have

$$u = \left(C_3 \cos \frac{\omega}{c} x + C_4 \sin \frac{\omega}{c} x \right)(C_1 \cos \omega t + C_2 \sin \omega t) \tag{6-365}$$

and

$$\frac{\partial u}{\partial x} = \frac{\omega}{c} \left(-C_3 \sin \frac{\omega}{c} x + C_4 \cos \frac{\omega}{c} x \right)(C_1 \cos \omega t + C_2 \sin \omega t) \tag{6-366}$$

Substituting the boundary condition $u = 0$ at $x = 0$ into Eq. (6-365) gives

$$0 = (C_3 + 0)(C_1 \cos \omega t + C_2 \sin \omega t) \tag{6-367}$$

which is satisfied for all values of t only when $C_3 = 0$.

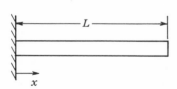

Fig. 6-49

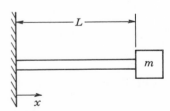

Fig. 6-50

Substituting $C_3 = 0$ and the boundary condition $\partial u/\partial x = 0$ at $x = L$ into Eq. (6-366) gives

$$0 = \frac{\omega}{c}\left(C_4 \cos \frac{\omega}{c} L\right)(C_1 \cos \omega t + C_2 \sin \omega t) \tag{6-368}$$

which is satisfied for all values of t only when $\cos(\omega L/c) = 0$. Therefore,

$$\frac{\omega L}{c} = \frac{\pi}{2}, \frac{3\pi}{2}, \frac{5\pi}{2}, \cdots = (\tfrac{1}{2} + n)\pi \tag{6-369}$$

where $n = 0, 1, 2, 3, \ldots$. The values of ω satisfying Eq. (6-369) are the natural frequencies. From Eqs. (6-369) and (6-357) we find

$$\omega_n = (\tfrac{1}{2} + n)\pi \frac{c}{L} = (\tfrac{1}{2} + n)\pi \sqrt{\frac{Eg}{\gamma L^2}} \tag{6-370}$$

where $n = 0, 1, 2, 3, \ldots$.

(b) The system is shown in Fig. 6-50, and the boundary conditions are that the displacement u is zero at $x = 0$ and the strain at $x = L$ will be that required to provide the force to accelerate the mass m.

Using Eq. (6-352), as in (a), we find that $C_3 = 0$. Thus, the equation we are interested in for the boundary conditions at $x = L$ is

$$u = C_4\left(\sin \frac{\omega}{c} x\right)(C_1 \cos \omega t + C_2 \sin \omega t) \tag{6-371}$$

The force required to accelerate the mass is

$$F = m \frac{\partial^2 u}{\partial t^2}\bigg|_{x=L} = -m\omega^2 C_4\left(\sin \frac{\omega}{c} L\right)(C_1 \cos \omega t + C_2 \sin \omega t) \tag{6-372}$$

and the force on m due to the strain will be

$$F = -AE \frac{\partial u}{\partial x}\bigg|_{x=L} = -AE \frac{\omega}{c} C_4\left(\cos \frac{\omega}{c} L\right)(C_1 \cos \omega t + C_2 \sin \omega t) \tag{6-373}$$

The forces in Eqs. (6-372) and (6-373) must be equal. Equating the two equations and simplifying gives

$$m\omega \sin \frac{\omega}{c} L = \frac{AE}{c} \cos \frac{\omega}{c} L \tag{6-374}$$

which can be rearranged to give

$$\frac{\omega}{c} L \tan \frac{\omega}{c} L = \frac{AEL}{mc^2} = \frac{AL\gamma}{mg} = \frac{m_{\text{rod}}}{m} \tag{6-375}$$

Equation (6-375) is of the form

$$\beta \tan \beta = K_1 \tag{6-376}$$

and the values of β can be found graphically or by referring to a table of functions.[1] Equation (6-374) can also be rearranged to give

$$\cot \frac{\omega}{c} L = \frac{mg}{AL} \frac{\omega}{c} L = \frac{m}{m_{\text{rod}}} \frac{\omega}{c} L \tag{6-377}$$

[1] E. Jahnke and F. Emde, "Tables of Functions," 4th ed., table V, addenda, pp. 32–35, Dover Publications, Inc., New York, 1945.

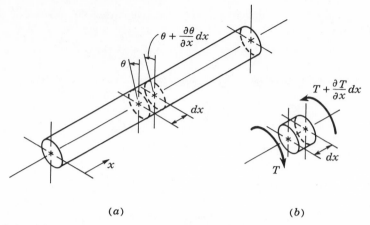

(a)   *(b)*

Fig. 6-51   *(a) Displacement of an element of a uniform rod in torsional vibrations; (b) free-body diagram showing torques acting on an element of the rod.*

which is of the form

$$\cot \beta = K_2 \beta \tag{6-378}$$

and tables are available[1] for finding roots 1 through 9 of this form.

The natural frequencies are then

$$\omega_n = \frac{c}{L} \beta = \beta \sqrt{\frac{Eg}{\gamma L^2}} \tag{6-379}$$

6-12. Torsional Vibrations of Uniform Rods   Figure 6-51 shows an element of a uniform rod that is vibrating in a torsional mode of free vibration.   As shown, the shaft is twisted and, in the absence of external forces, the unbalanced torque results in an acceleration of the element. From strength of materials the shaft torque is

$$T = JG \frac{\partial \theta}{\partial x} \tag{6-380}$$

where J = polar moment of inertia of the cross section

G = torsional modulus of elasticity

The rate of change of torque along the shaft will be

$$\frac{\partial T}{\partial x} = JG \frac{\partial^2 \theta}{\partial x^2} \tag{6-381}$$

and the net torque acting on the element will be

$$\frac{\partial T}{\partial x} dx = JG \frac{\partial^2 \theta}{\partial x^2} dx \tag{6-382}$$

[1] M. Abromowitz and I. A. Stegun (eds.), "Handbook of Mathematical Functions," National Bureau of Standards Applied Mathematics Series 55, Government Printing Office, Washington, D.C., 1964.

Applying Newton's second law to the element, we find

$$\Sigma T = I\alpha$$

$$\frac{\partial T}{\partial x}\, dx = \frac{dI}{dx}\, dx\, \frac{\partial^2\theta}{\partial t^2} = \frac{J\gamma}{g}\, dx\, \frac{\partial^2\theta}{\partial t^2} \tag{6-383}$$

Equating Eqs. (6-382) and (6-383) and rearranging gives

$$\frac{\partial^2\theta}{\partial t^2} = \frac{Gg}{\gamma}\, \frac{\partial^2\theta}{\partial x^2} \tag{6-384}$$

Equation (6-384) is identical in form with Eq. (6-346); thus, by direct analogy to the case of longitudinal vibrations, we rewrite Eq. (6-352) as

$$\theta = \left(C_3 \cos \omega \sqrt{\frac{\gamma}{Gg}}\, x + C_4 \sin \omega \sqrt{\frac{\gamma}{Gg}}\, x\right)(C_1 \cos \omega t + C_2 \sin \omega t) \tag{6-385}$$

The constants are determined by the boundary conditions, and the values of natural frequencies given in Table 6-3 can be applied to this case by simply replacing E with its torsional equivalent G.

chapter 7

CONTROL SYSTEMS

The systems considered in the first six chapters have one characteristic in common: they consist of *passive* elements that are not sources of energy and can only react to forces, torques, and displacements created by an external source of energy. That is, we were concerned solely with the response of the system to an external forcing function that in itself is unaffected by the action of the elements of the system. Borrowing a term from electrical engineering, we can say that in effect the source of energy has zero output impedance and, therefore, it is completely decoupled from the rest of the system. In reality, such a system does not exist, although many systems involving only passive elements of inertia, energy storage, and energy dissipation are so close to it that for practical purposes we can ignore—as we have ignored—any effects the system may have on the forcing function.

Systems containing only passive elements are passive systems. For example, masses, springs, dashpots, etc., are passive elements. Although external to, and thus not a part of, the systems we have previously considered, the element which supplies the energy for the forcing function is an *active* element.

A system designed to control a source of energy is called a control system. The energy may be in any form—chemical, mechanical, electrical, or even the paperwork output of human beings; and the desired output also may be almost anything—a new chemical, the indication of a position, the manufacture of a part, the generation of electric power, the reproduction of music with maximum fidelity, or the manufacture, distribution, and sale of some article. In general, the control system must cope with changes, particularly in the desired output, and we must consider the behavior of the system as a dynamic system.

In many respects the mathematics of control systems will be quite

248

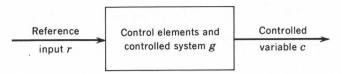

Fig. 7-1   *General control system.*

similar to that already presented in relation to vibrations. The major difference is that we shall now be working with active systems, and the problem becomes considerably more complicated because one or more sources of energy to act on the system will be internal, not external, to the system. As a consequence, in most cases, our analyses must be extended to determining whether or not the system is stable and under what conditions it can become unstable.

Although the following discussion will be related most closely to mechanical systems, the reader should keep in mind that the mathematics and concepts apply equally well to every kind of control system. There may, however, be some disconcerting differences in terminology.

7-1. Open-loop and Closed-loop Systems   The simplest approach to controlling an input to produce some desired output can be represented by the block diagram in Fig. 7-1. In this case the input *r* is in effect a simple command to the system *g* and the output *c* is directly related to the input *r* by the characteristic or properties built into the control elements and controlled system *g*. As discussed above, the output can be almost anything, for example, the position of an object, the temperature of a house or a heat-treating furnace, the speed of an automobile, the product of a manufacturing plant, or the supply of weapons and other matériel to an army.

The control elements consist of whatever is required to convert something from outside the system into the desired output of the controlled system. In the examples above, the controlled system may be the feed mechanism of a machine tool with electricity as the outside source of energy; a furnace with natural gas as the outside source of energy; the engine and drive train of an automobile with gasoline as the outside source of energy; a manufacturing plant supplied with many kinds of materials and different types of energy, including manpower; or a complete distribution system including warehouses, trucks, railroads, etc., with various outside sources of energy.

The input is a command—to make the diameter of a part 5.500 in., to make the temperature in the house be 72°F, to make the automobile speed be 65 mph, to manufacture 10,000 items per day, or to supply every army unit in the field with the exact quantities of everything needed to fight a war.

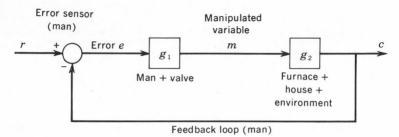

Fig. 7-2 *Feedback control system with man as a major element.*

One way in which a heating system for a residence could operate would be to have the setting of a valve control the flow of fuel to the furnace and, thus, the rate at which heat would be supplied to the house.   The reference input level could be indicated by the position of a pointer on a scale marked off in degrees Fahrenheit.   Based on one's personal experience, it becomes apparent that such a system would not be satisfactory, because it ignores such factors as the outside temperature, wind velocity, and how frequently doors are opened.   In general, the valve setting would give satisfactory results only for particular combinations of conditions, and most of the time the temperature in the house would be higher or lower than the number at which the valve indicator points.

Undoubtedly, the reader has already said to himself that no one would try to heat a house in this manner and that, at least, someone could walk over to the valve and turn it off when the temperature is too high and then turn it back on when the temperature drops below the desired level.   The result would be a more nearly uniform control of the temperature, but we no longer have the same control system; another element has been added—man has become part of the control system.   The person *senses* (reads) the output (temperature in the house), *feeds back* this information to the input, where he calculates the *error* (the input, i.e., the desired output, minus the actual output), and then takes action by closing or opening the valve, depending upon whether the error is negative (too hot) or positive (too cold).[1]   Although it is difficult to represent accurately a human being in a block diagram, the functions performed by him can be represented, as in the block diagram in Fig. 7-2.

It should be noted that a human being is a nonlinear element and any system in which he is a part will be a nonlinear system.   If, as is usually the case, a simple thermostat is used to open and close contacts in an electrical circuit that in turn opens and closes the fuel supply valve, the control system will be a two-position control system and, as such, still

[1] Defining the error as the desired value minus the actual value of the output is peculiar to control systems and is opposite to the normal definition of error.

nonlinear. Although this type is one of the most important systems, because of its widespread use in many different situations, we shall not consider it in detail until Chap. 10, because it is desirable first to gain a thorough understanding of linear systems.

Even at this early point in our discussion one can see the possibility of a system becoming unstable. For example, if the man becomes confused and opens the valve when he should have closed it, the temperature will continue to rise rather than fall toward the desired level. This could happen in several ways, but probably the simplest would be for the man to subtract the desired value from the actual value and thereby think the error is positive, when it is really negative, and then take the action appropriate to a positive error.

Although we shall not consider in detail control systems in which man plays a part, it should be noted that he is by far the most highly developed control system or element in a control system in existence. In varying degrees he possesses highly desirable characteristics, such as an extensive memory, versatility, adaptability, and judgment; and, most important of all, he can look ahead, or anticipate, and thus make allowances for changes or disturbances that have yet to exert a direct influence on the system. On the other hand, man can deliver very little power, is relatively slow in reacting, is subject to fatigue, error, and emotional problems, is highly variable in performance from day to day, and is often too expensive in relation to the value of the end result.

A control system in which there is feedback from the output to the input and which, on the basis of a comparison of the two, takes action by itself to make the output agree more closely with the input is called an *automatic control system*, a *self-regulating system*, or a *closed-loop control system*.[1] We shall be concerned with this type of system throughout the remainder of the text.

7-2. Transfer Functions Each of the blocks in Figs. 7-1 and 7-2 receives information and acts upon it in accordance with the characteristics built into the element or elements represented by the block. In general, we shall be interested in dynamic behavior and we shall find that a differential equation, with time as the independent variable, will be required to relate the output to the input.

For example, let us consider the simplified position control system, or *servomechanism*, in Fig. 7-3. In this case, the purpose of the system is to make the angular position of the rotor, the controlled variable θ_o, coincide with the command θ_i. The command can appear as almost anything—the position of a dial, the position of a lever, a pressure, or a voltage. In the last two cases, devices called *transducers* will have been used to con-

---

[1] A control system without feedback is called an open-loop control system. An example is the elementary heating system discussed above.

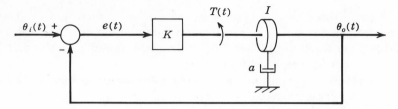

$T(t)$ I

$\theta_i(t)$ + $e(t)$ K $\theta_o(t)$

a

Fig. 7-3   *Schematic representation of a servomechanism.*

vert, or transduce, a position into a pressure or a voltage. Some information on transducers will be introduced from time to time as the need arises, but for detailed information the reader should consult manufacturers' catalogs or more specialized books.[1]

For the present, we shall assume that $\theta_i(t)$ and $\theta_o(t)$ have been converted from positions into compatible signals. Keeping in mind that the variables are functions of time, we can write the error signal as

$$e = \theta_i - \theta_o \tag{7-1}$$

The block labeled K is a major control element. It may be electrical, hydraulic, pneumatic, or mechanical; but its function is to provide a torque output that is directly proportional to its input, which in this case is the error signal. This block is the *active* element and K is called the *gain constant.*[2] Thus, we can write

$$T = Ke \tag{7-2}$$

The rotor has inertia I and the damping coefficient is a.* Applying Newton's second law to the rotor, we find

$$\Sigma T = I\alpha$$

or $$I\,D^2\theta_o + a\,D\theta_o = T \tag{7-3}$$

Substituting from Eq. (7-2) into Eq. (7-3) gives

$$I\,D^2\theta_o + a\,D\theta_o = Ke \tag{7-4}$$

[1] J. G. Truxal (ed.), "Control Engineers' Handbook," McGraw-Hill Book Company, New York, 1958.

J. E. Gibson and F. B. Tuteur, "Control System Components," McGraw-Hill Book Company, New York, 1958.

W. R. Ahrendt and C. J. Savant, "Servomechanism Practice," 2d ed., McGraw-Hill Book Company, New York, 1960.

[2] It should be noted that K can have unusual dimensions and units. For example, if e is an electrical signal, K probably will be expressed as pound-inches per volt.

* To avoid confusion with the controlled variable, we shall use a rather than c for the viscous damping coefficient throughout the remainder of the book.

and substituting from Eq. (7-1) for e in Eq. (7-4) gives

$$I\,D^2\theta_o + a\,D\theta_o = K(\theta_i - \theta_o) \tag{7-5}$$

Rearranging Eq. (7-5) results in a linear second-order differential equation with constant coefficients:

$$\frac{I}{K}\,D^2\theta_o + \frac{a}{K}\,D\theta_o + \theta_o = \theta_i \tag{7-6}$$

or, more usefully,

$$D^2\theta_o + \frac{a}{I}\,D\theta_o + \frac{K}{I}\,\theta_o = \frac{K}{I}\,\theta_i \tag{7-7}$$

Equation (7-7) is quite similar to the equations encountered in Chaps. 3 through 6, and the methods of solution discussed in Chap. 4 can be applied directly.   Considering the complementary solution of Eq. (7-7), we can write, by direct analogy with Eqs. (4-22) and (4-27), the characteristic equation

$$s^2 + \frac{a}{I}\,s + \frac{K}{I} = 0 \tag{7-8}$$

which, in turn by analogy with Eq. (4-91), can be written as

$$s^2 + 2\zeta\omega_n s + \omega_n{}^2 = 0 \tag{7-9}$$

where

$$\omega_n = \sqrt{\frac{K}{I}} \tag{7-10}$$

and

$$\zeta = \frac{a}{2\sqrt{KI}} = \frac{a}{2I\omega_n} \tag{7-11}$$

Thus, the concepts of undamped natural frequency and damping ratio are just as important and useful in the study of control systems as in the study of vibrations.

Stability considerations are of major importance in relation to control systems, and, as discussed in Sec. 4-23, stability can be defined in terms of the characteristics of the free vibration.   It should be recalled that *for a linear system* stability is determined completely by the roots of the characteristic equation.

Although systems can be studied by working with schematic diagrams, such as in Fig. 7-3, it becomes important (almost necessary) when studying any but the simplest systems to work with diagrams in which all elements are represented by blocks, except for summation points which are represented by circles.   To do this we must label each block in such a manner that the relationship of the output to the input is clearly evident.   A convenient, and *the accepted*, way is to place in the box the expression that gives the output when multiplied by the input; or, in

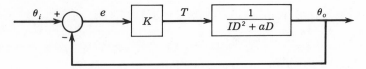

Fig. 7-4   *Block-diagram representation of a feedback control system in terms of transfer functions in operator notation.*

other words, the expression within the block defines the ratio of the output to the input.   For example, this has already been done for the amplifier in Fig. 7-3.   Another way of looking at this is to consider the effect the block has on a signal being transferred through it, and it becomes logical to call the expression in the block a *transfer function*.   Thus, for the amplifier, the transfer function is, from Eq. (7-2),

$$\frac{\text{Output}}{\text{Input}} = \frac{T}{e} = K \tag{7-12}$$

If we consider the rotor as a block, we find that the relationship between the output θ_o and the input T is expressed by Eq. (7-3).   However, it is more conveniently written as

$$(ID^2 + aD)\theta_o = T \tag{7-13}$$

from which the transfer function is

$$\frac{\text{Output}}{\text{Input}} = \frac{\theta_o}{T} = G(t) = \frac{1}{ID^2 + aD} \tag{7-14}$$

where, as before, D and D^2 are operators signifying d/dt and d^2/dt^2, respectively.   In terms of Eqs. (7-12) and (7-14), the schematic diagram in Fig. 7-3 becomes the block diagram in Fig. 7-4.

The concept of a transfer function is very useful, because it permits the reduction of the most complicated control system into a relatively simple diagram that presents complete information in the form in which it can be most readily used in analyzing the behavior of the system.   Several different ways of designating transfer functions will be presented below, but all are basically the same.

In every case the reader must keep in mind that an element or group of elements can be represented perfectly by a block with a transfer function only when the relationship between the output from and the input to the block is independent of anything that can happen anywhere else in the system.   This condition will be met whenever the block has an infinite input impedance and zero output impedance[1] or whenever the input to

---

[1] The term impedance is used here in a general way.   For example, an infinite input impedance means that zero energy passes from the input signal into the block and zero output impedance means that the block can deliver unlimited power without any change in its characteristics.

the block comes from an element with zero output impedance and the output from the block goes to an element with an infinite input impedance. Since there are no real elements with infinite input or zero output impedances, all block diagrams with transfer functions are only approximations of the actual systems. However, many elements, such as the amplifier and the error sensor in Figs. 7-3 and 7-4, can be designed to have very high input impedances and relatively low output impedances, and the resulting approximation will be quite satisfactory for most purposes.[1]

Transfer functions of the form in Fig. 7-4 contain all of the information required for simulating the system on an analog computer, deriving the characteristic equation for stability analysis, and deriving the differential equation for determining the response to any input or change in load. However, for many purposes it is convenient, even if not necessary, to go further into the subject of operational mathematics—in particular, to consider in some detail the Laplace transformation and its applications in the study of control systems.

7-3. The Laplace Transformation   The direct Laplace transformation is defined as

$$\mathcal{L}[f(t)] = \int_0^\infty e^{-st}f(t)\ dt = F(s) \qquad (7\text{-}15)$$

where $s = \sigma + j\omega$ and $F(s)$ is the Laplace transform of $f(t)$. A function $f(t)$ is considered to be in the time domain and the transform $F(s)$ is in the s domain. Since s is a complex number, the s domain becomes a complex plane called the s plane.[2]

By definition, we are concerned only with $t > 0$, and a function meeting the following two conditions will be transformable:[3] The function (1) must be continuous in every finite interval and (2) must be of exponential order as $t \to \infty$. These conditions are met by all functions encountered in control systems and, in general, we can assume that the transform exists.

The major advantages of the Laplace transformation, as well as other transformations used in operational mathematics, are that the operations

---

[1] The reader will find it worthwhile to prove to himself that the concept of a transfer function is not very useful in relation to systems made up of passive elements, such as discussed in earlier chapters, because the block must then include the entire system.

[2] The complex plane as used here is not the same as that used in Chaps. 4 and 6, where simple harmonic motion was represented by a rotating vector in a complex plane.

[3] For detailed proofs and related discussion, see R. V. Churchill, "Operational Mathematics," 2d ed., pp. 1–7, McGraw-Hill Book Company, New York, 1958, or W. T. Thomson, "Laplace Transformation," 2d ed., pp. 1–3, Prentice-Hall, Inc., Englewood Cliffs, N.J., 1960.

of differentiation and integration in the time domain are replaced by relatively simple algebraic operations in the s domain and that a considerable amount of information can be obtained from the transformed equation without actually solving the differential equation of the system.

Although we shall not be using the Laplace transformation to obtain complete solutions of differential equations, it should be noted that after the algebraic operations are performed in the s domain the result must be transformed back into the time domain. This process is called inverse transformation, and it may be most readily accomplished by making use of extensive tables of transform pairs.[1]

Of the numerous theorems related to the Laplace transformation, those given below will be most useful in our study of control systems.

(a) *Linearity theorem*   The Laplace transformation is a linear transformation. Thus,

$$\mathcal{L}[C_1f_1(t) + C_2f_2(t)] = C_1F_1(s) + C_2F_2(s) \qquad (7\text{-}16)$$

The proof of this theorem follows directly from the definition of the Laplace transformation, Eq. (7-15). Superposition applies, and obtaining the transform of a complicated function can often be simplified by breaking it down into a sum of functions and then adding up the transforms of the parts.

(b) *Final-value theorem*   If the function $f(t)$ and its first derivative $Df(t)$ are Laplace-transformable and the limit of $f(t)$ as $t \to \infty$ exists, then

$$\lim_{t \to \infty} f(t) = \lim_{s \to 0} sF(s) \qquad (7\text{-}17)$$

Since periodic functions, such as sin ωt, do not have a definite value when $t = \infty$, the limit does not exist and the final-value theorem cannot be applied. However, in many other important cases, such as for a step change in the command or load, the theorem can be quite useful in finding the steady-state solution without having to find the inverse transform to enter the time domain again.

(c) *Initial-value theorem*   If the function $f(t)$ and its first derivative $Df(t)$ are Laplace-transformable, then

$$\lim_{t \to 0} f(t) = \lim_{s \to \infty} sF(s) \qquad (7\text{-}18)$$

Although there will be less use for the initial-value theorem than for the final-value theorem, it can be used effectively when there is a need for investigating the behavior of a system at $t = 0$, or shortly after, $t = 0+$,

---

[1] Churchill, *op. cit.*, pp. 324–331.

Thomson, *op. cit.*, pp. 239–248.

F. E. Nixon, "Handbook of Laplace Transformation," pp. 66–112, Prentice-Hall, Inc., Englewood Cliffs, N.J., 1960.

and when one does not wish to find the inverse transform in order to enter again into the time domain.

(*d*) **Differentiation theorem**   If the function $f(t)$ and its first derivative $Df(t)$ are Laplace-transformable, and if the denominator of $sF(s)$ is of higher degree than the numerator,

$$sF(s) = \mathcal{L}[Df(t)] \tag{7-19}$$

As can be shown,

$$\mathcal{L}[Df(t)] = sF(s) - f(0) \tag{7-20}$$

where $f(0)$ is the value of the function at $t = 0$.

Thus, for Eq. (7-19) to be valid, $f(0) = 0$. When only the transform is available, the initial-value theorem can be used to determine whether or not $f(0) = 0$ without finding the inverse transform. It will be found that $f(0) = 0$ whenever the denominator of $sF(s)$ is of higher degree than the numerator.

(*e*) **Integration theorem**   If the function $f(t)$ is Laplace-transformable,

$$\frac{1}{s} F(s) = \mathcal{L}\left[\int_0^t f(t) \, dt \right] \tag{7-21}$$

7-4. Laplace Transforms and Transfer Functions   As derived in Sec. 7-2, the differential equation relating the output to the input for the rotor of the system in Fig. 7-3 is

$$I\,D^2\theta_o + a\,D\theta_o = T \tag{7-22}$$

where θ_o and T are functions of time.

By definition,

$$\theta_o(s) = \mathcal{L}[\theta_o(t)] \tag{7-23}$$

and
$$T(s) = \mathcal{L}[T(t)] \tag{7-24}$$

If we now limit ourselves to the case where the rotor is at rest at $t = 0$, that is, $\theta_o(0) = 0$ and $D\theta_o(0) = 0$, we can use the differentiation theorem to write

$$s\theta_o(s) = \mathcal{L}[D\theta_o(t)] \tag{7-25}$$

and
$$s^2\theta_o(s) = \mathcal{L}[D^2\theta_o(t)] \tag{7-26}$$

By use of the linearity theorem and the transforms in Eqs. (7-24) through (7-26), we find

$$\mathcal{L}[I\,D^2\theta_o + a\,D\theta_o] = \mathcal{L}[T(t)] \tag{7-27}$$

and
$$(Is^2 + as)\theta_o(s) = T(s) \tag{7-28}$$

Solving Eq. (7-28) for the ratio of the output to the input, we find, for the transfer function in terms of the Laplace transformation,

$$\frac{\theta_o(s)}{T(s)} = G(s) = \frac{1}{Is^2 + as} \tag{7-29}$$

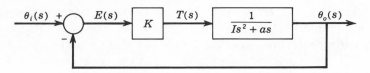

Fig. 7-5   *Block-diagram representation of a feedback control system in terms of transfer functions derived as Laplace transforms.*

Comparing the transfer function in terms of transforms in Eq. (7-29) with that in terms of the operator D in Eq. (7-14) shows that they are identical except that the former has s^2 and s where the latter has D^2 and D.   In terms of Laplace transforms, the block diagram in Fig. 7-4 becomes that in Fig. 7-5.

The remaining type of transfer function that needs to be considered is that related to steady-state sinusoidal response.   In Chaps. 4 and 6 the complex number $e^{j\omega t}$ was used extensively to represent sinusoidal motion. On the basis of the discussion in those chapters we know that, for the rotor in the system in Fig. 7-3, we can write

$$T = T_0 e^{j\omega t} \tag{7-30}$$

where T_0 is the amplitude of the torque, and

$$\theta_o = \Theta_o e^{j(\omega t + \phi)} \tag{7-31}$$

where ϕ is the phase angle of Θ_o relative to T_0.

Differentiating Eq. (7-31) twice with respect to time and substituting the values into Eq. (7-22) leads to

$$(-I\omega^2 + ja\omega)\Theta_o e^{j(\omega t + \phi)} = T_0 e^{j\omega t} \tag{7-32}$$

When Eq. (7-32) is solved for the ratio of the output to the input, we find

$$\frac{\text{Output}}{\text{Input}} = \frac{\theta_o(j\omega)}{T(j\omega)} = G(j\omega) = \frac{1}{-I\omega^2 + ja\omega} \tag{7-33}$$

in which

$$\theta_o(j\omega) = \Theta_o e^{j(\omega t + \phi)} \tag{7-34}$$

and

$$T(j\omega) = T_0 e^{j\omega t} \tag{7-35}$$

Thus, we can also write

$$G(j\omega) = \frac{\Theta_o e^{j\phi}}{T_0} = \frac{1}{\sqrt{(I\omega^2)^2 + (a\omega)^2}}\underline{/\phi} \tag{7-36}$$

where

$$\phi = -\tan^{-1}\frac{a\omega}{-I\omega^2} = -\tan^{-1}\frac{a}{-I\omega} \tag{7-37}$$

Comparison of Eq. (7-33) with Eq. (7-29) shows that the transfer function for sinusoidal response, $G(j\omega)$, can be obtained directly from

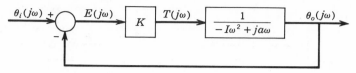

Fig. 7-6   *Block-diagram representation of a feedback control system in terms of transfer functions for sinusoidal response.*

$G(s)$ by simply substituting $j\omega$ for s.   In terms of sinusoidal response, the block diagram in Fig. 7-5 becomes that shown in Fig. 7-6.

The three types of transfer functions used in the block diagrams in Figs. 7-4 through 7-6 have much in common, but each has its advantages and limitations, as follows:

1   The operator-notation transfer function (Fig. 7-4) is in the time domain and there are no restrictions on initial conditions.   This form is most convenient when using an analog computer to investigate the behavior of the system.

2   The Laplace-transform transfer function (Fig. 7-5) is in the s domain and is restricted to use in situations where the initial conditions are such that all values are zero up through the derivative just below that corresponding to the highest degree of s.   For example, for the system in Fig. 7-5, $\theta_i(0)$ and $D\theta_i(0)$ must equal zero.   In practically all cases this limitation on initial conditions is not important.   As will be shown in Chap. 8, this form is closely related to the characteristic equation for the system, and, since stability can be defined completely in terms of the real parts of the roots of the characteristic equation, this form is the most widely used of all.

3   The $j\omega$ transfer function is limited to use in determining the steady-state sinusoidal response, which is of major interest in its own right and becomes essential in several methods of stability analysis.

Since all of the forms of transfer function can be obtained from each other by simply replacing s, D, or $j\omega$ by s, D, or $j\omega$—as the case may be— we shall simply use capital letters to represent signals and blocks, such as R, C, E, G, etc., and insert s, D, or $j\omega$ as the need arises.   In line with this, the block diagram for the system in Fig. 7-3 will appear in general form as shown in Fig. 7-7.   The symbol M stands for a manipulated variable.

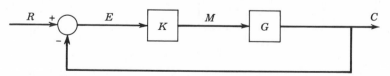

Fig. 7-7   *Block-diagram representation of a feedback control system in terms of generalized transfer functions.*

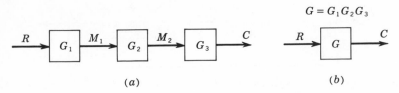

$$G = G_1G_2G_3$$

(a) *(b)*

Fig. 7-8 *Simplification of block diagrams by multiplication of transfer functions when blocks are connected in series.*

7-5. Algebra of Block Diagrams

Figure 7-8a shows a simple open-loop system containing three blocks. As discussed in Sec. 7-2, the input and output impedances of the blocks must be such that the ratio of the output to the input is unaffected by anything else in the system. Under this condition, the output of block 1 is

$$M_1 = RG_1 \tag{7-38}$$

the output of block 2 is

$$M_2 = M_1G_2 \tag{7-39}$$

and the output of block 3 is

$$C = M_2G_3 \tag{7-40}$$

Equations (7-38) through (7-40) can be combined to give

$$C = RG_1G_2G_3 \tag{7-41}$$

which can be rewritten as

$$C = RG \tag{7-42}$$

where $G = G_1G_2G_3$.

Since Eqs. (7-41) and (7-42) are really the same, the three blocks in Fig. 7-8a can be combined into one block and the diagram can be simplified to that in Fig. 7-8b. Being able to combine transfer functions by simple multiplication is a major feature of the transfer-function–block-diagram concept, and it can be used to advantage in simplifying complicated diagrams.

For many purposes the most useful reduced, or simplified, form of the system is that in which only one block remains, for example, the system in Fig. 7-9b, derived from that in Fig. 7-9a. The transfer function Y of

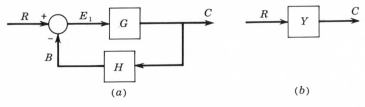

(a) *(b)*

Fig. 7-9 *(a) Complete block diagram; (b) reduced to a single block containing the closed-loop transfer function Y.*

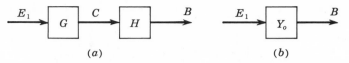

Fig. 7-10 (a) *Open-loop system;* (b) *reduced to a single block containing the open-loop transfer function Y_0.*

the single block is known as the *system transfer function,* the *control ratio,* or, for closed-loop systems, the *closed-loop transfer function.* In Fig. 7-9a, G and H are the combined transfer functions of all of the elements in the *forward loop* and *feedback loop,* respectively, and B is the *feedback signal.* The output of the reference, or summing, junction has been designated as E_1 rather than E to indicate that, in general, it is no longer the error. Under these circumstances, the term *actuating signal* is appropriate.

In terms of the operations indicated in the diagram and the properties of transfer functions, we can write the following equations for the system in Fig. 7-9a:

$$B = HC \tag{7-43}$$
and
$$E_1 = R - B \tag{7-44}$$
and
$$C = GE_1 \tag{7-45}$$

Substituting from Eq. (7-43) for B in Eq. (7-44), and then substituting the modified expression for E_1 in Eq. (7-45) and solving for C/R, results in

$$\frac{C}{R} = Y = \frac{G}{1 + GH} \tag{7-46}$$

The other reduced diagram that will be important to us is that corresponding to the system when the feedback loop is opened at the summing junction. For example, Fig. 7-10a and b are the *open-loop* equivalents of Fig. 7-9a and b, respectively. In consideration of the properties of transfer functions, we can write

$$B = GHE_1 \tag{7-47}$$
from which
$$\frac{B}{E_1} = Y_o = GH \tag{7-48}$$

The ratio B/E_1 is known as the *loop ratio* and as the *open-loop transfer function.*

Many control systems involve feedback loops within feedback loops and are subjected to disturbances that can be introduced at any point within the system, as illustrated in Fig. 7-11. The general procedure to follow in reducing, or trying to reduce, a block diagram to a single block is to start with the inside loop and reduce it to a single block in the next loop, and so on, until the entire system is represented by one block.

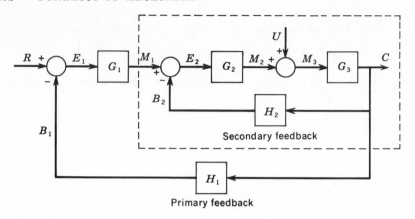

Fig. 7-11 *Control system with primary and secondary feedback loops and subjected to a disturbance U.*

Considering the secondary feedback loop, within the dashed lines in Fig. 7-11, we have the system in Fig. 7-12.   To find the output C as a function of the inputs M_1 and U, we can write the following equations:

$$B_2 = H_2C \tag{7-49}$$
$$E_2 = M_1 - B_2 = M_1 - H_2C \tag{7-50}$$
$$M_2 = G_2E_2 = G_2(M_1 - H_2C) \tag{7-51}$$
$$M_3 = M_2 + U = G_2(M_1 - H_2C) + U \tag{7-52}$$

and, finally, $$C = G_3M_3 = G_3[G_2(M_1 - H_2C) + U] \tag{7-53}$$

From Eq. (7-53) we find

$$C = \frac{G_2G_3}{1 + G_2G_3H_2}\left(M_1 + \frac{U}{G_2}\right) \tag{7-54}$$

which can be put into the form of a transfer function *only* when one of the inputs is zero.

Simpler versions of this secondary or inner-loop system will be considered in Chap. 8, but it can be noted at this time that if the magnitude of G_2 is large, the effect of U becomes small relative to that of M_1.

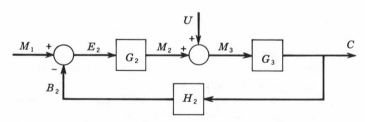

Fig. 7-12 *Secondary-feedback-loop subsystem of the system in Fig. 7-11.*

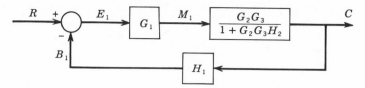

Fig. 7-13 *System in Fig. 7-11 with secondary-feedback-loop subsystem reduced to a single block.*

Considering now only the case where $U = 0$, Eq. (7-54) can be rewritten as

$$\frac{C}{M_1} = \frac{G_2 G_3}{1 + G_2 G_3 H_2} \tag{7-55}$$

Equation (7-55) is a transfer function, and in terms of it the system in Fig. 7-11 becomes that in Fig. 7-13, for which the following equations apply:

$$B_1 = H_1 C \tag{7-56}$$
$$E_1 = R - B_1 = R - H_1 C \tag{7-57}$$
$$M_1 = G_1 E_1 = G_1 (R - H_1 C) \tag{7-58}$$

and

$$C = \frac{G_2 G_3}{1 + G_2 G_3 H_2} M_1 = \frac{G_2 G_3}{1 + G_2 G_3 H_2} G_1 R - \frac{G_2 G_3}{1 + G_2 G_3 H_2} G_1 H_1 C \tag{7-59}$$

From Eq. (7-59) we find

$$\frac{C}{R} = \frac{G_1 G_2 G_3}{1 + G_2 G_3 H_2 + G_1 G_2 G_3 H_1} \tag{7-60}$$

and the final reduced form of block diagram for the system in Fig. 7-11 becomes, *for $U = 0$,* that in Fig. 7-14.

The remaining major considerations are related to system behavior. In general, we shall be interested in questions of stability, response to a change in reference input or load, and how the system can be modified to improve its performance. Although stability is actually the primary consideration, the insight gained by first considering in some detail several simple systems whose functional characteristics make them very useful as basic building blocks in control systems design will add meaning to stability analysis and will, therefore, be considered next.

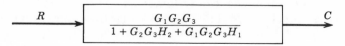

Fig. 7-14 *System in Fig. 7-11 reduced to a single block.*

chapter 8

RESPONSE
OF BASIC LINEAR CONTROL
SYSTEMS

The fundamental characteristic of all feedback control systems is that the corrective action is a function of the error signal. Since the error signal will itself be a function of time, we can choose from its time derivatives or integrals the one or the combination that will result in the simplest system that meets the performance specifications. The other main possibility is to feed back the time derivative of the output in addition to the output itself.

For many years control systems were designed and used in different fields of engineering with apparently little recognition of the extent to which the systems were similar in principle. The resulting variations in terminology are slowly dying out, but in any case there should be little confusion, because the terms are sufficiently descriptive that one can do his own translating.

8-1. Introduction In many respects the understanding of behavior and the methods of solving problems carry over directly from vibrations to control systems. In particular, both are concerned with dynamic systems, and we must solve differential equations. However, there are basic differences in the types of elements and components making up the systems and in the types of inputs for which response characteristics become important.

With respect to the elements and components, the major difference is that, as discussed above, control systems contain active as well as passive elements. Another important difference is that there can be no vibration problem without inertia, or mass, in the system and, therefore, the simplest case requires the solution of a second-order differential equation;

in other words, the simplest vibration system is a second-order system. Many elements in control systems, for example, many types of chemical reactions or processes, fluid-flow and heat-transfer problems, the response of valves and actuators, and electrical networks involving resistance and capacitance or resistance and inductance, have no inertia, or in reality so little that it can be neglected without introducing significant error and can be represented by first-order differential equations. The result is that, whereas in vibrations we considered only systems with even orders (2, 4, etc.),[1] we shall now be interested in systems of any order.

With respect to response criteria, in vibrations we were interested almost exclusively in the steady-state response of a system to periodic forcing functions. Most of our time was spent in discussing the sinusoidal response of systems, and we studied in detail the significance of terms such as magnification factor, transmissibility, resonance, critical frequency, damping ratio, and phase angle. The response of a control system to a sinusoidal disturbance, whether in reference level or load, will again be of major interest; but most control systems must perform satisfactorily when subjected to other types of disturbances. These disturbances are most easily classified with respect to their variation with time, and those we shall consider are as follows: (1) a disturbance that changes from one magnitude to another instantaneously, or almost instantaneously, known as a *step function*, (2) a disturbance that increases in magnitude linearly with time, known as a *ramp function* or a *constant-velocity input*, and (3) a disturbance that increases in magnitude with the square of time, known as a *parabolic* or *constant-acceleration input*. In addition, the concept of sinusoidal response will be applied to systems which must operate over frequencies ranging from a few cycles a day (or lower) to many cycles per second, with process control systems at the lower end and electronic feedback amplifiers at the high end of the frequency range.

Several additional, and possibly unfamiliar, terms will become important in our vocabulary. For example, overshoot, settling time, rise time, time constant, steady-state error, gain, decibel, octave, and decade will all be useful terms in describing quantitatively the response of a system to a given input.

The utility of the analog computer (electronic differential analyzer) for studying vibrating systems has been pointed out many times. Its utility is even greater in the case of control systems, for several reasons: (1) We must often be concerned with the response to inputs that are not sinusoidal. (2) The complexity of all except the simplest types is such

---

[1] Although not so stated at the time (Sec. 6-1), the determination of natural frequencies for systems with two degrees of freedom involved, in effect, converting the simultaneous second-order differential equations into a single fourth-order equation.

that a solution in closed form is extremely laborious. (3) The transfer function of one or more elements in the system can often be changed with little trouble and the effect of a change on system behavior can be investigated with relative ease by simply changing values of resistance and capacitance in the computer circuit. (4) The system can be simulated on the computer so that blocks in the system diagram are represented directly by equivalent blocks on the computer diagram. Thus, real components and simulated components (on the computer) can often be combined to permit the study of the effect of a variation in a real component on the behavior of the complete system without having to construct the entire actual system.

8-2. Proportional Control When the *corrective action is directly proportional to the error*, we have what is called *proportional control*. Figure 7-3 shows such a system and the appropriate block diagram was given in Fig. 7-7. If, however, we extend the discussion related to those figures to include the more general case where an external, or load, torque T_L acts on the rotor, we find the total torque acting on the rotor is the sum of that from the control element and the load. Thus, Eq. (7-3) becomes

$$I\,D^2\theta_o + a\,D\theta_o = T + T_L \tag{8-1}$$

which, in terms of the general inputs R and L and output C, can be written as

$$I\,D^2c + a\,Dc = T + L \tag{8-2}$$

and the block diagram for the system becomes that in Fig. 8-1. Following the procedure presented in Sec. 7-5, we find for the output of the system in Fig. 8-1

$$C = \frac{(KR + L)G}{1 + KG} \tag{8-3}$$

or, more usefully,

$$C = \frac{KG}{1 + KG}\,R + \frac{G}{1 + KG}\,L \tag{8-4}$$

In Eq. (8-4) it can be seen that the output is the sum of the responses to R and L. Therefore, we can investigate the response to each input

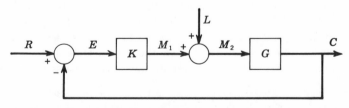

Fig. 8-1 *Proportional control system with load disturbance.*

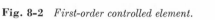
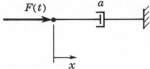

Fig. 8-2 *First-order controlled element.*

separately and then apply the principle of superposition when necessary to find the output when both inputs are acting.

Considering first the response to R alone, the output of the system in Fig. 8-1 is

$$C = \frac{KG}{1 + KG} R \qquad (8\text{-}5)$$

First-order system   If the inertia of a controlled system is negligible in relation to the damping term, we can represent the element as in Fig. 8-2.   For this figure,

$$a\,Dx = F(t) \qquad (8\text{-}6)$$

and the transfer function is

$$G = \frac{x}{F(t)} = \frac{1}{aD} \qquad (8\text{-}7)$$

and Eq. (8-5) becomes

$$C = \frac{K/aD}{1 + K/aD} R = \frac{1}{1 + (a/K)D} R \qquad (8\text{-}8)$$

Equation (8-8) can be rearranged as

$$\left(1 + \frac{a}{K} D\right) c = c + \frac{a}{K} Dc = r \qquad (8\text{-}9)$$

which is a first-order linear differential equation with constant coefficients and can be solved by use of classical methods without difficulty.

For the complementary solution, we have

$$c + \frac{a}{K} Dc = 0 \qquad (8\text{-}10)$$

and we assume

$$c = e^{st} \qquad (8\text{-}11)$$

from which

$$Dc = se^{st} \qquad (8\text{-}12)$$

Substituting from Eqs. (8-11) and (8-12) into Eq. (8-10) and simplifying gives the characteristic equation,

$$1 + \frac{a}{K} s = 0$$

from which

$$s = -\frac{K}{a} \qquad (8\text{-}13)$$

Therefore, the complementary solution is

$$c = C_1 e^{-(K/a)t} \tag{8-14}$$

Since the particular solution depends upon the forcing function, we must now select the type of function (r) for which the response is desired.

First-order system—unit-step-function input   The unit-step function is defined by:

For $t \leq 0$

$$r = 0 \tag{8-15}$$

For $t > 0$

$$r = 1.0 \tag{8-16}$$

For the particular solution, we assume, for $t > 0$, a solution in the form

$$c = C_2 r = C_2 \times 1.0 \tag{8-17}$$

Substituting from Eq. (8-17) into Eq. (8-9) and simplifying results in

$$C_2 = 1.0 \tag{8-18}$$

Therefore, the particular solution is

$$c = 1 \tag{8-19}$$

The response of the system described by Eq. (8-9) to a unit-step-function input is the sum of Eqs. (8-14) and (8-19).   Thus,

$$c = C_1 e^{-(K/a)t} + 1 \tag{8-20}$$

Considering the system to be at rest with $c = 0$ at $t \leq 0$, we find by substituting the initial conditions into Eq. (8-20)

$$C_1 = -1 \tag{8-21}$$

for which Eq. (8-20) becomes

$$c = 1 - e^{-(K/a)t} \tag{8-22}$$

The variation of c and the variation of r with time are shown in Fig. 8-3.   The time at which $e^{-(K/a)t} = e^{-1}$ is called the *time constant*, and it is

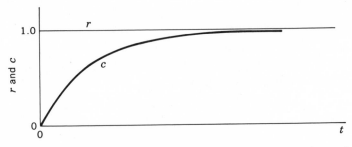

Fig. 8-3   *Response of a first-order proportional control system to a unit-step change in reference level.*

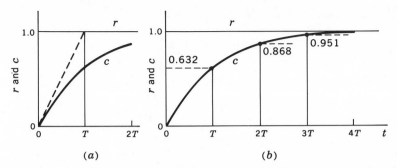

Fig. 8-4 *Response of a first-order proportional control system to a unit-step change in reference level in terms of the time constant.*

quite useful because it defines completely the shape of the curve. The time constant is

$$T = \frac{a}{K} \tag{8-23}$$

and Eq. (8-22) becomes

$$c = 1 - e^{-t/T} \tag{8-24}$$

In terms of the response curve, there are two ways of using the concept of the time constant: (1) As shown in Fig. 8-4a, it is the time interval during which the response would rise to its final value ($t = \infty$) if it increases linearly at its initial rate and (2) as shown in Fig. 8-4b, $e^{-t/T}$ decreases from 1.0 to 0.368 as the time increases from 0 to T; thus $1 - e^{-t/T}$ increases to 0.632 of its final value in the time interval 0 to T.

The latter concept is the more useful. For example, in four time constants from $t = 0$,

$$e^{-t/T} = e^{-4T/T} = e^{-4} = 0.368^4 = 0.018 \tag{8-25}$$

and

$$1 - e^{-t/T} = 1 - 0.018 = 0.982 \tag{8-26}$$

Considering the steady-state response to be that reached when $t = \infty$, it can be seen in Fig. 8-3 and from Eq. (8-24) that the steady-state response of the first-order system to a unit-step function is unity and, therefore, the steady-state error is zero. Although not really necessary in this simple situation, let us make use of the final-value theorem for the Laplace transformation (Sec. 7-3) to determine the steady-state error. From Fig. 8-1 for $L = 0$,

$$E = R - C \tag{8-27}$$

Substituting from Eq. (8-5) for C in Eq. (8-27) gives

$$E = R - \frac{KG}{1 + KG} R = \frac{1}{1 + KG} R \tag{8-28}$$

The final-value theorem, Eq. (7-17), is

$$\lim_{t \to \infty} f(t) = \lim_{s \to 0} sF(s) \tag{8-29}$$

which in this case becomes

$$e_{ss} = \lim_{t \to \infty} e(t) = \lim_{s \to 0} s \left[\frac{1}{1 + KG(s)} \right] R(s) \tag{8-30}$$

Substituting s for D in Eq. (8-7) gives

$$G(s) = \frac{1}{as} \tag{8-31}$$

We must now find $R(s)$ by use of Eq. (7-15):

$$\mathcal{L}[r(t)] = \int_0^\infty e^{-st} r(t) \, dt = R(s) \tag{8-32}$$

For the unit-step function, $r(t) = 1$; thus,

$$R(s) = \int_0^\infty e^{-st}(1) \, dt = -\frac{1}{s} e^{-st} \Big]_0^\infty = \frac{1}{s} \tag{8-33}$$

Substituting from Eqs. (8-31) and (8-33) into Eq. (8-30) gives

$$e_{ss} = \lim_{t \to \infty} e = \lim_{s \to 0} s \left(\frac{1}{1 + K/as} \right) \frac{1}{s} = \lim_{s \to 0} \frac{as}{as + K} = 0 \tag{8-34}$$

which agrees with the above conclusion based on the solution of the differential equation.

As a practical matter, a first-order system is usually considered to have reached its final value at $t = 4T$, when the response is 98.2 percent of the ultimate final value. It should also be noted that in most cases the designer will have already made a, the damping coefficient, as small as possible. Therefore, if faster response is desired, a larger value of K, the gain constant, must be specified.

First-order system—unit-ramp-function input The unit-ramp function is defined by:

For $t \le 0$

$$r = 0 \tag{8-35}$$

For $t > 0$

$$r = t \tag{8-36}$$

For the particular solution to Eq. (8-9), we assume

$$c = C_2 t + C_3 \tag{8-37}$$

from which

$$Dc = C_2 \tag{8-38}$$

Substituting from Eqs. (8-37) and (8-38) into Eq. (8-9) gives

$$C_2 t + C_3 + \frac{a}{K} C_2 = t \tag{8-39}$$

Equating coefficients of like powers of t, we find, from Eq. (8-39),

$$C_2 t = t \tag{8-40}$$

$$C_3 + \frac{a}{K} C_2 = 0 \tag{8-41}$$

From Eq. (8-40), we have

$$C_2 = 1 \tag{8-42}$$

and substituting from Eq. (8-42) for C_2 in Eq. (8-41) leads to

$$C_3 = - \frac{a}{K} \tag{8-43}$$

Thus, the particular solution becomes, from Eqs. (8-37), (8-42), and (8-43),

$$c = t - \frac{a}{K} \tag{8-44}$$

The complete solution for the response of the single-order system to a unit-ramp input is the sum of Eqs. (8-14) and (8-44).   Thus,

$$c = C_1 e^{-(K/a)t} + t - \frac{a}{K} \tag{8-45}$$

or, in terms of the time constant $T = a/K$,

$$c = C_1 e^{-t/T} + t - T \tag{8-46}$$

Considering the case where the system starts from $c = 0$ at $t = 0$, we find for these initial conditions

$$0 = C_1 + 0 - T \tag{8-47}$$

from which

$$C_1 = T \tag{8-48}$$

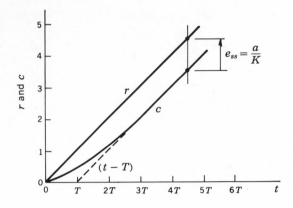

Fig. 8-5 *Response of a first-order proportional control system to a unit-ramp change in reference level.*

Substituting from Eq. (8-48) for C_1 in Eq. (8-46) and rearranging gives

$$c = t - T(1 - e^{-t/T}) \qquad (8\text{-}49)$$

Curves of r and c as functions of time are presented in Fig. 8-5. As can be seen, the response never catches up with the input and c approaches the line $t - T$ as $t \to \infty$. Thus, there is a steady-state error of T or a/K. Considering that a first-order controlled system with a transfer function $1/aD$ acts like a viscous damper, one would expect to find a steady-state error because a constant force will be required if the moving end is to move at a constant velocity.

The steady-state error can even more readily be found by using the final-value theorem of the Laplace transformation. In this case we can use Eq. (8-30), which was derived for this system. However, we must first determine $R(s)$ by use of Eq. (7-15). For $r = t$, we have

$$R(s) = \int_0^\infty e^{-st} t \, dt \qquad (8\text{-}50)$$

which can be integrated by parts, making use of the relationship

$$\int u \, dv = uv - \int v \, du \qquad (8\text{-}51)$$

In this case we shall let

$$u = t \qquad (8\text{-}52)$$

and

$$dv = e^{-st} \, dt \qquad (8\text{-}53)$$

From Eq. (8-52), we find

$$du = dt \qquad (8\text{-}54)$$

and from Eq. (8-53),

$$v = \int e^{-st} \, dt = -\frac{1}{s} e^{-st} \qquad (8\text{-}55)$$

Substituting from Eqs. (8-52) through (8-55) into Eq. (8-51) gives

$$R(s) = \int_0^\infty e^{-st}t\,dt = -\frac{te^{-st}}{s}\Big]_0^\infty + \frac{1}{s}\int_0^\infty e^{-st}\,dt$$

$$= 0 - \frac{1}{s^2}e^{-st}\Big]_0^\infty = \frac{1}{s^2} \quad (8\text{-}56)*$$

Substituting from Eqs. (8-31) and (8-56) for $G(s)$ and $R(s)$, respectively, into Eq. (8-30) gives

$$e_{ss} = \lim_{t\to\infty} e(t) = \lim_{s\to 0} s\left(\frac{1}{1 + K/as}\right)\frac{1}{s^2} = \frac{a}{K} \quad (8\text{-}63)$$

which agrees with the conclusion above based on the solution of the differential equation.

---

* An alternative, actually more convenient, way to find the Laplace transform of a unit-ramp-function input is to consider that it is the time integral of a unit-step function of velocity and apply the integration theorem, Eq. (7-21). For example, for the unit-ramp function

For $t \leq 0$

$$r = 0 \quad \text{and} \quad \frac{dr}{dt} = 0$$

For $t > 0$

$$r = t \quad \text{and} \quad \frac{dr}{dt} = 1 \quad (8\text{-}57)$$

From Eq. (7-21) we can write

$$\mathcal{L}[r(t)] = \mathcal{L}\left[\int_0^t \frac{dr}{dt}\,dt\right] = \frac{1}{s}\mathcal{L}\left[\frac{dr}{dt}\right] \quad (8\text{-}58)$$

Substituting from Eq. (8-57) into Eq. (7-15) gives

$$\mathcal{L}\left[\frac{dr}{dt}\right] = \mathcal{L}[1] = \frac{1}{s} \quad (8\text{-}59)$$

and substituting from Eqs. (8-57) and (8-59) into Eq. (8-58) results in

$$\mathcal{L}[r(t)] = \mathcal{L}[t] = \frac{1}{s^2} \quad (8\text{-}60)$$

Continuing this procedure, we find the Laplace transform for a unit parabolic input to be

$$\mathcal{L}[t^2] = 2\mathcal{L}\left[\int t\,dt\right] = \frac{2}{s}\mathcal{L}[t] = \frac{2}{s^3} \quad (8\text{-}61)$$

and, in general, for $m = 0, 1, 2, \ldots$

$$\mathcal{L}[t^m] = \frac{m!}{s^{m+1}} \quad (8\text{-}62)$$

First-order system—parabolic-input response   If the response were to be calculated for a parabolic (constant-acceleration) input, we would find that c would fall farther behind r as time increased and the steady-state error would be infinite.   A simple proportional control system should not be used if the input is to be parabolic, and further discussion is not warranted.[1]

First-order system—sinusoidal response   In the study of vibrations we found that the steady-state sinusoidal response could be characterized most conveniently in terms of the ratio of the response amplitude to the input amplitude and the phase angle of the response relative to the input.   This is also true for control systems, and in this case we shall be interested in rearranging Eq. (8-5) to give

$$\frac{C(j\omega)}{R(j\omega)} = \frac{KG(j\omega)}{1 + KG(j\omega)} \tag{8-64}$$

which can be seen to be the control ratio or closed-loop transfer function, as discussed in Sec. 7-4.   For the system under consideration we find $G(j\omega)$ by simply substituting $j\omega$ for D in Eq. (8-7) to give

$$G(j\omega) = \frac{1}{j\omega a} \tag{8-65}$$

Substituting from Eq. (8-65) into Eq. (8-64) and simplifying gives

$$\frac{C(j\omega)}{R(j\omega)} = \frac{K}{K + j\omega a} \tag{8-66}$$

Although the concept of a time constant may not appear to bear any relationship to sinusoidal response, we shall see shortly that a considerable simplification can be introduced in the handling of block diagrams and transfer functions if we write equations in terms of time constants whenever possible.   Since we have seen above that, for this system, the time constant is $T = a/K$, we can rewrite Eq. (8-66) as

$$\frac{C(j\omega)}{R(j\omega)} = \frac{1}{1 + j\omega a/K} = \frac{1}{1 + jT\omega} \tag{8-67}$$

Equation (8-67) can be rewritten in terms of magnitude and phase

---

[1] Noting that the steady-state error for the response of a first-order system is zero for a step-function input, a constant for a ramp-function input, and infinite for a parabolic input leads to the useful observation that the steady-state error will be (1) zero for all inputs of a lower degree than that resulting in a constant steady-state error and (2) infinite for all inputs of a higher degree than that resulting in a constant steady-state error.   This type of observation will be considered in a more general and more useful manner in Sec. 8-7.

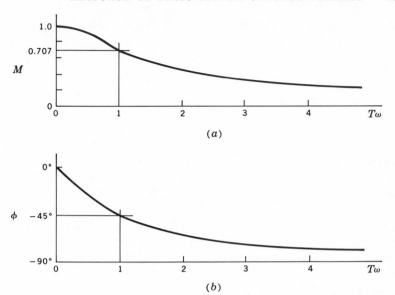

Fig. 8-6 *Steady-state response of a first-order proportional control system to a sinusoidally varying reference level. (a) Ratio of magnitudes of output to input; (b) phase angle φ of output relative to input.*

angle as

$$\frac{C(j\omega)}{R(j\omega)} = \frac{1}{\sqrt{1 + (T\omega)^2}} e^{j\phi} \tag{8-68}$$

or, as is more commonly done for control systems,

$$\frac{C(j\omega)}{R(j\omega)} = M\underline{/\phi} \tag{8-69}$$

where M is the magnitude of the ratio of the amplitudes and ϕ is the phase angle of C relative to R; that is,

$$M = \left| \frac{C(j\omega)}{R(j\omega)} \right| = \frac{1}{\sqrt{1 + (T\omega)^2}} \tag{8-70}$$

and

$$\phi = -\tan^{-1} T\omega \tag{8-71}$$

The relationships in Eqs. (8-70) and (8-71) are presented in Fig. 8-6 in nondimensional form as functions of $T\omega$. Comparing the curves in Fig. 8-6 with those in Fig. 4-20 for a second-order system (consisting of a mass, a spring, and a damper) leads to two important observations:

1 The amplitude ratio of a first-order system will never be greater than 1, whereas that of the second-order system can become infinite.

2 The phase angle of the first-order system approaches $-90°$, whereas that of the second-order system approaches $-180°$ as ω approaches infinity.

The significance of observation 1 is that in the absence of an inertial element, or one analogous to it, there can be no storage of energy within the system in the form of kinetic energy and thus the output can never "overshoot" (exceed) the input. Another consequence is that the terms critical frequency and natural frequency have no meaning and the dimensionless form of the independent variable must be $T\omega$ rather than ω/ω_n. The significance of observation 2 is that *each order* will contribute a maximum of 90° to the total phase angle for the system.

As mentioned in Sec. 8-1, we shall be interested in the sinusoidal response of systems for which there will be a wide range in values of amplitude ratio over a wide range of frequencies. We have already found it convenient to plot the ratio of the amplitudes on a logarithmic scale, and we now find it will be desirable, sometimes necessary, to plot the frequency also on a logarithmic scale. Such plots are called, for obvious reasons, log-magnitude–log-frequency plots, and three different ways are used to present the information: (1) using a logarithmic scale for the ordinate, as in Figs. 4-20, 4-22, etc., (2) using a linear scale for the ordinate and plotting the logarithm of the magnitude, and (3) plotting the magnitude in decibels.

When presentation of response data is the sole function of the plot, the straightforward presentation in method 1 makes it the most desirable. However, when the plot is used in design, the effect of introducing additional blocks or modifying existing blocks can be studied more easily if either method 2 or 3 is used, because multiplication becomes only a matter of addition of logarithms or decibels. Since the reader will encounter far wider usage of the magnitude in decibels in the literature, we shall follow that convention here.

The decibel (db) is defined as one-tenth of a bel, which in turn is defined as the logarithm to the base 10 of the ratio of two magnitudes of *power*. Thus, by definition,

$$\text{Number of decibels} = \text{db} = 10 \log \frac{P_2}{P_1} \qquad (8\text{-}72)$$

The decibel has been particularly useful in acoustics, where we must work with a range of powers of about 1 billion billion to 1 $(10^{18}:1)$ or, from Eq. (8-72), 180 db. It has also found considerable use in electronics, where ratios of 10^5 are not uncommon, and a natural extension of its use has been to the study of control systems.

The definition in terms of power is useful in many situations, particularly where the signal is not periodic, such as in the analysis of noise when

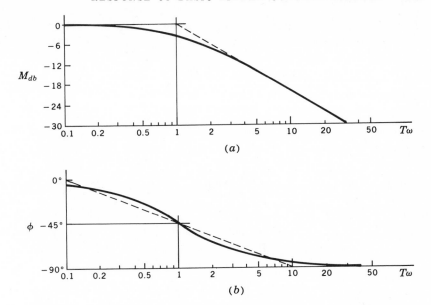

Fig. 8-7 *Replot of Fig. 8-6 in terms of M in decibels and log frequency.*

the total is made up of sounds emitted from several unrelated sources; but for most purposes in control systems and electronics we are more interested in the ratio of amplitudes than we are in the ratio of powers. Since for a sinusoidally varying electrical signal the power is proportional to the square of the amplitude of the voltage, we can rewrite Eq. (8-72) as

$$\text{db} = 10 \log \left(\frac{E_2}{E_1}\right)^2 = 20 \log \frac{E_2}{E_1} \tag{8-73}$$

where E_2 and E_1 are amplitudes.

Several points worth noting about Eq. (8-73) are (1) when $E_2/E_1 = 1.0$, we have 0 db; (2) when $E_2/E_1 = 2$, we have 6 db; (3) when $E_2/E_1 = 0.5$, we have -6 db; and (4) when $(E_2/E_1)^2 = 0.5$ or $E_2/E_1 = 0.707$, the power ratio is one-half and we have -3 db.

Referring to Eq. (8-70), we note that M is the ratio of amplitudes and, therefore, we can write

$$M_{\text{db}} = 20 \log \frac{1}{\sqrt{1 + (T\omega)^2}} \tag{8-74}$$

In terms of decibels, from Eq. (8-74), and log frequency, the curves in Fig. 8-6 become those in Fig. 8-7. The curves in Fig. 8-7 have several interesting properties. For example, the log-magnitude curve approaches a straight horizontal line (0 db) for frequencies below $\omega = 1/T$ and a straight line with a negative slope at frequencies above $\omega = 1/T$. In fact, the maximum error is only 3 db (at $T\omega = 1$) if the straight lines are used as the curve.

For values of $T\omega \gg 1$, Eq. (8-70) can be written as

$$M \approx \frac{1}{T\omega} \qquad (8\text{-}75)$$

and from Eq. (8-75) it can be seen that every time the frequency is doubled, the magnitude is halved. Similarly, every time the frequency is multiplied by 10, the magnitude is divided by 10. From sound and music, an octave corresponds to doubling the frequency, and in general terms a factor of 10 becomes a decade. From Eq. (8-73) we find that a ratio of one-half corresponds to -6 db and a ratio of one-tenth corresponds to -20 db. If we combine the concepts of an octave and a decade with the corresponding values of decibels, we can express the slope of the straight-line approximation for $T\omega \gg 1$ as -6 db/octave or -20 db/decade.

As shown by the dashed lines in Fig. 8-7a, the 0-db straight-line approximation for $T\omega \ll 1$ and the -6 db/octave straight-line approximation for $T\omega \gg 1$ intersect at $T\omega = 1$. Combining this observation with the fact that at $T\omega = 1$ the curve is 3 db below the value for $T\omega = 1$, as calculated for either approximation, permits one to sketch, quickly and accurately, the sinusoidal response curve for any first-order system on the basis of knowing only the time constant T for the system.

In Fig. 8-7 or from Eq. (8-71), it can be seen that at $T\omega = 1$ the phase angle will be $-45°$. The dashed straight-line approximation runs from $0°$ at the frequency one decade below to $-90°$ at the frequency one decade above the frequency at which $\phi = -45°$. The difference between the approximate and true curves will be less than $6°$. The frequency $\omega = 1/T$ is called the *break* or *corner frequency* because it corresponds to a break or a corner in the straight-line approximation of the response curve.

In later sections we shall be concerned with relatively complex systems in which a number of blocks are cascaded, i.e., in series in the block diagram. In these cases the transfer functions multiply, as discussed in Sec. 7-5, and the resultant or overall transfer function will have a magnitude equal to the product of the individual magnitudes and a phase angle equal to the sum of the individual phase angles. For example,

$$G_1(j\omega)G_2(j\omega) = |G_1|e^{j\phi_1}|G_2|e^{j\phi_2} = M_1e^{j\phi_1}M_2e^{j\phi_2} = M_1M_2e^{j(\phi_1+\phi_2)}$$
$$= M_1M_2\underline{/\phi_1 + \phi_2} = M\underline{/\phi} \qquad (8\text{-}76)$$

In terms of logarithms the magnitudes become

$$\log M = \log M_1 + \log M_2 \qquad (8\text{-}77)$$

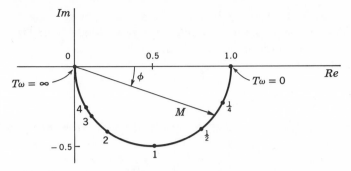

Fig. 8-8 *Polar plot of the steady-state response of a first-order proportional control system to a sinusoidally varying reference level.*

from which

$$20 \log M = 20 \log M_1 + 20 \log M_2$$

and
$$M_{\mathrm{db}} = M_{1_{\mathrm{db}}} + M_{2_{\mathrm{db}}} \tag{8-78}$$

Consequently, by sketching the M_{db} and ϕ versus log-frequency curves for the separate blocks, we can construct the response curves of the entire system by using only simple addition.

In Fig. 8-6 two curves were required to present the complete sinusoidal response of the system. For most purposes this method, or the equivalent in Fig. 8-7, is best. However, as we shall see in Chap. 9, there are situations in which it is desirable to combine the magnitude and phase-angle information into a polar plot, as in Fig. 8-8. The curve can be plotted directly by use of Eqs. (8-70) and (8-71) without drawing the curves in Fig. 8-6, or it can be plotted in terms of imaginary and real components by rationalizing Eq. (8-67). To rationalize a complex fraction, we multiply both the denominator and numerator by the complex conjugate of the denominator. Since the complex conjugate of $1 + jT\omega$ is $1 - jT\omega$, Eq. (8-67) becomes

$$\frac{C(j\omega)}{R(j\omega)} = \frac{1}{1 + jT\omega} = \frac{1 - jT\omega}{(1 + jT\omega)(1 - jT\omega)} = \frac{1 - jT\omega}{1 + (T\omega)^2} \tag{8-79}$$

For convenience in plotting the polar diagram, Eq. (8-79) can be split up into its real and imaginary parts to give

$$\frac{C(j\omega)}{R(j\omega)} = \frac{1}{1 + (T\omega)^2} - j\frac{T\omega}{1 + (T\omega)^2} \tag{8-80}$$

First-order system—step change in load   If we consider the case where the reference input is held at zero and only the load changes, from Eq. (8-4) we have

$$C = \frac{G}{1 + KG}L \tag{8-81}$$

The response in the time domain can be calculated by using differential equations, as above; but let us limit our efforts to determining the steady-state error introduced by a step change in load.

By definition,

$$E = R - C \tag{8-82}$$

which for $R = 0$ becomes

$$E = -C \tag{8-83}$$

and for the first-order system under consideration, from Eq. (8-81),

$$E = - \frac{G}{1 + KG} L \tag{8-84}$$

By use of the final-value theorem of the Laplace transformation, Eq. (7-17),

$$\lim_{t \to \infty} f(t) = \lim_{s \to 0} sF(s) \tag{8-85}$$

In this case, from Eq. (8-31),

$$G(s) = \frac{1}{as} \tag{8-86}$$

and from Eq. (8-33), for a unit-step change in load with zero initial conditions,

$$L(s) = \frac{1}{s} \tag{8-87}$$

Substituting from Eqs. (8-86) and (8-87) into Eq. (8-84) and simplifying results in

$$E(s) = - \frac{1}{as + K} \frac{1}{s} \tag{8-88}$$

Then substituting from Eq. (8-88) into Eq. (8-85) gives

$$e_{ss} = \lim_{t \to \infty} e(t) = \lim_{s \to 0} s \left(- \frac{1}{as + K} \right) \frac{1}{s} = - \frac{1}{K} \tag{8-89}$$

In Eq. (8-89) it can be seen that a steady-state error will result from a step change in load and that the magnitude can be reduced by increasing the value of the gain constant K. These results could have been deduced by just looking at the diagram and noting that for equilibrium (steady state) the output from box K must be equal in magnitude and opposite in sign to the load so that the sum equals zero. The steady-state error will be found to be infinite for higher-degree changes, such as ramp and parabolic, in load.

Second-order system   In Sec. 7-2 the transfer function for a second-order controlled system with inertia I and damping a was shown to be

$$G = \frac{1}{ID^2 + aD} \tag{8-90}$$

If we consider the system in Fig. 8-1 with $L = 0$, we find, by substituting from Eq. (8-90) into Eq. (8-5) and simplifying,

$$C = \frac{K}{ID^2 + aD + K} R \qquad (8\text{-}91)$$

Dividing the numerator and the denominator of Eq. (8-91) by K gives

$$C = \frac{1}{(I/K)D^2 + (a/K)D + 1} R \qquad (8\text{-}92)$$

in which the denominator is recognized as a second-order linear differential equation in operator notation—one that was considered in detail in Chap. 4 and in which we recognize

$$\frac{a}{K} = \frac{2\zeta}{\omega_n} \qquad (8\text{-}93)$$

and

$$\frac{I}{K} = \frac{1}{\omega_n{}^2} \qquad (8\text{-}94)$$

Substituting from Eqs. (8-93) and (8-94) into Eq. (8-92) gives

$$C = \frac{1}{(1/\omega_n{}^2)D^2 + (2\zeta/\omega_n)D + 1} R \qquad (8\text{-}95)$$

which can be written as a differential equation in the time domain as

$$\left(\frac{1}{\omega_n{}^2} D^2 + \frac{2\zeta}{\omega_n} D + 1 \right) c = r \qquad (8\text{-}96)$$

or

$$\frac{1}{\omega_n{}^2} D^2 c + \frac{2\zeta}{\omega_n} Dc + c = r \qquad (8\text{-}97)$$

From Chap. 4 we recognize the necessity for considering three different complementary solutions of Eq. (8-97)—one for $\zeta < 1$, one for $\zeta = 1$, and one for $\zeta > 1$:

For $\zeta < 1$, from Eq. (4-98)

$$c = e^{-\zeta\omega_n t}(C_1 \cos \omega_n \sqrt{1 - \zeta^2}\, t + C_2 \sin \omega_n \sqrt{1 - \zeta^2}\, t) \qquad (8\text{-}98)$$

For $\zeta = 1$, from Eq. (4-115)

$$c = C_1 e^{-\omega_n t} + C_2 t e^{-\omega_n t} \qquad (8\text{-}99)$$

For $\zeta > 1$, from Eq. (4-114)

$$c = C_1 e^{(-\zeta\omega_n + \omega_n\sqrt{\zeta^2 - 1})t} + C_2 e^{(-\zeta\omega_n - \omega_n\sqrt{\zeta^2 - 1})t} \qquad (8\text{-}100)$$

The particular solution cannot be determined until the input is specified.

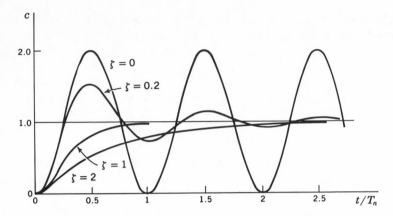

Fig. 8-9 *Response of underdamped, critically damped, and overdamped second-order proportional control systems to a unit-step change in reference level.*

Second-order system—unit-step-function input For a unit-step-function input, Eq. (8-97) becomes

$$\frac{1}{\omega_n{}^2} D^2 c + \frac{2\zeta}{\omega_n} Dc + c = 1 \tag{8-101}$$

for which the particular solution is

$$C = 1 \tag{8-102}$$

The complete solution is the sum of the appropriate complementary solution, Eq. (8-98), (8-99), or (8-100), and Eq. (8-102). As in vibrations, the underdamped case is the most important and for it we find, for zero initial conditions,

$$c = 1 - e^{-\zeta \omega_n t}\left(\cos \omega_n \sqrt{1 - \zeta^2}\, t + \frac{\zeta}{\sqrt{1 - \zeta^2}} \sin \omega_n \sqrt{1 - \zeta^2}\, t\right) \tag{8-103}$$

The response to a unit-step function has already been considered in Example 4-3 for a particular system consisting of a mass, a spring, and a damper. In that case the mass was released from an initial deflection at $t = 0$. However, we are now interested in the general behavior of second-order systems and we shall refer to the curves in Fig. 8-9, in which time has been nondimensionalized as the ratio of the time to the period of the undamped natural frequency. Thus,

$$T_n = \frac{2\pi}{\omega_n} \tag{8-104}$$

Although the observations based on Fig. 8-9 must be identical with those made in Secs. 4-8 through 4-10 about the significance of the damping ratio in relation to the mass-spring-damper system, we can combine the important points into one statement. The *lowest damping ratio that will result in a response to a step function without overshoot is the critical damping ratio* $\zeta = 1.0$. Thus, for $\zeta < 1.0$ there will always be overshoot and for $\zeta > 1.0$ there will be no overshoot. Since, in general, we want a control system to reach the steady-state value in the shortest possible time, it is apparent that we shall not normally want to use $\zeta > 1.0$.

It can also be concluded from Fig. 8-9 or Eq. (8-103) and its equivalent for $\zeta = 1.0$ that an infinite length of time will be required for the system response to reach the steady-state value of 1.0. Consequently, if we wish to compare different systems on the basis of rapidity of response, we must do so in a manner similar to that above for a first-order element of system, where it was found convenient to describe the response in terms of the time required for the response to reach a prescribed fraction of its steady-state value. The main difference is that the response of systems having orders greater than 1 can be oscillatory, and we can better describe the response in terms of the time required for the response *to enter and remain within* a band of response levels about the steady-state value. This time is called the *settling* time and the band is usually specified as ± 2, 5, or 10 percent of the steady-state value.[1]

For second-order systems the first overshoot is the maximum. As discussed in Sec. 4-8, the time interval between successive peaks, occurring when the response curve is tangent to the envelope $(1 - e^{-\zeta\omega_n t})$ on one side of the steady-state value, is

$$T = \frac{2\pi}{\omega_n \sqrt{1 - \zeta^2}} \tag{8-105}$$

In the present case we are interested in the time from zero to the first point of tangency with the envelope reflected about the steady-state value, as shown by the dashed curve labeled $1 + e^{-\zeta\omega_n t}$ in Fig. 8-10. Thus, the time interval from zero to the first peak overshoot is

$$T_{P1} = \frac{\pi}{\omega_n \sqrt{1 - \zeta^2}} \tag{8-106}$$

and the ratio of the magnitude of the peak overshoot to the steady-state

---

[1] Another term used to describe the response to a step function is *rise time*, which is the time required for the output to increase from one specified percentage of the final value to another. Usually the rise time is the time between the 10 and 90 percent levels.

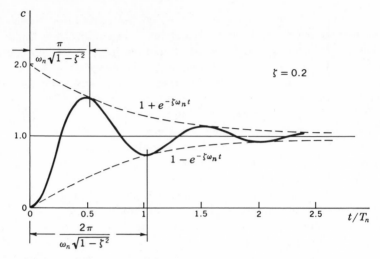

Fig. 8-10 *Response of an underdamped, second-order proportional control system to a unit-step change in reference level.*

value is, from Fig. 8-10 and Eq. (8-106),

$$\left.\frac{c}{c}\right|_{P1} \Big/ \left. c \right|_{ss} = M_{P1} = 1 + \exp\left(-\zeta\,\frac{\pi}{\sqrt{1-\zeta^2}}\right) \qquad (8\text{-}107)$$

The relationship in Eq. (8-107) is presented in Fig. 8-11, in which it can be seen that if the maximum first overshoot is to be less than 5 percent of the steady-state response, the damping ratio must equal or exceed about 0.69. As shown in Fig. 8-12, the settling time for a ± 5 percent tolerance

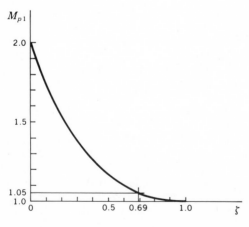

Fig. 8-11 *Ratio of magnitudes of first peak overshoot to steady-state value as a function of the damping ratio for response of a second-order proportional control system to a step change in reference level.*

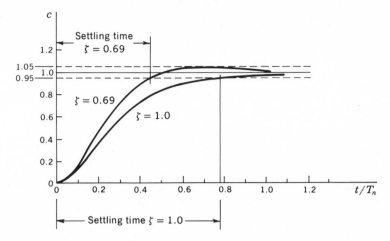

Fig. 8-12 *Settling time for the response of a second-order proportional control system to a unit-step change in reference level.*

band is much less when the damping ratio is 0.69 than when it is 1.0. Thus, from a practical viewpoint, most systems are designed for considerably less than critical damping.

Second-order system—unit-ramp-function input   The response in the time domain to a unit-ramp-function input can be found readily by determining the particular solution for Eq. (8-97) with $r = t$ and then adding it to the appropriate complementary solution from Eqs. (8-98) through (8-100).   In consideration of the oscillatory nature of the complementary solution in Eq. (8-98), we can expect an oscillatory response if $\zeta < 1.0$.   We are now most interested in determining the value of the steady-state error, if any.

To use the final-value theorem of the Laplace transformation, we need first to rewrite Eq. (8-92) as

$$C(s) = \frac{1}{(I/K)s^2 + (a/K)s + 1} R(s) \tag{8-108}$$

From Fig. 8-1 for $L = 0$,

$$E(s) = R(s) - C(s) \tag{8-109}$$

Substituting from Eq. (8-108) into Eq. (8-109) and simplifying gives

$$E(s) = \frac{(I/K)s^2 + (a/K)s}{(I/K)s^2 + (a/K)s + 1} R(s) \tag{8-110}$$

The final-value theorem, Eq. (7-17), is

$$\lim_{t \to \infty} f(t) = \lim_{s \to 0} sF(s) \tag{8-111}$$

$R(s)$ for a unit-ramp function was found [Eq. (8-56)] to be $1/s^2$. Substituting this value into Eq. (8-110) and then substituting $E(s)$ into Eq. (8-111) results in

$$e_{ss} = \lim_{t \to \infty} e(t) = \lim_{s \to 0} s \left[\frac{(I/K)s^2 + (a/K)s}{(I/K)s^2 + (a/K)s + 1} \right] \frac{1}{s^2} = \frac{a}{K} \tag{8-112}$$

Comparing the value in Eq. (8-112) with that in Eq. (8-63) for the first-order system shows that they are identical. In this case, as for most simple systems, the steady-state characteristics can be deduced by considering the relationship of the physical properties of the elements to the type of input. For example, the ramp function is in effect a constant-velocity input and, after the starting transient dies out, the effect of inertia disappears and the block K must supply only the force required to overcome the damping force.

Second-order system—sinusoidal response For a sinusoidal variation in reference level, Eq. (8-95) can be rewritten as

$$\frac{C(j\omega)}{R(j\omega)} = \frac{1}{(j\omega)^2/\omega_n^2 + j2\zeta\omega/\omega_n + 1} = \frac{1}{1 - (\omega/\omega_n)^2 + j2\zeta\omega/\omega_n} \tag{8-113}$$

from which

$$\frac{C(j\omega)}{R(j\omega)} = M\underline{/\phi} \tag{8-114}$$

where

$$M = \frac{1}{\sqrt{[1 - (\omega/\omega_n)^2]^2 + (2\zeta\omega/\omega_n)^2}} \tag{8-115}$$

and

$$\phi = -\tan^{-1} \frac{2\zeta\omega/\omega_n}{1 - (\omega/\omega_n)^2} \tag{8-116}$$

Equations (8-115) and (8-116) are identical with Eqs. (4-167) and (4-166), respectively. Consequently, the curves in Fig. 4-20 and the conclusions and observations in Secs. 4-12 and 4-13 apply to second-order control systems just as well as to a vibration system with a single degree of freedom. However, as for the first-order control system, it will now be more convenient to work with log-magnitude vs. log-frequency and phase-angle vs. log-frequency curves, as presented in Fig. 8-13.

In comparison with the curves for the first-order system in Fig. 8-7, it should be noted that (1) the second-order system can also be approximated by straight lines, although the approximation can be seriously in error near the critical frequency, i.e., near the corner frequency at

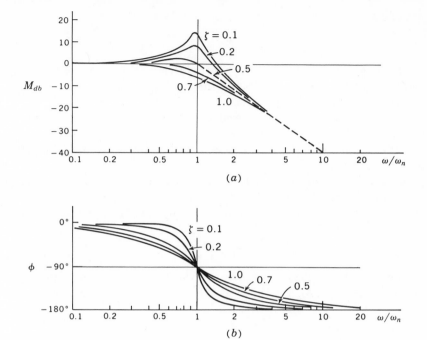

Fig. 8-13 M_{db} *and phase angle vs. log frequency for the steady-state response of a second-order proportional control system to a sinusoidally varying reference level.*

$\omega/\omega_n = 1.0$, and (2) the slope at $\omega/\omega_n \gg 1.0$ is -12 db/octave or -40 db/decade.   Curves for correcting the straight-line approximation are available in many books[1] but, from Eq. (8-115), we see that at $\omega/\omega_n = 1$,

$$M = \frac{1}{2\zeta} \tag{8-117}$$

or

$$M_{db} = 20 \log \frac{1}{2\zeta} \tag{8-118}$$

and for most purposes the complete response curve can be drawn accurately enough by combining this one calculated point with the straight-line approximation.

The slope of -12 db/octave results because Eq. (8-115) becomes, for

[1] H. M. James, N. B. Nichols, and R. S. Phillips, "Theory of Servomechanisms," p. 176, McGraw-Hill Book Company, New York, 1947.

J. J. D'Azzo and C. H. Houpis, "Feedback Control System Analysis and Synthesis," 2d ed., p. 294, McGraw-Hill Book Company, New York, 1966.

J. G. Truxal (ed.), "Control Engineers' Handbook," pp. 2-25, 2-26, McGraw-Hill Book Company, New York, 1958.

$\omega/\omega_n \gg 1$,

$$M \approx \frac{1}{(\omega/\omega_n)^2} \tag{8-119}$$

As previously noted, each order introduces a phase angle of 90°. We can now extend this by concluding that each $j\omega$ in the transfer function contributes a slope of ± 6 db/octave and a phase angle of $\pm 90°$ to the system response at frequencies well above the corner or break frequency for the term.

The polar diagram of the sinusoidal response, Eqs. (8-115) and (8-116), is shown in Fig. 8-14.

Second-order system—step change in load When the reference input is held at zero and only the load is changed, we find that Eq. (8-84) becomes, for the second-order system we have been considering,

$$E = -\frac{1}{ID^2 + aD + K}L \tag{8-120}$$

Applying the final-value theorem of the Laplace transformation, Eq. (7-17), to Eq. (8-120) for the case of a unit-step change in load, where,

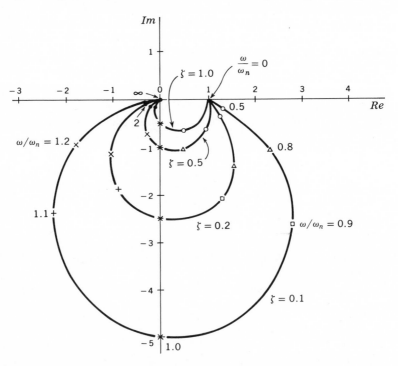

Fig. 8-14 *Polar plot of the steady-state response of a second-order proportional control system to a sinusoidally varying reference level.*

from Eq. (8-33), $L(s) = 1/s$, we find

$$e_{ss} = \lim_{t \to \infty} e(t) = \lim_{s \to 0} s \left(- \frac{1}{Is^2 + as + K} \right) \frac{1}{s} = - \frac{1}{K} \qquad (8\text{-}121)$$

which, again, is the same as the value of steady-state error found for the first-order system, Eq. (8-89).

Higher-order systems   As discussed in Chaps. 3 through 6, irrespective of the number of degrees of freedom, all lumped-parameter systems containing inertial, spring, and damping elements can be treated as a family of second-order systems.   The reason is simply that a second-order differential equation completely defines the forces acting on a single inertial element.   Although control systems may or may not involve force in a literal sense, elements analogous to force, inertia, springs, and dampers will be present, and every system can be shown to be made up of first- and second-order elements or systems.

In Chap. 6 it became evident that if a system required more than a single second-order differential equation, the number of variables would be too large to permit the presentation of the general solution in the form of curves, as in Figs. 8-7 and 8-13.   Consequently, we can expect to have to consider each higher-order system as a special case with respect to response and stability characteristics.   This will be done in Chap. 9.

As shown above, proportional control will have (1) a zero steady-state error only for a step change in input reference level, (2) a constant steady-state error for a ramp-function change in input reference level or a step change in load, and (3) an infinite steady-state error for higher-degree functions of reference-level input or load change.

For many applications these characteristics are satisfactory, but in many others the system will be required to follow a rapidly changing reference level or maintain a constant output when subjected to a rapidly changing load, as, for example, in control of the position of the cutting tool by letting a tracing point follow a template or curve in contour machining, control of the position of the antenna of a tracking radar that must follow rapidly moving objects while being subjected to large varying forces from wind gusts, and control of the output of a manufacturing plant when management changes its mind about production schedules and/or the consumers change their minds about how many articles they will buy and when they will buy them.   Under these conditions, control systems of greater complexity than simple proportional control must be used if the results are to be satisfactory.

The next several sections will be given to discussion of simple modifications that will greatly increase the usefulness of the control system by providing improved response to ramp function and sinusoidal reference

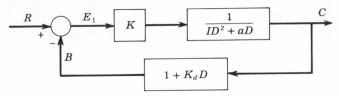

Fig. 8-15 *Control system with a second-order controlled system and proportional plus derivative-feedback control.*

inputs and step changes in load. The principles introduced here will also find direct application in more complicated systems.

8-3. Proportional plus Derivative-feedback Control As shown in Fig. 8-13, the sinusoidal response of a second-order system with zero (or very little) damping will be unsatisfactory, if not disastrous, at or near $\omega/\omega_n = 1.0$. Therefore, in general, appreciable damping will be desirable. There is no best value of damping—the choice must be based on a compromise between response time and overshoot, and values between 0.4 and 0.7 are usually specified.

The main problem is that most systems, particularly mechanical systems, contain very little damping. In fact, we generally go to considerable trouble to minimize friction to decrease the power required and to minimize the problems of wear and heat dissipation. In many cases, it is difficult to introduce sufficient damping and/or to make certain it does not change with time. Thus, just introducing more damping is not a good approach, and we need to look further.

If we consider the effect of damping on the time variation of the response, we find that damping always opposes any change and, in effect, acts like an output from the control element that has a magnitude proportional to the rate of change of the controlled variable and a sign opposite to that of the rate of change. Thus, we can achieve the same effect by supplying the control element with a signal, e_1, that is a function of both the difference between the reference input r and the controlled variable c, and the rate of change of the controlled variable Dc. This can be accomplished in different ways, but for mechanical systems where position is the controlled variable, i.e., a servomechanism, the derivative is a velocity and the output of a tachometer or any other type of velocity transducer, such as that shown in Fig. 4-41, can be used to provide this part of the signal. In any case, the block diagram becomes that shown in Fig. 8-15, where K_d is a multiplying factor (usually <1), and the control ratio is

$$\frac{C}{R} = \frac{K}{ID^2 + (a + KK_d)D + K} \tag{8-122}$$

in which we recognize the coefficient of D as the effective damping coefficient.   Thus,

$$a_{eff} = a + KK_d \tag{8-123}$$

Equation (8-122) can be written as

$$\frac{C}{R} = \frac{1}{(1/\omega_n{}^2)D^2 + (2\zeta/\omega_n)D + 1} \tag{8-124}$$

where

$$\omega_n{}^2 = \frac{K}{I}$$

and

$$\zeta = \frac{1}{2}\left(\frac{a}{K} + K_d\right)\omega_n \tag{8-125}$$

Therefore, as can be seen in Eq. (8-123) or Eq. (8-125), we can effectively provide any desired degree of damping, even if the original system had none, by simply adjusting the value of K_d.   A major advantage of using derivative feedback over increasing the actual damping is that there is no energy dissipated in that part of the effective damping due to derivative feedback.   Since for sinusoidal response the velocity leads the displacement, this method of modifying the performance characteristics is often called *lead compensation*.

Although it is difficult to imagine a situation in which it would be desired, the effective damping can be decreased by making the transfer function of the block in the feedback loop $1 - K_dD$, resulting in, from Eq. (8-123),

$$a_{eff} = a - KK_d \tag{8-126}$$

In Eq. (8-126), if $KK_d > a$, the effective damping becomes negative and, as discussed in Sec. 4-23, the system will be unstable.

8-4. Proportional plus Error-rate Control   As discussed above, the effective damping can be increased by adding the time derivative of the response of the controlled variable to the response in the feedback signal.   From the viewpoint of response alone, comparison of Eq. (8-122) with Eq. (8-91) shows that the effect is exactly the same as increasing the damping coefficient.   This was shown to be beneficial with respect to sinusoidal response, but the increased damping will result in a slower response to a step-function input and a greater steady-state error with a ramp-function input.   These undesirable effects arise because the increased damping acts as long as the response is changing, even though the system has reached steady-state operation.

Since the purpose of the system is to reduce the error, a logical approach is to try to prevent the error from arising in the first place.   Thus, a corrective signal that is a function of the rate of change of error can have a significant effect before the error itself has become appreciable.   In addition, the effect becomes zero when the error reaches a constant (steady-state) value.   Therefore, if we can add effective damping while,

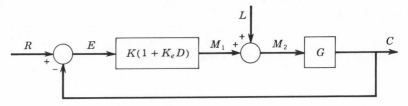

Fig. 8-16 *Control system with proportional plus error-rate control.*

and only while, the error is changing, we can have a system that behaves as a heavily damped system during transient and sinusoidal conditions and as a lightly damped system during steady-state operation. The result is a system with proportional plus error-rate control,[1] as shown in Fig. 8-16.

In reality, true error-rate control cannot be achieved. For example, an infinite signal would be required for a step change in reference input. In addition, differentiation is inherently an undesirable process, whether accomplished numerically, graphically, electrically, or pneumatically, because the effects of minor variations, particularly those due to unavoidable errors or noise in the signal, are magnified rather than minimized. As a result, the devices that are used give only approximate differentiation. Most provide a good approximation at low frequency with heavy filtering of the high frequencies. We shall assume here that the differentiation is exact, as indicated by the K_eD term in the control-element transfer function.

Considering first the case when the load L is zero, the control ratio for the system in Fig. 8-16 is

$$\frac{C}{R} = \frac{K(1 + K_eD)G}{1 + K(1 + K_eD)G} \tag{8-127}$$

and the error is

$$E = R - C = \frac{1}{1 + K(1 + K_eD)G} R \tag{8-128}$$

Unit-ramp-function input Since the steady-state error in the response to a ramp-function input was found in Sec. 8-2 to be the same for first- and second-order controlled systems and since we are more interested in the sinusoidal response of second-order systems, only the latter will be considered further in this section.

Substituting for the second-order controlled system, from Eq. (8-90),

$$G = \frac{1}{ID^2 + aD} \tag{8-129}$$

---

[1] *Error-rate* control is commonly called *rate* control and *derivative* control in the process industry.

into Eq. (8-127) and rearranging gives

$$\frac{C}{R} = \frac{1 + K_e D}{(I/K)D^2 + (a/K + K_e)D + 1} \tag{8-130}$$

which we recognize can be written as

$$\frac{C}{R} = \frac{1 + K_e D}{(1/\omega_n{}^2)D^2 + (2\zeta/\omega_n)D + 1} \tag{8-131}$$

where

$$\omega_n{}^2 = \frac{K}{I}$$

and

$$\zeta = \frac{1}{2}\left(\frac{a}{K} + K_e\right)\omega_n \tag{8-132}$$

Therefore as can be seen in Eq. (8-132), we can have any desired damping ratio by adjusting the value of K_e. Comparing Eq. (8-132) with Eq. (8-125) shows that as far as the damping ratio is concerned error-rate and derivative feedback have the same effect.

Substituting from Eq. (8-129) into Eq. (8-128) and simplifying results in

$$E = \frac{ID^2 + aD}{ID^2 + (a + KK_e)D + K} R \tag{8-133}$$

Applying the final-value theorem of the Laplace transformation, Eq. (7-17), to Eq. (8-133) for the case of a unit-ramp-function input, where, from Eq. (8-56), $R(s) = 1/s^2$, we find

$$e_{ss} = \lim_{t \to \infty} e(t) = \lim_{s \to 0} s\left[\frac{Is^2 + as}{Is^2 + (a + KK_e)s + K}\right]\frac{1}{s^2} = \frac{a}{K} \tag{8-134}$$

Comparing Eq. (8-134) with Eq. (8-112) shows that the steady-state error for ramp-function response is unchanged by adding error-rate control. This is the major advantage of error-rate control over derivative feedback control, which results in an increased steady-state error.

Sinusoidal response   For sinusoidal response, Eq. (8-131) becomes

$$\frac{C(j\omega)}{R(j\omega)} = \frac{1 + jK_e\omega}{1 - (\omega/\omega_n)^2 + j2\zeta\omega/\omega_n} \tag{8-135}$$

where, from Eq. (8-132), $\zeta = \frac{1}{2}(a/K + K_e)\omega_n$. Equation (8-135) is a function of ω as well as ω/ω_n and cannot be treated in nondimensional terms. However, in most cases the designer will try to minimize the actual damping, and we shall find $a/K \ll K_e$. In this case,

$$\zeta \approx \frac{K_e}{2}\omega_n \tag{8-136}$$

and Eq. (8-135) becomes

$$\frac{C(j\omega)}{R(j\omega)} = \frac{1 + j2\zeta\omega/\omega_n}{1 - (\omega/\omega_n)^2 + j2\zeta\omega/\omega_n} \tag{8-137}$$

which is nondimensional.   The magnitude of $C(j\omega)/R(j\omega)$ is

$$M = \frac{\sqrt{1 + (2\zeta\omega/\omega_n)^2}}{\sqrt{[1 - (\omega/\omega_n)^2]^2 + (2\zeta\omega/\omega_n)^2}} \tag{8-138}$$

which is identical with Eq. (4-189) for transmissibility ratio of a mass-spring-damper system, and, consequently, the curves in Fig. 4-25 apply also to Eq. (8-138).

As noted in Sec. 8-2, the curves in Fig. 4-20 apply to the system with proportional control. Comparing Fig. 4-25 with Fig. 4-20 shows that for the same damping ratio, we can expect a higher peak amplitude with proportional plus error-rate control than with proportional control only. It can also be seen that the phase angle for proportional plus error-rate control approaches $-90°$ rather than $-180°$ at $\omega/\omega_n \gg 1$. The first point is of little consequence because we can increase the error-rate damping simply and at no expense in power consumption. The significance of the phase-angle behavior will be discussed in Sec. 9-4.

Another major difference between the curves in Figs. 4-20 and 4-25 is that at $\omega/\omega_n \gg 1$ the magnitude decreases at a lower rate for proportional plus error-rate control. At $\omega/\omega_n \gg 1$, Eq. (8-138) becomes

$$M \approx \frac{2\zeta}{\omega/\omega_n} \tag{8-139}$$

and the slope is -6 db/octave rather than -12 db/octave, as found for proportional control alone.

Unit-step change in load   If we consider the system in Fig. 8-16 for the case when $R = 0$ and there is a change in load, we find

$$C = \frac{G}{1 + K(1 + K_eD)G} L \tag{8-140}$$

For the second-order controlled element, substituting for G from Eq. (8-129), Eq. (8-140) becomes

$$C = \frac{1}{ID^2 + (a + KK_e)D + K} L \tag{8-141}$$

and from Fig. 8-16,

$$E = -C = -\frac{1}{ID^2 + (a + KK_e)D + K} L \tag{8-142}$$

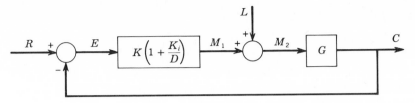

Fig. 8-17   *Control system with proportional plus integral control.*

Applying the final-value theorem of the Laplace transformation, Eq. (7-17), to Eq. (8-142) for the case of unit-step change in load, where, from Eq. (8-33), $L(s) = 1/s$, we have

$$e_{ss} = \lim_{t \to \infty} e(t) = \lim_{s \to 0} s \left[- \frac{1}{Is^2 + (a + KK_e)s + K} \right] \frac{1}{s} = - \frac{1}{K} \quad (8\text{-}143)$$

Comparing Eq. (8-143) with Eq. (8-121) leads, again, to the conclusion that the addition of error-rate control has no effect on the steady-state error resulting from a step change in load.

8-5. Proportional plus Integral Control   A major limitation of all of the systems discussed thus far is the presence of a steady-state error in their responses to a ramp change in reference level or a step change in load.   Since the derivative of the error disappears when the steady-state error becomes constant, it is apparent that there is little to be gained by trying further to reduce the error by use of any sort of derivative control.

In fact, the only way in which the presence of a steady signal can be utilized is to integrate the signal, thus providing another signal that increases in magnitude as long as the original signal does not change in sign.   This increasing signal can then be the input to a control element. In the case at hand the obvious application is to use the integral of the error signal to drive the output in the direction to reduce the error.

This type of control is commonly called *integral* or *reset control*.   The latter term is being used less frequently as time goes by, and we shall not use it further in this text.   The logical reason for the term was that the integral of the error effectively resets the system for a new set of steady-state operating conditions.

When integral control is added to the system in Fig. 8-1, we have the proportional plus integral control system in Fig. 8-17, where $1/D$ is the operator notation for integration and K_i is a multiplying factor.

Considering first the case with $L = 0$, we find for the system in Fig. 8-17 the control ratio

$$\frac{C}{R} = \frac{K(D + K_i)G}{D + K(D + K_i)G} \quad (8\text{-}144)$$

and the error

$$E = R - C = \frac{D}{D + K(D + K_i)G} R \qquad (8\text{-}145)$$

First-order controlled element From Eq. (8-7), for a first-order controlled element,

$$G = \frac{1}{aD} \qquad (8\text{-}146)$$

Substituting from Eq. (8-146) into Eq. (8-144) gives

$$\frac{C}{R} = \frac{(1/K_i)D + 1}{(a/KK_i)D^2 + (1/K_i)D + 1} \qquad (8\text{-}147)$$

which can be rearranged as

$$\left(\frac{a}{KK_i}D^2 + \frac{1}{K_i}D + 1\right)c = \left(\frac{1}{K_i}D + 1\right)r \qquad (8\text{-}148)$$

or

$$\left(\frac{1}{\omega_n{}^2}D^2 + \frac{2\zeta}{\omega_n}D + 1\right)c = \left(\frac{2\zeta}{\omega_n}D + 1\right)r \qquad (8\text{-}149)$$

where

$$\omega_n{}^2 = \frac{KK_i}{a} \qquad (8\text{-}150)$$

and

$$\zeta = \frac{1}{2K_i}\omega_n = \frac{1}{2}\sqrt{\frac{K}{aK_i}} \qquad (8\text{-}151)$$

In Eq. (8-149) we see the interesting fact that the addition of integral control converts a first-order proportional control system with no possibility of overshoot or oscillation into a second-order system that can overshoot and oscillate.

Substituting from Eq. (8-146) into Eq. (8-145) gives

$$E = \frac{aD^2}{aD^2 + KD + KK_i} R \qquad (8\text{-}152)$$

First-order controlled element—unit-ramp-function input
Applying the final-value theorem of the Laplace transformation, Eq. (7-17), to Eq. (8-152) for the case of a unit-ramp-function input, where $R(s) = 1/s^2$, from Eq. (8-56), we find

$$e_{ss} = \lim_{t \to \infty} e(t) = \lim_{s \to 0} s \left(\frac{as^2}{as^2 + Ks + KK_i}\right)\frac{1}{s^2} = 0 \qquad (8\text{-}153)$$

Therefore, as expected, the steady-state error for a ramp input is zero.

Second-order controlled system From Eq. (8-90), we have for a second-order controlled system

$$G = \frac{1}{ID^2 + aD} \qquad (8\text{-}154)$$

Substituting from Eq. (8-154) into Eqs. (8-144) and (8-145) gives, respectively,

$$\frac{C}{R} = \frac{K(D + K_i)}{ID^3 + aD^2 + KD + KK_i} \tag{8-155}$$

and

$$E = \frac{ID^3 + aD^2}{ID^3 + aD^2 + KD + KK_i} R \tag{8-156}$$

As for the first-order controlled element, the addition of integral control results in a system order one greater than that of the element. In this case, we have a third-order system, for which the characteristic equation is

$$Is^3 + as^2 + Ks + KK_i = 0 \tag{8-157}$$

In terms of previous discussions, two observations become important:

1   The concepts of natural frequency and damping ratio that were so useful in relation to second-order systems no longer apply.
2   We can no longer look at the characteristic equation, as in Sec. 4-23, and say that since all coefficients are positive, the system is stable.

Consequently, from observation 1, each set of parameters becomes a separate case and the use of an analog computer (electronic differential analyzer) is indicated. From observation 2, we see the necessity for going deeper into the subject of stability analysis,[1] which will be done in Chap. 9.

Second-order controlled system—unit-ramp-function input Applying the final-value theorem of the Laplace transformation, Eq. (7-17), to Eq. (8-156) for a unit-ramp-function input, where $R(s) = 1/s^2$ from Eq. (8-56), we find

$$e_{ss} = \lim_{t \to \infty} e(t) = \lim_{s \to 0} s \left(\frac{Is^3 + as^2}{Is^3 + as^2 + Ks + KK_i} \right) \frac{1}{s^2} = 0 \tag{8-158}$$

and, as for the first-order controlled system, the steady-state error in the response to a ramp-function input is zero.

Second-order controlled system—unit-step change in load Considering only the effect of the load, when $R = 0$, we find for the system in Fig. 8-17

$$C = \frac{DG}{D + K(D + K_i)G} L \tag{8-159}$$

---

[1] Actually, the third-order system can be analyzed in a relatively simple manner by use of Routh's criterion. It can be shown, as in Sec. 9-1, that if the characteristic equation is expressed as

$$As^3 + Bs^2 + Cs + D = 0$$

the system will be stable if A, B, C, and D are all positive numbers and $BC > AD$.

which, for a second-order controlled system, from Eq. (8-154) becomes

$$C = \frac{D}{ID^3 + aD^2 + KD + KK_i}L \qquad (8\text{-}160)$$

From Fig. 8-17,

$$E = -C = -\frac{D}{ID^3 + aD^2 + KD + KK_i}L \qquad (8\text{-}161)$$

Applying the final-value theorem of the Laplace transformation, Eq. (7-17), to Eq. (8-161) for the case of a unit-step change in load, where $L(s) = 1/s$ from Eq. (8-33), we find

$$e_{ss} = \lim_{t \to \infty} e(t) = \lim_{s \to 0} s\left(-\frac{s}{Is^3 + as^2 + Ks + KK_i}\right)\frac{1}{s} = 0 \qquad (8\text{-}162)$$

and we see that the addition of integral control has eliminated the steady-state error in the response to a step change in load.

8-6. Proportional plus Error-rate plus Integral Control   As discussed in the sections immediately above, the effective damping in a proportional control system can be increased readily by adding error-rate control and the steady-state error in the response to a ramp-function input or step change in load can be eliminated by adding integral control. Since error-rate control acts only when the error is changing and integral control is effective mainly when the error is constant, it is apparent that neither will have appreciable effect on the other in their respective regions of major effect.   Therefore, one can expect the proportional plus error-rate plus integral control system in Fig. 8-18 to combine the best features of the individual actions.   Such is the case, and this type of control system finds widespread use in many different situations.

8-7. System Type Numbers   The importance of the steady-state error as a basis for comparing the performance of systems has been demonstrated in the preceding sections.   All systems can be treated in exactly the same way, but the method can be generalized and simplified *for those systems with unity feedback*, i.e., with no blocks in the feedback

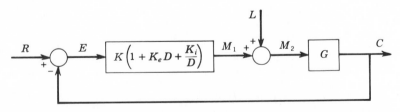

Fig. 8-18   *Control system with proportional plus error-rate plus integral control.*

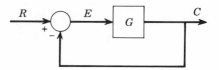

Fig. 8-19 *Control system with unity feedback.*

loop. For example, if we consider the system in Fig. 8-19 as the general system, we find

$$\frac{C}{R} = \frac{G}{1 + G} \qquad (8\text{-}163)$$

and

$$E = R - C = \frac{1}{1 + G} R \qquad (8\text{-}164)$$

In general, the open-loop transfer function can be written in the form

$$G(s) = \frac{K'(a_k s^k + a_{k-1} s^{k-1} + \cdots + a_1 s + 1)}{s^n(b_i s^i + b_{i-1} s^{i-1} + \cdots + b_1 s + 1)} \qquad (8\text{-}165)$$

where $K' =$ gain constant.

As before, we can find the steady-state error by applying the final-value theorem of the Laplace transformation, Eq. (7-17), to Eq. (8-164). Therefore, we can write

$$e_{ss} = \lim_{t \to \infty} e(t) = \lim_{s \to 0} \frac{s}{1 + G(s)} R(s) \qquad (8\text{-}166)$$

We could substitute from Eq. (8-165) at this time, but it will be simpler if we first consider the limit of $G(s)$ as $s \to 0$. Doing this, we find, from Eq. (8-165),

$$\lim_{s \to 0} G(s) = \lim_{s \to 0} \frac{K'}{s^n} \qquad (8\text{-}167)$$

which we now substitute in Eq. (8-166) to give

$$e_{ss} = \lim_{s \to 0} s \left(\frac{1}{1 + K'/s^n} \right) R(s) = \lim_{s \to 0} \frac{s^{n+1}}{s^n + K'} R(s) \qquad (8\text{-}168)$$

If, as is usually the case, $n \neq 0$ Eq. (8-168) becomes

$$e_{ss} = \lim_{s \to 0} \frac{s^{n+1}}{K'} R(s) \qquad (8\text{-}169)$$

If we now consider input functions of the form

$$r = 0$$

for $t \leq 0$, and

$$r = t^m \qquad (8\text{-}170)$$

for $t > 0$, we find for $t > 0$, from Eq. (8-62),

$$R(s) = \frac{m!}{s^{m+1}} \qquad (8\text{-}171)$$

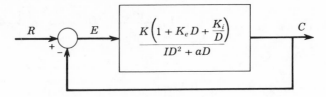

Fig. 8-20 *Control system with a second-order controlled system and proportional plus error-rate plus integral control.*

Substituting from Eq. (8-171) into Eq. (8-169) gives

$$e_{ss} = \lim_{s \to 0} \frac{m!}{K'} \frac{s^{n+1}}{s^{m+1}} = \lim_{s \to 0} \frac{m!}{K'} \frac{s^n}{s^m} \tag{8-172}$$

From Eq. (8-172) we can now conclude the following:

When $n > m$

$$e_{ss} = 0 \tag{8-173}$$

When $n = m \neq 0$

$$e_{ss} = \frac{m!}{K'} \tag{8-174}$$

When $n < m$

$$e_{ss} = \infty \tag{8-175}$$

Therefore, we need to know only n and K' from the open-loop transfer function $G(s)$ in order to tell what the steady-state error will be for any input that is expressed as a simple power function of time.

As is apparent, the value of n by itself can be used to classify systems in terms of their steady-state characteristics. It is called the system *type* number. For example, when $n = 3$, we have a type-3 system and it will have a zero steady-state error for step-, ramp-, and parabolic-function inputs and an error of $6/K'$ for a unit-cubic input.

Example 8-1 Determine (*a*) the type number and (*b*) expected steady-state errors for the input functions $r = 5$, $r = 5t$, and $r = 5t^2$ for a proportional plus error-rate plus integral control system when the controlled element has a transfer function

$$G = \frac{1}{ID^2 + aD} \tag{8-176}$$

(*a*) *System Type.* The system in Fig. 8-17 becomes that in Fig. 8-20 when $G = 1/(ID^2 + aD)$. The transfer function of the resultant single block can be written as

$$\begin{aligned}
G(s) &= \frac{K(1 + K_e s + K_i/s)}{Is^2 + as} \\
&= \frac{KK_i}{as^2} \left[\frac{(K_e/K_i)s^2 + (1/K_i)s + 1}{(I/a)s + 1} \right] \\
&= \frac{K'}{s^n} \left[\frac{(K_e/K_i)s^2 + (1/K_i)s + 1}{(I/a)s + 1} \right]
\end{aligned} \tag{8-177}$$

where
$$K' = \frac{KK_i}{a} \tag{8-178}$$

and
$$n = 2 \tag{8-179}$$

From Eq. (8-179), $n = 2$, and we see that we have a type-2 system.

(b) *Steady-state Errors.* From Eq. (8-173), a type-2 system will have a zero steady-state error for inputs of the form Ct^m when $m < 2$. Therefore, the steady-state error will be zero for $r = 5$ and $r = 5t$. For $m = 2$, the steady-state error for a *unit*-parabolic input will be, from Eq. (8-174),

$$e_{ss} = \frac{2!}{K'} = \frac{2}{K'} \tag{8-180}$$

which, from Eq. (8-178), becomes

$$e_{ss} = \frac{2a}{KK_i} \tag{8-181}$$

Since the system is linear, the error for $r = 5t^2$ will be five times the error for $r = t^2$. Thus, for $r = 5t^2$,

$$e_{ss} = \frac{10a}{KK_i}$$

chapter 9

STABILITY ANALYSIS
OF LINEAR SYSTEMS

As discussed in Sec. 4-23, the question of whether or not a linear system is stable is determined completely by the roots of the characteristic equation. In fact, one might be tempted to dismiss the entire subject by simply stating that a system will be unstable if the real part of any root of the characteristic equation is positive. However, this still leaves us with the problems of finding the roots and relating them to the performance of the system, particularly with respect to how the roots should be changed to improve the performance.

In many cases the designer is interested in modifying an existing system and an analytical expression for the characteristic equation is not known. He must base his analysis on data obtained during operation of the existing system, in part or as a whole.

From a purely practical viewpoint, all performance characteristics can be found by simulating the system on an analog computer. Then, if improvement is required, we can usually achieve the desired results by combining a basic understanding of the physical behavior of the blocks with a trial-and-error adjustment of parameters and the introduction of additional elements to the computer model.

Although in many cases the understanding gained in Chap. 8 will be adequate for this procedure, there is still much to be gained from considering stability in more general, as well as more specific, terms than we have thus far. Of the numerous methods and variations of methods for studying stability and overall system performance, the four that can be considered basic and will be emphasized in this chapter are:

1 Routh's criterion
2 The root locus

3   Conformal mapping

4   Use of attenuation-phase plots

9-1. Stability Analysis by Inspection   In general, the characteristic equation for a system can be expressed as either

$$(s - s_1)(s - s_2) \cdot \cdot \cdot (s - s_n) = 0 \qquad (9\text{-}1)$$

or $$b_n s^n + b_{n-1} s^{n-1} + \cdot \cdot \cdot + b_1 s + b_0 = 0 \qquad (9\text{-}2)$$

where $s_1 \cdot \cdot \cdot s_n$ are the roots, real and/or complex, of the characteristic
equation and the b's are real coefficients.

If the equation exists in the form of Eq. (9-1), we can immediately
state whether or not the system will be stable.   For example, let us consider the characteristic equation

$$(s + 1)(s + 1 + j2)(s + 1 - j2)(s - 2) = 0 \qquad (9\text{-}3)$$

where $s_1 = -1$

$s_2 = -1 - j2$

$s_3 = -1 + j2$

$s_4 = +2$

Since at least one root (s_4) has a positive real part, we conclude that
the system will be unstable.   Although not related to stability, it should
be noted that s_2 and s_3 are complex conjugates and it should be recalled
from Sec. 4-8 that complex roots can only occur as conjugate pairs.

If, as will be the usual case, the characteristic equation is available only
in the form of Eq. (9-2), the matter may not be quite so simple.   However, even for the most complex system there are two conditions under
which we can state that the system will be unstable.   These conditions
are (1) if any term in Eq. (9-2) is missing, i.e., any $b = 0$, and/or (2) if the
coefficients do not all have the same sign.

For example, Eq. (9-3) can be written as

$$s^4 + s^3 + s^2 - 9s - 10 = 0 \qquad (9\text{-}4)$$

and on the basis of condition 2 we would conclude, again, that the system
is unstable.

The remaining problem is that the two conditions above *are only
sufficient*—they are not necessary—for instability and the existence of all
terms with coefficients of the same sign does not guarantee that the
system will be stable.

9-2. Routh's Criterion   Routh's criterion provides a straightforward
procedure for determining the number of roots of a characteristic equation
with positive real parts.   It has limited usefulness because it gives no
information about the location of the roots.   However, it can be quite
useful if we are interested only in whether or not the system is stable and
the characteristic equation is of the form of Eq. (9-2), with all terms

present and with all coefficients having the same sign, so that simple inspection alone is not adequate.

The method involves constructing an array by cross-multiplying the coefficients in a prescribed manner. To illustrate the method, let us repeat Eq. (9-2):

$$b_n s^n + b_{n-1} s^{n-1} + \cdots + b_1 s + b_0 = 0 \tag{9-5}$$

The array is then

s^n	b_n	b_{n-2}	b_{n-4}	$\cdots$
s^{n-1}	b_{n-1}	b_{n-3}	b_{n-5}	$\cdots$
s^{n-2}	c_1	c_3	c_5	$\cdots$
s^{n-3}	d_1	d_3	d_5	$\cdots$
s^{n-4}	e_1	e_3	e_5	$\cdots$
$\cdot$	$\cdot$			
$\cdot$	$\cdot$			
$\cdot$	$\cdot$			
s^0	$\cdots$			

where

$$c_1 = \frac{-1}{b_{n-1}} \begin{vmatrix} b_n & b_{n-2} \\ b_{n-1} & b_{n-3} \end{vmatrix} \qquad c_3 = \frac{-1}{b_{n-1}} \begin{vmatrix} b_n & b_{n-4} \\ b_{n-1} & b_{n-5} \end{vmatrix} \qquad \cdots \tag{9-6}$$

$$d_1 = \frac{-1}{c_1} \begin{vmatrix} b_{n-1} & b_{n-3} \\ c_1 & c_3 \end{vmatrix} \qquad d_3 = \frac{-1}{c_1} \begin{vmatrix} b_{n-1} & b_{n-5} \\ c_1 & c_5 \end{vmatrix} \qquad \cdots \tag{9-7}$$

$$e_1 = \frac{-1}{d_1} \begin{vmatrix} c_1 & c_3 \\ d_1 & d_3 \end{vmatrix} \qquad e_3 = \frac{-1}{d_1} \begin{vmatrix} c_1 & c_5 \\ d_1 & d_5 \end{vmatrix} \qquad \cdots \tag{9-8}$$

. . . .

In Eqs. (9-6) through (9-8) the vertical bars signify the determinant of the enclosed array. The final array will be roughly triangular, and there will be $n + 1$ rows. *The number of roots with positive real parts will equal the number of changes of sign of the coefficients in the first column of coefficients in the final array* (column headed by b_n).

When the first term in a row is zero or when all terms in a row are zero, special procedures must be followed. The significance of the case where all terms in a row are zero will be considered briefly in Sec. 9-4, but for detailed discussions of these and other related topics, the reader is referred to the literature.[1]

[1] J. G. Truxal (ed.), "Control Engineers' Handbook," pp. 2-9–2-11, McGraw-Hill Book Company, New York, 1958.

J. J. D'Azzo and C. H. Houpis, "Feedback Control System Analysis and Synthesis," 2d ed., pp. 121–127, McGraw-Hill Book Company, New York, 1966.

M. F. Gardner and J. L. Barnes, "Transients in Linear Systems," vol. 1, pp. 197–201, John Wiley & Sons, Inc., New York, 1942.

F. H. Raven, "Automatic Control Engineering," pp. 106–108, McGraw-Hill Book Company, New York, 1961.

Another method that is quite similar to Routh's criterion is the Hurwitz criterion.[1]   This method requires finding certain determinants related to a special array of the coefficients of the characteristic equation.   Consequently, it is more laborious than Routh's criterion when solved manually, but the simplicity of describing the method of solution makes it particularly adaptable for programming it for solution by use of a digital computer.

Example 9-1   Apply Routh's criterion to determine the number of roots with positive real parts of the characteristic equation in Eq. (9-4).

From Eq. (9-4), $n = 4$; thus, the Routh array will have $n + 1 = 5$ rows, and the first two will be

$$
\begin{array}{c|ccc}
s^4 & 1 & 1 & -10 \\
s^3 & 1 & -9 &
\end{array}
$$

From Eq. (9-6),

$$
c_1 = \frac{-1}{b_{n-1}} \begin{vmatrix} b_n & b_{n-2} \\ b_{n-1} & b_{n-3} \end{vmatrix} = \frac{-1}{1} \begin{vmatrix} 1 & 1 \\ 1 & -9 \end{vmatrix} = -1(-9 - 1) = 10
$$

$$
c_3 = \frac{-1}{b_{n-1}} \begin{vmatrix} b_n & b_{n-4} \\ b_{n-1} & b_{n-5} \end{vmatrix} = \frac{-1}{1} \begin{vmatrix} 1 & -10 \\ 1 & 0 \end{vmatrix} = -1(0 + 10) = -10
$$

$$
c_5 = \frac{-1}{b_{n-1}} \begin{vmatrix} b_n & b_{n-6} \\ b_{n-1} & b_{n-7} \end{vmatrix} = \frac{-1}{1} \begin{vmatrix} 1 & 0 \\ 1 & 0 \end{vmatrix} = 0
$$

from Eq. (9-7),

$$
d_1 = \frac{-1}{c_1} \begin{vmatrix} b_{n-1} & b_{n-3} \\ c_1 & c_3 \end{vmatrix} = \frac{-1}{10} \begin{vmatrix} 1 & -9 \\ 10 & -10 \end{vmatrix} = -\tfrac{1}{10}(-10 + 90) = -8
$$

$$
d_3 = \frac{-1}{c_1} \begin{vmatrix} b_{n-1} & b_{n-5} \\ c_1 & c_5 \end{vmatrix} = \frac{-1}{10} \begin{vmatrix} 1 & 0 \\ 10 & 0 \end{vmatrix} = 0
$$

and from Eq. (9-8),

$$
e_1 = \frac{-1}{d_1} \begin{vmatrix} c_1 & c_3 \\ d_1 & d_3 \end{vmatrix} = \frac{-1}{-8} \begin{vmatrix} 10 & -10 \\ -8 & 0 \end{vmatrix} = \tfrac{1}{8}(0 - 80) = -10
$$

Therefore, the complete Routh array is

$$
\begin{array}{c|ccc}
s^4 & 1 & 1 & -10 \\
s^3 & 1 & -9 & \\
s^2 & 10 & -10 \cdot & \\
s^1 & -8 & & \\
s^0 & -10 & &
\end{array}
$$

and since there is one change of sign, from 10 to -8, in the first column of coefficients, there is one root with a positive real part and the system will be unstable.

9-3. Root Locus   The preceding sections have been concerned with methods that are quite useful when one is interested in determining

[1] Raven, *op. cit.*, pp. 108–109.

simply whether or not a system is stable. However, in most cases the designer is also interested in the response characteristics that can be expected and how they can be modified. As discussed in Chaps. 4 and 8, the response characteristics, such as critical frequency, logarithmic decrement, response time, settling time, magnification factor, and transmissibility, are all directly related to the roots of the characteristic equation.

For example, we found for second-order systems that (1) if the roots are real negative numbers, the system is overdamped and the response to a disturbance will be a rapid return to the equilibrium position without oscillation, (2) if the roots are complex with zero real parts, the system has zero damping and the response to a disturbance will be an oscillation that continues indefinitely, and (3) if the roots are complex with negative real parts, the system is underdamped and the response to a disturbance will be a decaying oscillation, with the rate of decay increasing as the real parts of the roots become increasingly negative.

Although the response of a complex system cannot really be described in such simple terms, it will be found that the response will be quite similar to that of a simple system whose roots are identical with those of the complex system having the least negative parts. For example, if one root of the characteristic equation has a positive real part, the system is unstable, and if a pair of roots are complex with zero real parts, we can expect the system to oscillate indefinitely even though all of the other roots are negative real numbers.

Some of the roots of the characteristic equation can be expected to be complex; and if we wish to present them graphically, we must use the complex s plane. For example, if the characteristic equation is

$$s^2 + 2\zeta\omega_n s + \omega_n{}^2 = 0 \tag{9-9}$$

we know that for $\zeta < 1$ the roots are

$$s_1 = -\zeta\omega_n + j\omega_n \sqrt{1 - \zeta^2} \tag{9-10}$$

and
$$s_2 = -\zeta\omega_n - j\omega_n \sqrt{1 - \zeta^2} \tag{9-11}$$

Equations (9-10) and (9-11) are complex conjugates and can be expressed in general form as

$$s = \sigma \pm j\omega \tag{9-12}$$

where, for Eq. (9-10), $\sigma = -\zeta\omega_n$, the real part of the root, and

$$j\omega = j\omega_n \sqrt{1 - \zeta^2}$$

the imaginary part of the root. If we designate the axes of the complex s plane by σ and $j\omega$, the roots in Eqs. (9-10) and (9-11) will be located as shown in Fig. 9-1.

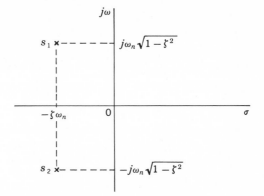

Fig. 9-1 *Complex conjugate roots on the s plane.*

Previously, we have found it convenient to express complex terms as a vector with a magnitude and an angle. This procedure is just as useful when considering the roots of the characteristic equation as when determining the ratio of amplitudes and phase angles for the sinusoidal response of a system. In general,

$$s = \sigma + j\omega = |s|\underline{/s} = \sqrt{\sigma^2 + \omega^2} \underline{/\phi} \qquad (9\text{-}13)$$

where

$$\phi = \tan^{-1} \frac{\omega}{\sigma} \qquad (9\text{-}14)$$

Substituting from Eq. (9-10) into Eqs. (9-13) and (9-14), we find

$$s_1 = \omega_n \underline{/\phi_1} \qquad (9\text{-}15)$$

where

$$\phi_1 = \tan^{-1} \frac{\sqrt{1 - \zeta^2}}{-\zeta} = \cos^{-1}(-\zeta) \qquad (9\text{-}16)$$

In terms of Eq. (9-15), s_1 becomes the vector in Fig. 9-2.

A simple plot of the roots, as in Fig. 9-1, has limited usefulness, because it shows the location of the roots for only one combination of values of ζ and ω_n. The usefulness can be increased by plotting the roots calculated by permitting *one* parameter to vary. If we do this, we shall find that all

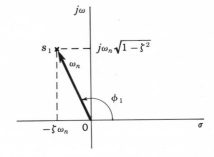

Fig. 9-2 *Relationship between the location of roots on the s plane and the natural frequency, the damping ratio, and the phase angle for the steady-state sinusoidal response of a second-order system.*

roots lie on smooth curves. These curves become the loci of the roots, and the plot is known as a *root-locus plot*.

In general, most of the parameters in the controlled system have already been fixed before the designer begins to select or design the control elements. Thus, the designer can most readily vary such factors as the gain and the magnitudes of coefficients for error-rate and integral control, and he will be most interested in the effect of these variations on the locus of the roots of the characteristic equation.

To illustrate the general procedure, let us develop the root-locus plot for the second-order proportional control system discussed in Sec. 8-2. From Eq. (8-91) we find the characteristic equation to be

$$Is^2 + as + K = 0 \qquad (9\text{-}17)$$

Assuming, as is usually the case, that a and I are beyond the control of the designer, K becomes the only parameter with which he can work. The roots of Eq. (9-17) are

$$s_1 = -\frac{a}{2I} + \sqrt{\left(\frac{a}{2I}\right)^2 - \frac{K}{I}} \qquad (9\text{-}18)$$

and

$$s_2 = -\frac{a}{2I} - \sqrt{\left(\frac{a}{2I}\right)^2 - \frac{K}{I}} \qquad (9\text{-}19)$$

The first step is to find the roots for the values of K for which the roots can be calculated almost by inspection, for example, for the limiting values of $K = 0$ and $K = \infty$. In this case we find

For $K = 0$

$$s_1 = 0 \qquad \text{and} \qquad s_2 = -\frac{a}{I}$$

For $K = \infty$

$$s_1 = -\frac{a}{2I} + j\infty \qquad \text{and} \qquad s_2 = -\frac{a}{2I} - j\infty$$

For $K = \dfrac{a^2}{4I}$

$$s_1 = -\frac{a}{2I} \qquad \text{and} \qquad s_2 = -\frac{a}{2I}$$

Also by inspection, we can see that as K increases from 0 to $a^2/4I$, s_1 decreases from 0 to $-a/2I$ and s_2 increases from $-a/I$ to $-a/2I$, with both remaining real numbers. Similarly, we can see that as K increases from $a^2/4I$ to ∞, the real parts of s_1 and s_2 remain equal to $-a/2I$ and the magnitudes of the imaginary parts increase from 0 to ∞. Based on the above observations, the root-locus plot for Eq. (9-17) can be drawn as shown in Fig. 9-3.

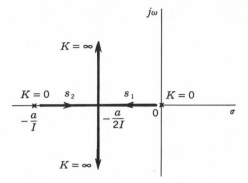

Fig. 9-3 *Root-locus plot for a second-order system.*

It should be apparent that for a simple system, for which the roots of the characteristic equation can be determined relatively easily, there is not much point in constructing a root-locus plot, because it contains no additional information.   The root-locus plot and the graphical techniques for constructing it become important when the complexity of the characteristic equation becomes so great that determining the roots becomes unduly laborious.

In the general case the system will be that in Fig. 9-4, for which the control ratio is

$$\frac{C}{R} = \frac{G}{1 + HG} \tag{9-20}$$

From Eq. (9-20) we find the characteristic equation to be the denominator of the closed-loop transfer function

$$1 + H(s)G(s) = 0 \tag{9-21}$$

where $H(s)G(s)$ is the open-loop transfer function.

Both $H(s)$ and $G(s)$ can be made up of numerous terms but, as discussed in Sec. 8-2, no matter how complicated the expression may be, it will consist of the product of constants and first- and second-order terms, such as $1/s$, s, $1/(Ts + 1)$, $Ts + 1$, $1/(s^2 + 2\zeta\omega_n s + \omega_n^2)$, and $s^2 + 2\zeta\omega_n s + \omega_n^2$. For example, we might find

$$H(s)G(s) = \frac{K(T_1 s + 1)}{s(T_2 s + 1)(s^2 + 2\zeta\omega_n s + \omega_n^2)} \tag{9-22}$$

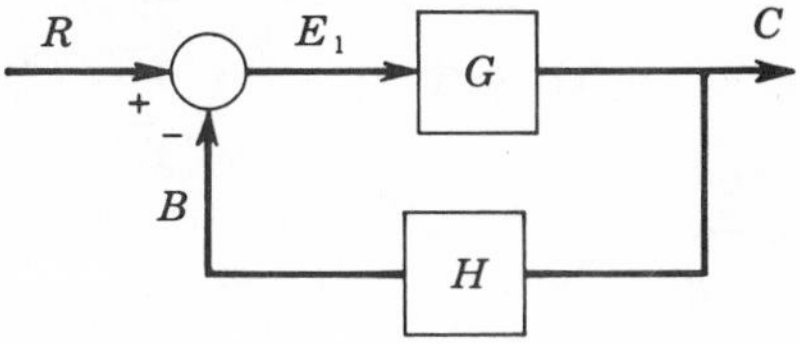

Fig. 9-4

for which the characteristic equation becomes

$$1 + H(s)G(s) = s(T_2s + 1)(s^2 + 2\zeta\omega_n s + \omega_n{}^2)$$
$$+ K(T_1s + 1) = 0 \quad (9\text{-}23)$$

In order to find the roots of Eq. (9-23) in the same manner as we did for Eq. (9-17), we would have to perform the multiplications indicated, collect terms, and then factor the resulting expression into the product of first- and second-order terms. In problems for which there is need for the root-locus plot this is usually impractical and we must find another approach.

Since, in general, we shall know the individual first- and second-order terms, the open-loop transfer function will be available in factored form, as in Eq. (9-22), and we can write Eq. (9-21) in a more useful form as

$$H(s)G(s) = -1 \qquad (9\text{-}24)$$

The values of s that fulfill the equality in Eq. (9-24) will be the roots of the characteristic equation. Thus, Eq. (9-24) permits us to work directly with the open-loop transfer function.

Since the roots of the characteristic equation can be complex, we shall actually use Eq. (9-24) in the form

$$H(s)G(s) = |H(s)G(s)|\underline{/H(s)G(s)} = 1\underline{/180° \pm m360°} \qquad (9\text{-}25)$$

where $m = 0, 1, 2, \ldots$

In the interest of simplifying the procedure for plotting the root loci, we shall find it desirable to rewrite the transfer function so that the coefficient of the highest power of s in each term is 1. For example, we shall rewrite Eq. (9-22) as

$$H(s)G(s) = \frac{s + 1/T_1}{s(s + 1/T_2)(s^2 + 2\zeta\omega_n s + \omega_n{}^2)} \frac{KT_1}{T_2} \qquad (9\text{-}26)$$

which in turn will be written as

$$H(s)G(s) = \frac{s - z_1}{s(s - p_1)(s - p_2)(s - p_3)} K' \qquad (9\text{-}27)$$

where $z_1 = -1/T_1$, a zero of $H(s)G(s)$
$p_1 = -1/T_2$, a pole of $H(s)G(s)$
$p_2 = -\zeta\omega_n + j\omega_n \sqrt{1 - \zeta^2}$, a pole of $H(s)G(s)$
$p_3 = -\zeta\omega_n - j\omega_n \sqrt{1 - \zeta^2}$, a pole of $H(s)G(s)$
$K' = KT_1/T_2$, the static loop sensitivity

The terms *zero* and *pole* will be encountered frequently throughout the remainder of this text. The terms are descriptive—zero meaning that when $s = z$, the magnitude of the function is zero, and pole meaning that when $s = p$, the magnitude of the function is infinite.

As a matter of interest it should be noted that the roots of the characteristic equation are the poles of the control ratio. Thus, the loci of the roots of the characteristic equation are the loci of the poles of the control ratio.

9-4. Constructing the Root Loci As discussed in the preceding section, the roots of the characteristic equation are those values of s corresponding to the conditions

$$|H(s)G(s)| = 1 \tag{9-28}$$

and $$\underline{/H(s)G(s)} = 180° \pm m360° \tag{9-29}$$

where $m = 0, 1, 2, \ldots$.

In general, the loci will be determined by consideration of the angular requirements, Eq. (9-29), and the roots, i.e., the points on the loci, will be determined for a given set of operating conditions by the value of the static loop sensitivity. Consequently, we shall consider first the conditions under which the angle of $H(s)G(s) = 180°$ or an odd multiple thereof.

The general form of $H(s)G(s)$ is

$$H(s)G(s) = K' \frac{(s - z_1)(s - z_2) \cdots (s - z_u)}{(s - p_1)(s - p_2) \cdots (s - p_v)} \tag{9-30}$$

where, for control systems, $v > u$.

In terms of the expression in Eq. (9-30), we find

$$\underline{/H(s)G(s)} = \underline{/s - z_1} + \underline{/s - z_2} + \cdots + \underline{/s - z_u} - \underline{/s - p_1}$$
$$- \underline{/s - p_2} - \cdots - \underline{/s - p_v} = 180° \pm m360° \tag{9-31}$$

where $m = 0, 1, 2, \ldots$.

Equation (9-31) can be simplified to become

$$\underline{/H(s)G(s)} = \Sigma\underline{/s - z_i} - \Sigma\underline{/s - p_i} = 180° \pm m360° \tag{9-32}$$

or $\underline{/H(s)G(s)} = \Sigma$ angles of numerator terms

$$- \Sigma \text{ angles of denominator terms} = 180° \pm m360° \tag{9-33}$$

where $m = 0, 1, 2, \ldots$.

Since each $s - z$ and $s - p$ term in Eq. (9-30) is a vector quantity, each can be drawn in the s plane. If we consider the case where the denominator of Eq. (9-30) is

$$(s - p_1)(s - p_2)(s - p_3)(s - p_4)$$

where p_1 and p_2 are real numbers and p_3 and p_4 are complex conjugates, the vectors for an arbitrary (trial) value of s would appear as in Fig. 9-5.

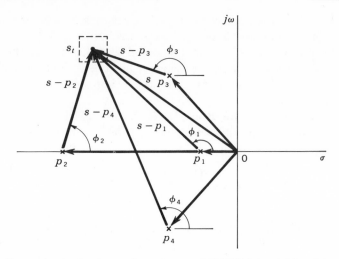

Fig. 9-5   *Angles from poles and zeros of $H(s)G(s)$ to a trial point s_t.*

The sum of the angles will be

$$\Sigma/s - p_i = \phi_1 + \phi_2 + \phi_3 + \phi_4 \tag{9-34}$$

It is important to note that if graphical constructions are to be used in finding the root loci, *the same scale must be used for the ordinate and abscissa;* otherwise, the angles, and therefore the results, will be meaningless.

It should be apparent that, for all but the simplest systems, trying to solve Eq. (9-32) graphically in a hit-or-miss fashion will be extremely laborious and could lead to serious error in that it would be easy to miss completely one or more loci. Fortunately, again, we can deduce much useful information by considering the behavior of $H(s)G(s)$ at, and in the vicinity of, limiting or special values of K' and s, such as $K' = 0$, $K' = \infty$, $s = \infty$, $s = $ a real number, and $s = j\omega$. As a result, the root loci for complicated systems can be drawn fairly quickly with a minimum number of trial-and-error or iteration steps. The basic method and procedures are due to Evans.[1]   We shall limit the discussion here to the more important principles and techniques, and the reader is referred to the literature for more detailed presentations.[2]

[1] W. R. Evans, Graphical Analysis of Control Systems, *Trans. AIEE*, vol. 67, pp. 547–551, 1948, and "Control-system Dynamics," McGraw-Hill Book Company, New York, 1954.

[2] J. G. Truxal, "Automatic Feedback Control System Synthesis," chap. 4, McGraw-Hill Book Company, New York, 1955.

D'Azzo and Houpis, *op. cit.*, chap. 7.

H. L. Harrison and J. G. Bollinger, "Introduction to Automatic Controls," chap. 12, International Textbook Company, Scranton, Pa., 1963.

Step 1. $K' = 0$ Considering the general form of $H(s)G(s)$ in Eq. (9-30), we can see that the condition $|H(s)G(s)| = 1$ can be satisfied for $K' = 0$ only when the magnitude of the remainder is infinite. Thus, for $K' = 0$, the roots of the characteristic equation correspond to the poles of the open-loop transfer function.

The root loci will be continuous curves with K' as the independent variable. Therefore, if we assume that the loci begin at $K' = 0$, we conclude that (1) *the initial points of the root loci are the poles of the open-loop transfer function, and* (2) *there will be as many loci as there are poles of the open-loop transfer function.*

Step 2. $K' = \infty$ Considering the general form of $H(s)G(s)$ in Eq. (9-30), we can see that the condition $|H(s)G(s)| = 1$ can be satisfied for $K' = \infty$ only when the magnitude of the remainder equals zero. This requires the numerator of the remainder to equal zero or the denominator to be infinite. Assuming that the root loci begin at $K' = 0$ and end at $K' = \infty$, we can conclude that *the loci terminate either* (1) *at the zeros of the open-loop transfer function or* (2) *at $s = \infty$*.

Step 3.   Real-axis loci   As can be seen in Fig. 9-6, the sum of the angles of the vectors from complex conjugate poles (or zeros) to a point on the real axis is 360°. Thus, the sections on the real axis that are root loci will be determined only by the poles and zeros on the real axis. The angle of a vector from a pole or zero to a trial point s on the real axis will be 0° if the trial point is to the right of the pole or zero and $\pm 180°$ if the trial point is to the left of the pole or zero. Consequently, we need consider only the poles and zeros to the right of the trial point, and, since each pole or zero contributes 180° to the summation in Eq. (9-32), we can conclude that *the trial point on the real axis lies on a locus whenever the total number of poles and zeros on the real axis to the right of the trial point is an odd number.*

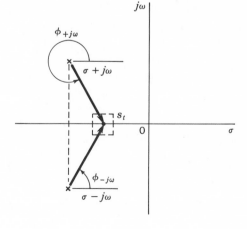

Fig. 9-6 *Angles from complex conjugate poles or zeros of $H(s)G(s)$ to a trial point on the real axis.*

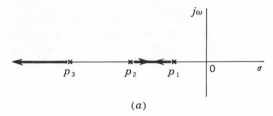

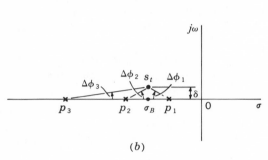

(a)

(b)

Fig. 9-7 *Relationships for determining the point at which a root locus breaks away from the real axis.*

Step 4. Breakaway from real axis   Since the summation of the angles of the vectors from the poles and zeros to a point on a locus will equal 180° or an odd multiple thereof, the sum of the changes in angles for a small change in position on a locus must equal zero.   If we consider the case in Fig. 9-7a, where there are three poles on the real axis, we find from step 3 that the root loci must lie on the real axis between p_1 and p_2 and must extend from p_3 on the real axis to the left to ∞.   Since a locus will start from each pole (at $K' = 0$), one locus will move from p_1 to the left and one locus will move from p_2 to the right.   To find the point σ_B of breakaway from the real axis between p_1 and p_2, we pick a trial point s_t close to the real axis.   As shown in Fig. 9-7b, when s moves from σ_B to s_t, the angles change by $\Delta\phi_1$, $\Delta\phi_2$, and $\Delta\phi_3$.   For the net change in angle to be zero, we must have

$$\Delta\phi_1 + \Delta\phi_2 + \Delta\phi_3 = 0 \qquad (9\text{-}35)$$

For small angles, we can let $\tan \Delta\phi = \Delta\phi$, and Eq. (9-35) becomes

$$\frac{\delta}{\sigma_B - p_1} + \frac{\delta}{\sigma_B - p_2} + \frac{\delta}{\sigma_B - p_3} = 0 \qquad (9\text{-}36)$$

which, after dividing through by δ, becomes

$$\frac{1}{\sigma_B - p_1} + \frac{1}{\sigma_B - p_2} + \frac{1}{\sigma_B - p_3} = 0 \qquad (9\text{-}37)$$

Equation (9-37) can usually be most quickly solved for σ_B by trial and error.

Considering that zeros have an effect on the angles just opposite to that of poles and that Eq. (9-37) can be written as a summation, we find that when *all of the poles and zeros are on the real axis*, the breakaway point will be that corresponding to

$$\sum \frac{1}{\sigma_B - p_n} - \sum \frac{1}{\sigma_B - z_m} = 0 \qquad (9\text{-}38)$$

where p_n are the open-loop real poles and z_m are the open-loop real zeros. The effect of complex poles and zeros on the location of the breakaway point is small unless they are located relatively close to the real axis, and it can usually be neglected without introducing appreciable error.

It should be noted that identical conclusions would have been reached if we had drawn δ downward in Fig. 9-7b. Therefore, there will be two loci from the breakaway point between p_1 and p_2. Actually, since we know that complex roots will appear as complex conjugates, we can conclude without further discussion that *the root loci must be mirror images reflected about the real axis.*

When the real axis between two zeros is a root locus, the loci must terminate at the zeros, and the point at which the loci join the real axis is a *break-in* point. Equation (9-38) applies also to this case.

Step 5. $s \to \infty$ For s very large, the poles and zeros appear as a point and all of the vectors from the poles and zeros to the trial point lie on a single straight line. The net angle of $H(s)G(s)$ is the angle of the straight line multiplied by the number of poles minus the number of zeros. The straight line becomes the root locus, asymptotically, whenever the net angle equals $180° \pm m360°$. Thus, the angle for the asymptote will be

$$\gamma = \frac{180° \pm m360°}{N_p - N_z} \qquad (9\text{-}39)$$

where $m = 0, 1, 2, \ldots$
$N_p = $ number of poles of $H(s)G(s)$
$N_z = $ number of zeros of $H(s)G(s)$

There will be $N_p - N_z$ asymptotes, and, in general, they will not intersect at the origin, even though because of symmetry we can say they will intersect on the real axis. To determine the point of intersection of the asymptotes, let us rewrite Eq. (9-30) as

$$H(s)G(s) = K' \frac{s^u + a_1 s^{u-1} + \cdots + a_u}{s^v + b_1 s^{v-1} + \cdots + b_v} \qquad (9\text{-}40)$$

where, from algebra, we know that

$$a_1 = \Sigma(-z) = -\Sigma z \qquad (9\text{-}41)$$
and
$$b_1 = \Sigma(-p) = -\Sigma p \qquad (9\text{-}42)$$

As a root locus, $H(s)G(s) = -1$, and Eq. (9-40) can be rearranged as

$$\frac{s^v + b_1 s^{v-1} + \cdots + b_v}{s^u + a_1 s^{u-1} + \cdots + a_u} = -K' \tag{9-43}$$

Performing the division indicated in Eq. (9-43) gives

$$s^{v-u} + (b_1 - a_1)s^{v-u-1} + \cdots = -K' \tag{9-44}$$

In Eq. (9-44) there will be $v - u$ roots, each of which will provide a vector from the pole to the trial point s. Since, *for $s \gg 1$*, all of the vectors are identical and since the coefficient of the second term in a poly-nomial is the negative sum of the roots, we conclude that the roots are equal and the asymptotes intersect the real axis at their common value. Thus,

$$s_a = -\frac{b_1 - a_1}{v - u} \tag{9-45}$$

where s_a is the point of intersection of the asymptotes with the real axis.

Equation (9-45) can be put into a more convenient form by noting that $v = $ number of poles and $u = $ number of finite zeros of $H(s)G(s)$ and then substituting from Eqs. (9-41) and (9-42) for a_1 and b_1, respectively, to give

$$s_a = \frac{\Sigma p - \Sigma z}{N_p - N_z} \tag{9-46}$$

Step 6. *Angles of departure and arrival* In many cases, partic-ularly when there are complex poles and zeros, it is helpful to know the angles at which the loci leave the poles or approach the zeros. As shown in Fig. 9-8, the angles contributed by the other poles and zeros are prac-tically independent of the position of the trial point, as long as it is close

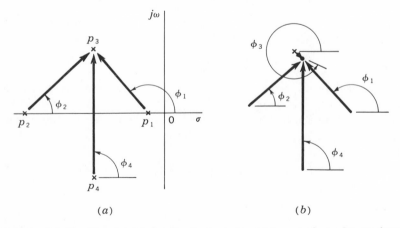

(a)   (b)

Fig. 9-8 *Relationships for determining the angle at which a root locus departs from a complex pole or arrives at a complex zero of $H(s)G(s)$.*

to the pole or zero of interest. Thus, the angles of departure and arrival can be found from, respectively,

$$\phi_d + \Sigma\phi_p - \Sigma\phi_z = 180° \pm m360° \tag{9-47}$$

and

$$-\phi_a + \Sigma\phi_p - \Sigma\phi_z = 180° \pm m360° \tag{9-48}$$

where $\Sigma\phi_p$ and $\Sigma\phi_z$ are the sums of the angles contributed by the vectors from the other poles and zeros, respectively, to the one for which the angle is being calculated.

Step 7. Determination of K' Once the root loci have been drawn on the basis of the six points given above, the loop sensitivity K' can be determined for any point on the locus by direct application of Eq. (9-28). For convenience, let us rewrite Eq. (9-28) in terms of Eq. (9-30) as

$$|H(s)G(s)| = K' \frac{|s - z_1|\,|s - z_2|\, \cdots \,|s - z_u|}{|s - p_1|\,|s - p_2|\, \cdots \,|s - p_v|} = 1 \tag{9-49}$$

from which

$$K' = \frac{|s - p_1|\,|s - p_2|\, \cdots \,|s - p_v|}{|s - z_1|\,|s - z_2|\, \cdots \,|s - z_u|} = \frac{\Pi|s - p_n|}{\Pi|s - z_m|} \tag{9-50}$$

where $\Pi|s - p_n|$ and $\Pi|s - z_m|$ are the continued products of the lengths of the vectors from the respective poles and zeros to a point on a root locus, i.e., the root s for which we wish to know the required value of K'.

Step 8. Intersection of the $j\omega$ axis Since complex roots can exist only as conjugates, loci of complex roots will always intersect the $j\omega$ axis as a pair of loci. The roots will be $+j\omega_i$ and $-j\omega_i$, and the case becomes that of zero damping with $\omega_i = \omega_n$. If K' is increased further, the loci enter the right half of the s plane and the real parts of the roots become positive. Thus, the intersection of the $j\omega$ axis is the dividing line between stability and instability.

The point of intersection and the loop sensitivity K' can be determined accurately enough for most purposes from a sketch of the root loci made by considering the seven points given just above. However, if increased accuracy is desired, the Routh criterion, as discussed in Sec. 9-2, can be used to calculate the exact value of K' at which the system becomes neutrally stable, and a slight extension of the discussion in Sec. 9-2 then permits determination of the corresponding value of $j\omega$.

Since a change in sign in the first column of coefficients in the Routh array indicates instability, the system will become unstable whenever K' becomes large enough to make a term in the first column have a sign different from that in the preceding rows. However, this does not in itself tell whether the instability is due to a root being a positive real number or a complex number with a positive real part, and we must investigate further.

The conditions that must be fulfilled in the Routh array if the roots are to be conjugate imaginary numbers, indicating zero damping, are[1] (1) all terms in a row are zeros and (2) the auxiliary polynomial formed from the terms in the last nonvanishing row has a pair of conjugate imaginary zeros. From the plot of the root loci we can determine by inspection the number of roots that are real positive numbers and the approximate frequencies at which the complex loci cross the imaginary axis. If root loci cross the imaginary axis at nonzero values of $j\omega$, we know that conditions 1 and 2 are satisfied and we can then utilize the Routh array to determine the exact values of K' and $j\omega$. For example, if the characteristic equation is

$$1 + H(s)G(s) = s^3 + b_2s^2 + b_1s + b_0K' = 0 \qquad (9\text{-}51)$$

the Routh array, from Sec. 9-2, is

$$
\begin{array}{c|cc}
s^3 & 1 & b_1 \\
s^2 & b_2 & b_0K' \\
s^1 & \dfrac{-b_0K' + b_1b_2}{b_2} & \\
s^0 & b_0K' &
\end{array}
$$

Since $b_0K' = 0$ is a trivial case, we find condition 1 is fulfilled when all terms in the s^1 row are zero. Consequently, we want

$$-b_0K' + b_1b_2 = 0 \qquad (9\text{-}52)$$

or

$$K' = \frac{b_1b_2}{b_0} \qquad (9\text{-}53)$$

The last nonvanishing row is the s^2 row, and the auxiliary polynomial for this row is

$$b_2s^2 + b_0K' = 0 \qquad (9\text{-}54)$$

which, for $K' = b_1b_2/b_0$, becomes

$$s^2 + b_1 = 0 \qquad (9\text{-}55)$$

The roots (zeros) of Eq. (9-55) are

$$s = \pm j \sqrt{b_1} \qquad (9\text{-}56)$$

Therefore, the root loci intersect the imaginary axis at $j\omega = +j \sqrt{b_1}$ and $j\omega = -j \sqrt{b_1}$ and we can say that the undamped natural frequency is $\omega_n = \sqrt{b_1}$.

[1] M. F. Gardner and J. L. Barnes, "Transients in Linear Systems," vol. 1, p. 200, John Wiley & Sons, Inc., New York, 1942.

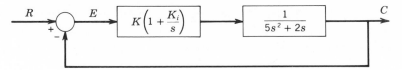

Fig. 9-9

Example 9-2 The block diagram for a second-order controlled system with proportional plus integral control is given in Fig. 9-9. We are asked (a) to draw the root loci for the case when $K_i = 0.2$, (b) to discuss the general behavior characteristics of the system, and (c) to discuss the effect of varying the magnitude of K_i.

(a) *Constructing the Root Loci.* For $K_i = 0.2$, we have

$$H(s)G(s) = K\left(1 + \frac{0.2}{s}\right)\left(\frac{1}{5s^2 + 2s}\right) = \frac{(s + 0.2)}{s^2(s + 0.4)}\frac{K}{5} = \frac{(s + 0.2)}{s^2(s + 0.4)}K' \quad (9\text{-}57)$$

where $K' = K/5$.

In Eq. (9-57) we see that the open-loop transfer function has three poles and one zero: $p_1 = 0$, $p_2 = 0$, $p_3 = -0.4$, and $z_1 = -0.2$. The procedure is to plot the poles and zeros, as in Fig. 9-10a, and then follow through steps 1 through 8 above.

Step 1. $K' = 0$. There are three poles; thus, there will be three loci and one will start from each of the poles.

Step 2. $K' = \infty$. There is one zero; thus, one locus will end at the zero and the other two will end at $s = \infty$.

Step 3. *Real-axis Loci.* The total number of poles and zeros to the right on the real axis is an odd number for that section of the real axis between p_3 and z_1; therefore, as shown in Fig. 9-10b, this section is a root locus that runs from p_3 to z_1.

Step 4. *Breakaway from Real Axis.* The loci originating at p_1 and p_2 at 0 must depart from the real axis immediately.

Step 5. $s = \infty$. $N_p - N_z = 2$; thus, there will be two asymptotes and the angles will be, from Eq. (9-39),

$$\gamma = \frac{180° \pm m360°}{N_p - N_z} = \frac{180° \pm m360°}{2} = 90° \pm m180°$$
$$= 90° \text{ and } 270°, \text{ or } 90° \text{ and } -90°$$

From Eq. (9-46), the asymptotes will intersect the real axis at

$$s_a = \frac{\Sigma p - \Sigma z}{N_p - N_z} = \frac{(0 + 0 - 0.4) - (-0.2)}{2} = \frac{-0.4 + 0.2}{2} = -0.1$$

Step 6. *Angles of Departure.* Two loci must leave from the double pole at zero and end up on the asymptotes. The loci must be symmetrical about the real axis. Thus, if we consider the loci from p_1 to be in the top half of the s plane and the loci from p_2 to be in the bottom half of the s plane, we have, from Eq. (9-47), for the angle of departure from p_1,

$$\phi_d + \Sigma\phi_p - \Sigma\phi_z = 180° \pm m360°$$
$$\phi_{d,p_1} + \phi_{p_2} + \phi_{p_3} - \phi_{z_1} = \phi_{d,p_1} + \phi_{p_2} + 0 - 0$$
$$\phi_{d,p_1} + \phi_{p_2} = 180°$$

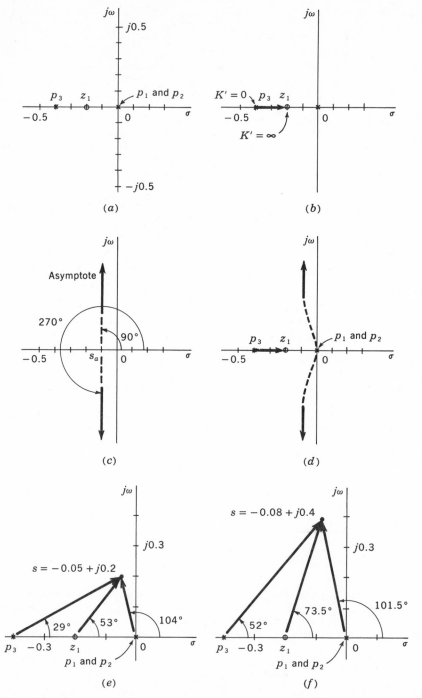

Fig. 9-10 *Steps in plotting the root loci for the system in Fig. 9-9.* (a) *Location of poles and zeros;* (b) *real-axis locus;* (c) *asymptotes as* $s \to \infty$; (d) *breakaway from real axis;* (e *and* f) *trial points on locus.*

Since p_1 and p_2 are at the same point, the angles from each to the trial point must be identical. Therefore,

$$\phi_{d,p_1} = \phi_{p_2} = \frac{180°}{2} = 90°$$

and, similarly, $\qquad \phi_{d,p_2} = 270°, \text{ or } -90°$

The general trends of the loci from zero to the asymptotes are shown as dashed lines in Fig. 9-10d. To obtain a better approximation in the transition region, we must resort to a trial-and-error solution based on the relationship between the angles as expressed in Eq. (9-32).

For a trial point $s = -0.05 + j0.2$, the angles from Fig. 9-10e are $\phi_{p_1} = \phi_{p_2} = 104°$, $\phi_{p_3} = 30°$, and $\phi_{z_1} = 53°$. For these angles,

$$\Sigma\underline{/s - z_i} - \Sigma\underline{/s - p_i} = 53° - (104° + 104° + 29°) = -184°$$

which is close enough to $-180°$ for most purposes. If we wish greater accuracy, we now must choose a new trial point. This choice does not have to be a blind choice, because we can, by inspection of the figure, determine the relative order of the contributions of the various vectors and can then deduce, at least qualitatively, the changes that should be made. For example, in this case we see that a slight shift in the real part of the trial root will have a greater effect on ϕ_{p_1} and ϕ_{p_2} than on any other angles. Since, from the angle summation above, ϕ_{p_1} and ϕ_{p_2} are too large, but only by a degree or so, we should choose a new trial point slightly to the right of the first one. For example, if we take a trial point $-0.04 + j0.2$, we shall find that the summation of the angles equals $-180°$. Therefore, $-0.04 \pm j0.2$ are points on the root loci.

Additional points on the loci can be located if desired. Although measuring the angles and performing the additions and subtractions indicated are neither particularly difficult nor time-consuming, anyone faced with drawing root loci for many complicated solutions should consider using a device called the Spirule[1] or using an analog computer to plot automatically the loci.[2]

Figure 9-10f shows the trial for $s = -0.08 + j0.4$. The angle summation yields $-181.5°$, which means that the loci will pass slightly to the right of that point and its conjugate.

Using the results from the steps above, the complete loci are drawn in Fig. 9-11. Values of K indicated were determined by using Eq. (9-50) and the relationship $K = 5K'$ for this particular case. The lengths of the vectors were measured from the plot in Fig. 9-11 for points on the loci beginning at 0. For example, at $s = -0.05 + j0.25$, Eq. (9-50) becomes

$$K' = \frac{0.253 \times 0.253 \times 0.43}{0.29} = 0.0948$$

and $\qquad K = 5K' = 5 \times 0.0948 = 0.474$

For points on the real axis, between p_3 and z_1,

$$K = 5K' = 5 \frac{|s - p_1|\,|s - p_2|\,|s - p_3|}{|s - z_1|}$$

$$= 5 \frac{|s - 0|\,|s - 0|\,|s + 0.4|}{|s + 0.2|} = 5 \frac{|s|\,|s|\,|s + 0.4|}{|s + 0.2|}$$

---

[1] Manufactured by the Spirule Company, Whittier, California.

[2] For example, see J. J. D'Azzo and C. H. Houpis, "Control System Analysis and Synthesis," 1st ed., pp. 536–542, McGraw-Hill Book Company, New York, 1960.

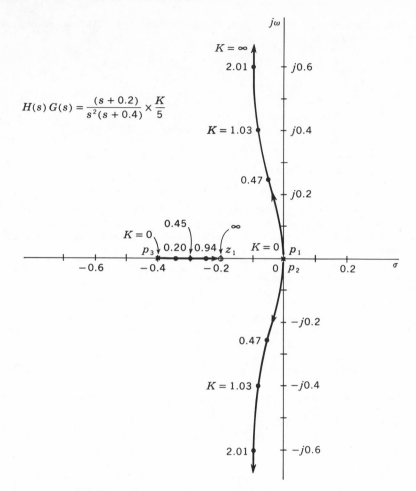

$$H(s)\,G(s) = \frac{(s + 0.2)}{s^2(s + 0.4)} \times \frac{K}{5}$$

Fig. 9-11   *Complete root-locus plot for system in Fig. 9-9.*

and, for example, for $s = -0.35$,

$$K = 5\,\frac{|-0.35|\,|-0.35|\,|-0.35 + 0.4|}{|-0.35 + 0.2|} = 5\,\frac{0.35 \times 0.35 \times 0.05}{0.15} = 0.202$$

(*b*) *Behavior Characteristics.*   The general behavior of a complicated system is determined largely by the roots with the most positive real parts.   For example, in Fig. 9-11, the locus on the real axis between p_3 and z_1 represents relatively heavy damping and the effect dies out quickly in comparison with that from the other two loci.   In this example, the roots with the most positive real parts lie on the two loci starting at 0.   Therefore, since there are two loci, we shall use a second-order system for comparison.   From Fig. 9-11 we can conclude that, since no locus crosses into the right half of the s plane, the system will be stable for all values of K.

If we compare this system with a second-order system, we can use Eq. (9-16) and the discussion relative to Fig. 9-2 to estimate the damping ratio for various values of K. For example, a line from zero tangent to the loci makes an angle of 101°, as shown in Fig. 9-12.   At the point of tangency we find, from Eq. (9-16),

$$\zeta = -\cos \phi = -\cos 101° = \cos 79° = 0.19$$

Consequently, since the angle is less to every other point on the locus, we can say that the damping ratio can never exceed this value of 0.19.   It should also be noted that the angle, and therefore the damping ratio, is relatively insensitive to the value of K as long as it is between about 0.5 and 1.2.

The damped natural frequency corresponds to the imaginary part of the root. Thus, if $K = 1$, the damped natural frequency will be about 0.4 rad/sec and the system will behave very much like a simple second-order system with $\zeta = 0.19$—except that, as shown in Sec. 8-5, the addition of integral control eliminates the steady-state error in the response to a ramp-function reference level and a step change in load.

(c) *Effect of Varying K_i.*   The zero occurs at $-K_i$; thus, from step 5, above, the asymptotes intersect the real axis at

$$s_a = \frac{+p_3 - z_1}{2} = \frac{-0.4 + K_i}{2}$$

As K_i decreases, z_1 moves to the right.   When $K_i = 0$, $z_1 = 0$ and one of the poles at 0 is canceled.   The system becomes a simple second-order system.   The net number of poles is still two, and there will still be two asymptotes, at 90° and 270°, that now intersect the real axis at -0.2.   Thus, one effect of decreasing K_i is to move the asymptotes to the left, thereby effectively increasing the damping ratio.

As K_i increases, z_1 moves to the left and the asymptotes move to the right.   When $K_i = 0.4$, $z_1 = -0.4$ and cancels the pole at -0.4.   The asymptotes are now the $j\omega$ axis and the system is neutrally stable, i.e., has zero damping, for all values of K. The only effect of K is to change the natural frequency.

When $K_i > 0.4$, the asymptotes and the root loci beginning at 0 lie in the right half of the s plane, as shown in Fig. 9-13, and the system is unstable for all values of $K > 0$.

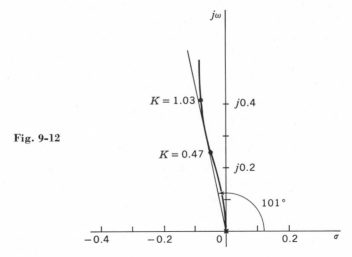

Fig. 9-12

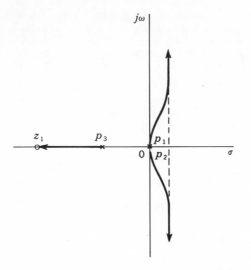

Fig. 9-13   *Root-locus plot for system in Fig.* 9-9 *when* $K_i > 0.4$.

9-5. Conformal Mapping   The root-locus method in Sec. 9-4 is probably the most useful method for investigating on paper the stability of a control system when the open-loop transfer function $H(s)G(s)$ is available in, or can readily be put into, the form of Eq. (9-30). Since both the real and imaginary parts of the roots are immediately available, we are not limited to considering the question of stability alone; but we can usually draw useful conclusions about the effective damping ratio and the damped natural frequency, and thus quickly estimate the response to transients as well as sinusoidal inputs. The root-locus plot is also useful to the designer, because he can often determine by inspection which poles and zeros should be shifted, and thus which elements in the system should be modified, to improve the performance.

About the only limitation to the root-locus method is that it cannot be applied unless $H(s)G(s)$ is available in the form of Eq. (9-30) or an equivalent. In some cases, the designer will be interested in a system that already exists and for which no equations are available. Numerous methods have been presented[1] for determining approximate transfer functions from experimental data; but in many cases it is relatively simple to excite the open-loop system with a sinusoidal forcing function, and more accurate information about system behavior will be obtained with less effort by using conformal mapping, Nyquist's criterion, or the attenuation-phase criterion. It should be noted that data from experiments with an open-loop system will have significance only when the open-loop system is stable. This will be true in the great majority of cases.

[1] Truxal, "Automatic Feedback Control System Synthesis," pp. 344–389.

The development of the Nyquist criterion[1] was a milestone in the development and understanding of the theory of feedback amplifiers and control systems. Nyquist relied upon the theory of complex variables, in particular the Cauchy integral theorem and the theorem of residues, in deriving his criterion. We shall not consider it further here for two main reasons: (1) A rigorous development of the method is beyond the scope of this book. (2) In practice, as far as the user is concerned, the method becomes a matter of simply following a series of steps for which only the end result has much significance. Identical answers can be obtained more simply and directly by use of conformal mapping. The major advantages of the conformal mapping approach are that (1) it is straightforward and (2) it applies directly and logically to all cases without the user having to keep track of any special situations or considerations.

The feature of conformal mapping that makes it so useful in the study of control systems is that it permits drawing conclusions about the behavior of a system over an entire region of operation rather than having to consider it point by point. For example, as discussed before, the stability of a linear control system is determined solely by the roots of the characteristic equation, with the system being unstable if any root or the real part of complex conjugate roots is positive. From Sec. 9-3, we know that for a general system of the form of Fig. 9-4, instability can result when

$$1 + H(s)G(s) = 0 \qquad (9\text{-}58)$$

or, more usefully,

$$H(s)G(s) = -1 \qquad (9\text{-}59)$$

Equation (9-59) can be used as it stands, but when the equation is available, as for the root-locus method, we can gain more information more readily if $H(s)G(s)$ is written in the form of Eq. (9-30), which is

$$H(s)G(s) = K' \frac{(s - z_1)(s - z_2) \cdots (s - z_u)}{(s - p_1)(s - p_2) \cdots (s - p_v)} \qquad (9\text{-}60)$$

Substituting from Eq. (9-60) into Eq. (9-59) gives

$$K' \frac{(s - z_1)(s - z_2) \cdots (s - z_u)}{(s - p_1)(s - p_2) \cdots (s - p_v)} = -1 \qquad (9\text{-}61)$$

which can be rearranged as

$$\frac{(s - z_1)(s - z_2) \cdots (s - z_u)}{(s - p_1)(s - p_2) \cdots (s - p_v)} = -\frac{1}{K'} \qquad (9\text{-}62)$$

or

$$\frac{H(s)G(s)}{K'} = -\frac{1}{K'} \qquad (9\text{-}63)$$

[1] H. Nyquist, Regeneration Theory, *Bell System Tech. J.*, vol. 11, pp. 126–147, January, 1932.

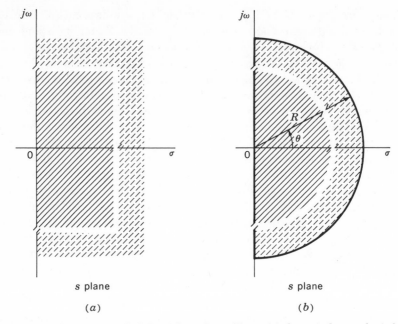

Fig. 9-14 (a) *Infinite right half of the s plane;* (b) *semicircle equivalent to the infinite right half of the s plane.*

In general, Eq. (9-63) will be used when $H(s)G(s)$ is available as an equation and Eq. (9-59) will be used when $H(s)G(s)$ is known only as experimental data from sinusoidal tests of the open-loop system.

Since we are concerned only with roots (zeros) of the characteristic equation for which σ is positive, knowing the behavior of $H(s)G(s)$ or $H(s)G(s)/K'$ *for all possible values of s in the right half of the s plane,* as indicated in Fig. 9-14a, will tell us whether or not the condition in Eq. (9-59), or in Eq. (9-63), is fulfilled. Therefore, we are interested in placing bounds on the value of $H(s)G(s)/K'$ or $H(s)G(s)$ for all values of s with positive values of σ and then seeing if the critical point $-1/K'$ or -1, respectively, is included within the bounds. As a matter of convenience, we shall consider the right half of the s plane lying within a large radius R, as shown in Fig. 9-14b, so that for all points on the arc we can write

$$s = Re^{j\theta} \tag{9-64}$$

Since R can be arbitrarily large, every possible finite value of s with positive σ can be included within the semicircle.

This can be accomplished most readily by using the principles and observations associated with conformal mapping. The topic of conformal mapping is of major importance in its own right and cannot be considered here in detail.[1] For our purposes, we need consider only those facets related to the investigation of linear systems.

In terms of the notation we have been using, *a mapping from the s plane into the* $H(s)G(s)/K'$ *plane will be conformal and sense-preserving at every point where* $H(s)G(s)/K'$ *is analytic*[2] *and its derivative with respect to s* $\neq 0$.

The term conformal means that the image of a small figure in the $H(s)G(s)/K'$ plane will conform to the original figure in the s plane, in that it will have approximately the same shape, and the transformation or mapping will be one-to-one. That is, for each value of s in the region of interest in the s plane, there will be one and only one value of $H(s)G(s)/K'$ in the $H(s)G(s)/K'$ plane.

When the mapping is sense-preserving, the region lying on the right-hand side of an observer traveling around a closed path in the s plane will appear on the right-hand side of an observer traveling around the corresponding closed path in the $H(s)G(s)/K'$ plane, and a change in direction angle at a point on the path in the s plane will appear as exactly the same angle in the same sense (clockwise or counterclockwise) at the corresponding point on the path in the $H(s)G(s)/K'$ plane. The usual practice, which we shall follow, is to use the term conformal to mean both conformal and sense-preserving.

From the discussion in Sec. 9-4, we see that we can always write $H(s)G(s)/K'$ in the form

$$\frac{H(s)G(s)}{K'} = \frac{(s - z_1)(s - z_2) \cdots (s - z_u)}{(s - p_1)(s - p_2) \cdots (s - p_v)} \tag{9-65}$$

where $v > u$, which will be analytic for all values of s *except those corresponding to the poles of the open-loop transfer function.* The mapping will be conformal for all other values of s, but points corresponding to the poles cannot be represented on the $H(s)G(s)/K'$ plane. Since

$$\frac{H(s)G(s)}{K'} = \infty$$

at a pole, it is obvious that $1 + H(s)G(s) \neq 0$, and, therefore, we need not worry about what happens at the poles of the *open-loop* transfer function.

The method can best be explained by considering several simple examples.

---

[1] See, for example, R. V. Churchill, "Introduction to Complex Variables and Applications," pp. 135ff., McGraw-Hill Book Company, New York, 1948, or W. Kaplan, "Advanced Calculus," pp. 580ff., Addison-Wesley Publishing Company, Inc., Reading, Mass., 1952.

[2] A function is analytic if the first derivative exists.

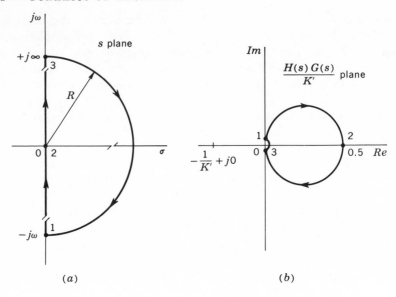

Fig. 9-15   *Conformal mapping for $H(s)G(s)/K' = 1/(s + 2)$.*

Example 9-3 $H(s)G(s) = K'/(s + 2)$, which has a pole at -2.   We shall now plot the curve of $H(s)G(s)/K'$ in the $H(s)G(s)/K'$ plane as s traverses the boundary of the right half of the s plane.   Since the boundary consists of the $j\omega$ axis and $s = Re^{j\theta}$, we need calculate values of $H(s)G(s)/K'$ only on these boundaries.

The starting point is immaterial, but, to be consistent, let us start at $s = 0 - j\infty$, as indicated by point 1 in Fig. 9-15a, and let s move up the $j\omega$ axis.   For $H(s)G(s) = K'/(s + 2)$, we have

$$\frac{H(s)G(s)}{K'} = \frac{1}{s + 2} = \frac{1}{\sigma + j\omega + 2} \tag{9-66}$$

which, for values of s along the imaginary axis, that is, $\sigma = 0$, becomes

$$\frac{H(s)G(s)}{K'} = \frac{H(j\omega)G(j\omega)}{K'} = \frac{1}{2 + j\omega} = \frac{2 - j\omega}{4 + \omega^2} = \frac{2}{4 + \omega^2} - j\frac{\omega}{4 + \omega^2} \tag{9-67}$$

When $j\omega = -j\infty$, $\omega = -\infty$ and Eq. (9-67) becomes

$$\frac{H(j\omega)G(j\omega)}{K'} = \frac{2}{\infty^2} + j\frac{1}{\infty} \tag{9-68}$$

which is not well defined and indicates that we need to use a limiting process.   To do this, let us consider what happens as $\omega \to -\infty$.   We see that the real part of Eq. (9-68) approaches zero faster than the imaginary part and that $H(j\omega)G(j\omega)/K'$ at $\omega \approx -\infty$ can be represented by point 1 on the $H(s)G(s)/K'$ plane in Fig. 9-15b.

Letting $j\omega = 0$, point 2 in Fig. 9-15a, we find, from Eq. (9-67),

$$\frac{H(j\omega)G(j\omega)}{K'} = \frac{2}{4} + j0 = 0.5 + j0 \tag{9-69}$$

which is labeled point 2 in Fig. 9-15b.

Additional points can be calculated for the path between 1 and 2, but, from the discussion related to Fig. 8-8, we know in this case the curve is a semicircle, as shown in Fig. 9-15b.

Continuing on up the $j\omega$ axis in the s plane, we find at point 3, for $j\omega = +\infty$,

$$\frac{H(j\omega)G(j\omega)}{K'} = \frac{2}{\infty^2} - j\,\frac{1}{\infty} \tag{9-70}$$

which is the complex conjugate of Eq. (9-68). The corresponding point in the $H(s)G(s)/K'$ plane is point 3 in Fig. 9-15b.

To complete the traverse of the boundary of the right half of the s plane, we shall use Eq. (9-64) with $R \gg 1$. For $s = Re^{j\theta}$,

$$\frac{H(s)G(s)}{K'} = \frac{1}{Re^{j\theta} + 2} \approx \frac{1}{R}\,e^{-j\theta} = [0(+)]e^{-j\theta} \tag{9-71}$$

where $0(+)$ signifies a positive value $\ll 1$.

From Eq. (9-71) we deduce that as we move from 3 along the arc with radius R to 1 in a clockwise direction in the s plane, we move from 3 along an arc with radius $1/R$ in a counterclockwise direction in the $H(s)G(s)/K'$ plane, with the angle being 180° in both planes. Since R can be made arbitrarily large, $1/R \to 0$ and the curve in the $H(s)G(s)/K'$ plane becomes a semicircle with a very small radius.

It should be noted that since we have turned clockwise through a 90° angle at 3 in the s plane, the sense-preserving feature of the mapping requires that we turn clockwise through 90° at 3 in the $H(s)G(s)/K'$ plane.

The remaining point of importance is determining the region in the $H(s)G(s)/K'$ plane that corresponds to the values of s in the right half of the s plane. This can be ascertained by (1) taking a trial point, such as $s = 1 + j0$, or (2) making use of the sense-preserving feature of the mapping. In general, particularly for complicated systems, it will be more convenient to do the latter. In this case, we note that *when traversing the curves in the directions of the arrows, the corresponding regions lie on the same hand of the observer.* In this example, the region of interest in the s plane lies to the right. Therefore, the region of interest, the critical region, in the $H(s)G(s)/K'$ plane must also lie to the right, which, as shown in Fig. 9-16, is the area enclosed by the path in the $H(s)G(s)/K'$ plane.

Since the critical value, or point, $-1/K' + j0$, is not in the critical region, the system is stable. It can further be concluded that the system will be stable for all values of K' and is, therefore, *inherently stable.*

The stability in this case can be checked by substituting $H(s)G(s)$ in Eq. (9-58) to give for the characteristic equation

$$1 + H(s)G(s) = s + 2 + K' = 0 \tag{9-72}$$

for which the root or zero is $-(2 + K')$, confirming that the system is stable for all values of K'.

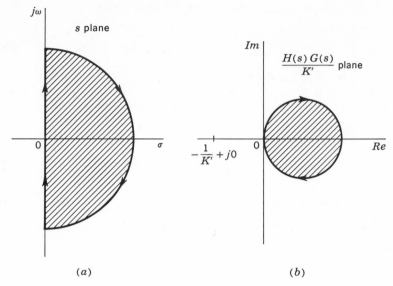

(a)   (b)

Fig. 9-16   *Corresponding regions in the s and $H(s)G(s)/K'$ planes for $H(s)G(s)/K' = 1/(s + 2)$.*

Example 9-4 $H(s)G(s) = K'/(s - 2)$, which has a pole at $+2$.   Following the procedure presented in Example 9-3, we find, for $s = 0 + j\omega$,

$$\frac{H(s)G(s)}{K'} = \frac{H(j\omega)G(j\omega)}{K'} = \frac{1}{j\omega - 2} = -\frac{2}{\omega^2 + 4} - j\frac{\omega}{\omega^2 + 4} \qquad (9\text{-}73)$$

and calculations for the several points lead to

Point	ω	$H(s)G(s)/K'$
1	$-\infty$	$j0(+)$
2	0	$-0.5 + j0$
3	$+\infty$	$-j0(+)$

As above, traversing in a clockwise direction the infinite semicircle in the s plane in Fig. 9-17a results in traversing in a counterclockwise direction a very small radius semicircle in the $H(s)G(s)/K'$ plane in Fig. 9-17b.

Comparing the closed path in the $H(s)G(s)/K'$ plane with that in the s plane in Fig. 9-17 leads to the observation that the sense of the direction of the paths is different.   As in Example 9-3, the region of interest in the s plane lies to the right when one is moving in the direction of the arrows in the s plane and the corresponding region in the $H(s)G(s)/K'$ plane will also lie to the right when moving along the path in the direction of the arrows.   However, we now note that the critical region in the $H(s)G(s)/K'$ plane is the region *outside* the closed path, as shown in Fig. 9-18b.

As can be seen, the critical point, $-1/K' + j0$, will lie in or out of the critical region, depending on the value of K'.   For example, if $1/K' > 0.5$, the point lies in the critical (crosshatched) region, and if $1/K' < 0.5$, the point lies inside the closed path, which is

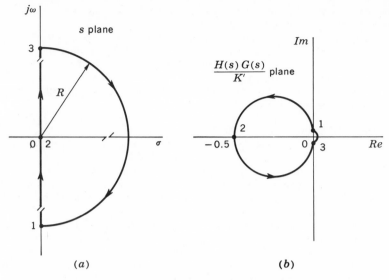

Fig. 9-17   *Conformal mapping for* $H(s)G(s)/K' = 1/(s-2)$.

outside the critical region.   Therefore, for $K' < 2$, the system will be unstable and, for $K' > 2$, the system will be stable.   In general terms, we shall classify this system as *conditionally stable*.

To check the condition for stability, let us substitute $H(s)G(s)$ into Eq. (9-58) to give

$$1 + H(s)G(s) = s - 2 + K' = 0 \qquad (9\text{-}74)$$

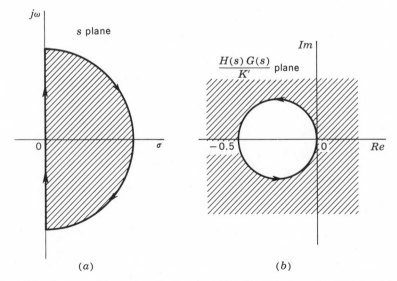

Fig. 9-18   *Corresponding regions in the s and* $H(s)G(s)/K'$ *planes for* $H(s)G(s)/K' = 1/(s-2)$.

The root is $-(K' - 2)$, which will be negative for $K' > 2$ and positive for $K' < 2$, confirming the conclusions reached above.

Noting that, when the path of $H(s)G(s)/K'$ passes to one side of $-1/K' + j0$, the root of the characteristic equation is a positive real number and, when the path passes on the other side of $-1/K' + j0$, the root is a negative real number leads to the conclusion that the root is zero when the path passes through the point $-1/K' + j0$. For this case, we have the condition of neutral stability. A system with a single real root equal to zero will behave as a first-order system and the condition of neutral stability will not be oscillatory, as for a second-order system. In effect, the system will remain wherever it is left. In this case, the path passes through $-1/K' + j0$ only once. The implications of multiple intersections will be considered in Example 9-6.

The difference in the location of the critical region relative to the closed path in the $H(s)G(s)/K'$ plane in Figs. 9-16b and 9-18b is due to the presence of a pole of the open-loop transfer function in the right half, i.e., within the bounded region, of the s plane in the latter case and not in the former.

Examination of the open-loop transfer function will show that the zeros of the denominator become the poles of the transfer function. Since the denominator of the open-loop transfer function is the characteristic equation for the *open-loop system*, we can conclude that poles with positive real parts result from zeros, or roots, of the characteristic equation with positive real parts and, thus, the existence of a pole of the open-loop transfer function in the right half of the s plane means that the open-loop system is unstable. Relatively few control systems are unstable when operated open loop, and, therefore, in most cases we shall find the critical area to lie within the closed path in the $H(s)G(s)/K'$ plane.

It should be apparent that significant experimental results can be obtained for open-loop systems only when they are stable and, thus, when working with experimental results, we can conclude that there are no poles of the open-loop transfer function in the right half of the s plane and that the critical region will lie inside the closed path in the $H(s)G(s)$ plane.

Example 9-5 $H(s)G(s) = K'/s$, which has a pole at 0. Following the procedure presented above, we find, for $s = 0 + j\omega$,

$$\frac{H(s)G(s)}{K'} = \frac{H(j\omega)G(j\omega)}{K'} = \frac{1}{j\omega} = -j\frac{1}{\omega} \tag{9-75}$$

Letting point 1 be at $j\omega \approx -j\infty$, or $\omega \approx -\infty$, we find

$$\frac{H(j\omega)G(j\omega)}{K'} = +j0(+) \tag{9-76}$$

Then, as shown in Fig. 9-19, when the value of s moves up the $j\omega$ axis from $-j\infty$ toward $j0$ in the s plane, $H(s)G(s)/K'$ moves up the imaginary axis from $+j0(+)$ toward $+j\infty$ in the $H(s)G(s)/K'$ plane. However, when $\omega = 0$ is substituted in Eq. (9-75), we find

$$\frac{H(j\omega)G(j\omega)}{K'} = -j\infty \tag{9-77}$$

To see what is happening, let us calculate $H(j\omega)G(j\omega)/K'$ for $\omega = -0(+)$ and $+0(+)$. Doing this, we find, at point 2,

$$\frac{H(j\omega)G(j\omega)}{K'}\bigg|_{\omega = -0(+)} \approx +j\infty \tag{9-78}$$

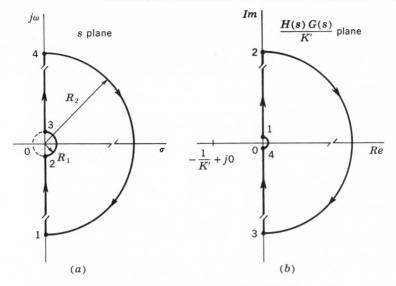

Fig. 9-19   *Conformal mapping for $H(s)G(s)/K' = 1/s$.*

and, at point 3,

$$\left. \frac{H(j\omega)G(j\omega)}{K'} \right|_{\omega \,=\, +0(+)} \approx -j\infty \tag{9-79}$$

From Eqs. (9-78) and (9-79), we conclude that as we try to go through $\omega = 0$, $H(s)G(s)/K'$ must jump from $+j\infty$ to $-j\infty$, and, since we are interested in the regions into which the closed curve divides the $H(s)G(s)/K'$ plane, we must know how the curve gets from $+j\infty$ to $-j\infty$.

The problem is that we have ignored the fact that $H(s)G(s) = K'/s$ has a pole at $s = 0$ and the function is not analytic at that point. Therefore, to ensure that $H(s)G(s)/K'$ is analytic everywhere on the boundary, we must choose the path in the right half of the s plane so that it does not pass through any poles of the open-loop transfer function.

To avoid the pole at 0, we can again let

$$s = Re^{j\theta} \tag{9-80}$$

and use a semicircular path around the pole. In this case, $R \ll 1$ and we need not worry about leaving out part of the s plane, because we can let the path come arbitrarily close to $s = 0$.

We must now choose whether to put the semicircle in the right half of the s plane, as shown by the solid arc between 2 and 3, or in the left half of the s plane, as shown by the dashed arc between 2 and 3 in Fig. 9-19a. If the former, we exclude the pole from, and if the latter, we include the pole in, the right half of the s plane. Since few systems will have open-loop transfer functions with poles in the right half of the s plane, it will be simpler to exclude all poles on the imaginary axis, and we shall, therefore, follow the solid-line semicircle in the right half of the s plane in Fig. 9-19a.

For the semicircle around the pole at $s = 0$, from Eq. (9-80),

$$\frac{H(s)G(s)}{K'} = \frac{1}{s} = \frac{1}{R_1 e^{j\theta}} = \frac{1}{R_1} e^{-j\theta} \tag{9-81}$$

which for $R_1 \ll 1$ becomes

$$\frac{H(s)G(s)}{K'} \approx \infty \, e^{-j\theta} \qquad (9\text{-}82)$$

Thus, as we traverse the very small radius semicircle in the counterclockwise direction from 2 to 3 in the s plane, we traverse a very large radius semicircle in the clockwise direction from 2 to 3 in the $H(s)G(s)/K'$ plane.

Continuing the mapping as before, we find that as s moves up the $j\omega$ axis from 3 to 4 in the s plane, $H(s)G(s)/K'$ moves up the imaginary axis from 3 to 4 in the $H(s)G(s)/K'$ plane.

To get from point 4 to point 1 in the s plane, we let

$$s = R_2 e^{j\theta} \qquad (9\text{-}83)$$

where $R_2 \gg 1$, and find that

$$\frac{H(s)G(s)}{K'} = \frac{1}{R_2} e^{-j\theta} = [0(+)]e^{-j\theta} \qquad (9\text{-}84)$$

Since θ changes by 180° in the clockwise direction in going from 4 to 1 in the s plane, $-\theta$ will change by 180° in the counterclockwise direction in going from 4 to 1 in the $H(s)G(s)/K'$ plane. Thus, the path is closed by the very small radius semicircle from 4 to 1 in Fig. 9-19b.

Because we selected the path around $s = 0$ to exclude the pole of the open-loop transfer function, there are no poles in the region bounded by the closed path in the s plane. Therefore, the critical region in the $H(s)G(s)/K'$ plane is enclosed by the curve in that plane, and, since the point $-1/K' + j0$ lies outside this region for all values of K', we conclude that the system is inherently stable, i.e., stable for all values of K'.

To check the stability, we substitute $H(s)G(s) = K'/s$ into Eq. (9-58) to give

$$1 + H(s)G(s) = s + K' = 0 \qquad (9\text{-}85)$$

for which the root is $-K'$, confirming that the system is indeed stable for all values of K'.

Example 9-6 $H(s)G(s) = K'/s^2$, which has a double pole at 0. Following the discussion in Example 9-5, we shall exclude the double pole at 0 by defining the right half of the s plane as the region within the closed curve in Fig. 9-20a.

For points on the $j\omega$ axis,

$$\frac{H(s)G(s)}{K'} = \frac{H(j\omega)G(j\omega)}{K'} = \frac{1}{(j\omega)^2} = -\frac{1}{\omega^2} \qquad (9\text{-}86)$$

and calculations for the several points give

Point	ω	$H(s)G(s)/K'$
1	$-\infty$	$-0(+) + j0$
2	$-0(+)$	$-\infty + j0$
3	$+0(+)$	$-\infty + j0$
4	$+\infty$	$-0(+) + j0$

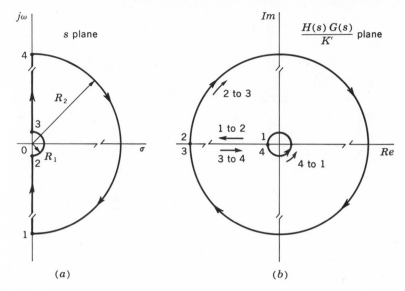

Fig. 9-20 *Conformal mapping for $H(s)G(s)/K' = 1/s^2$.*

As shown in Fig. 9-20, when s moves up the $j\omega$ axis from 1 to 2, $H(s)G(s)/K'$ moves to the left along the real axis from about 0 to $-\infty$, and as s moves up the $j\omega$ axis from 3 to 4, $H(s)G(s)/K'$ moves to the right along the real axis from $-\infty$ to almost 0.

Since in the $H(s)G(s)/K'$ plane point 2 coincides with point 3 and point 4 coincides with point 1, it may appear at first glance that the path is closed. However, such is not the case; we have not considered the semicircles in the s plane from 2 to 3 and from 4 to 1.

For the semicircle from 2 to 3, we let, as above,

$$s = R_1 e^{j\theta} \tag{9-87}$$

where $R_1 \ll 1$. In terms of Eq. (9-87), we find

$$\frac{H(s)G(s)}{K'} = \frac{1}{s^2} = \frac{1}{(R_1 e^{j\theta})^2} = \frac{1}{R_1^2} e^{-j2\theta} \tag{9-88}$$

From Eq. (9-88), we see that for $R_1 \ll 1$, $1/R^2 \approx \infty$, and that the point traverses the arc in the $H(s)G(s)/K'$ plane in the opposite direction to and through *twice the angle of* the traverse in the s plane. Therefore, we conclude that as s moves counterclockwise on an arc with $R \ll 1$ through an angle of 180° from 2 to 3 in the s plane, $H(s)G(s)/K'$ will move clockwise on an arc with $R \gg 1$ through an angle of 360° from 2 to 3 in the $H(s)G(s)/K'$ plane, as shown in Fig. 9-20b.

In a similar manner, we find that as s traverses in a clockwise direction the very large radius *semicircle* from 4 to 1 in the s plane, $H(s)G(s)/K'$ will traverse in a counterclockwise direction a very small radius *circle* from 4 to 1 in the $H(s)G(s)/K'$ plane.

Since we have excluded the poles at 0, there are no poles of the open-loop transfer function within the right half of the s plane and, therefore, the region enclosed by the curve in the $H(s)G(s)/K'$ plane corresponds to the right half of the s plane. However, the path from 1 to 2 and the path from 3 to 4 pass through the point $-1/K' + j0$. As discussed in Example 9-4, when a single path passes through the critical point, a

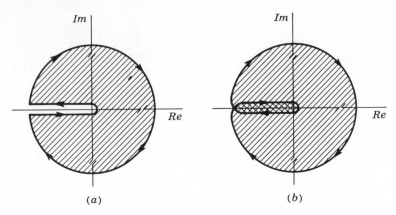

Fig. 9-21 *Critical regions for $H(s)G(s)/K' = 1/s^2$ when the paths on the real axis in $H(s)G(s)/K'$ plane are displaced arbitrarily small distances above and below the real axis to (a) exclude and (b) include the critical point.*

root of the characteristic equation equals zero and the system is neutrally stable. In this case, we have two paths passing through $-1/K' + j0$ and we need to know what it means.

Let us consider limiting cases, as in Fig. 9-21, where the paths 1 to 2 and 3 to 4 are displaced arbitrarily small distances above and below the real axis. We find that for Fig. 9-21a the point $-1/K' + j0$ is not in the critical region and that for Fig. 9-21b it is. Thus, we would conclude that the system in Fig. 9-21a is stable and that in Fig. 9-21b is unstable. As we let the displacement from the real axis → zero, the two conditions must also approach, and we conclude that the system becomes neutrally stable.

To determine the nature of the root or roots, let us substitute $H(s)G(s)$ into Eq. (9-58). Thus,

$$1 + H(s)G(s) = s^2 + K' = (s + j\sqrt{K'})(s - j\sqrt{K'}) = 0 \qquad (9\text{-}89)$$

and we see that the roots are a pair of complex conjugates with zero real parts. This corresponds to a second-order system with zero damping, or neutral stability, which becomes technically, from Sec. 4-23, an unstable system. Thus, we conclude that in this case Fig. 9-21b is more realistic than Fig. 9-21a, and, in general, *we shall consider points on the $H(s)G(s)/K'$ path as lying in the critical region.* It should also be noted that in Example 9-4 a single path passing through the point $-1/K' + j0$ corresponded to neutral stability of a first-order system, and in this example the two paths passing through the critical point correspond to neutral stability of a second-order system.

Example 9-7 $H(s)G(s) = K(4s + 1)^2/s^2(2s - 1)$, which has two poles at 0, a pole at $+0.5$, and two zeros at -0.25. Rewriting $H(s)G(s)$ in the form of Eq. (9-30), we find

$$H(s)G(s) = K\frac{(4s + 1)^2}{s^2(2s - 1)} = K'\frac{(s + 0.25)^2}{s^2(s - 0.5)} \qquad (9\text{-}90)$$

where $K' = 8K$. From Eq. (9-90), we find

$$\frac{H(s)G(s)}{K'} = \frac{(s + 0.25)^2}{s^2(s - 0.5)} = \frac{s^2 + 0.5s + 0.0625}{s^2(s - 0.5)} \qquad (9\text{-}91)$$

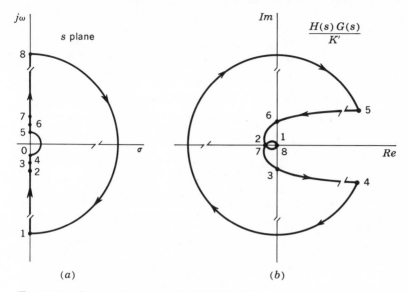

Fig. 9-22   *Conformal mapping for $H(s)G(s)/K' = (s + 0.25)^2/s^2(s - 0.5)$.*

The transfer function is now sufficiently complicated that the mapping of the right half of the s plane into the $H(s)G(s)/K'$ plane is somewhat more laborious than in the previous examples and several more calculations will be required. As before, we shall follow the path indicated in Fig. 9-22a, which excludes the two poles at 0 from the right half of the s plane. For values of $s = 0 + j\omega$, we find, from Eq. (9-91),

$$\frac{H(s)G(s)}{K'} = \frac{H(j\omega)G(j\omega)}{K'} = \frac{-\omega^2 + j0.5\omega + 0.0625}{-\omega^2(j\omega - 0.5)}$$

$$= -\frac{(\omega^2 - 0.0312)}{\omega^2(\omega^2 + 0.25)} - j\frac{\omega^2 - 0.312}{\omega(\omega^2 + 0.25)} \tag{9-92}$$

For $|\omega| \gg 1$

$$\frac{H(s)G(s)}{K'} \approx -\frac{1}{\omega^2} - j\frac{1}{\omega} \tag{9-93}$$

For $\omega \approx -\infty$, Eq. (9-93) becomes

$$\frac{H(s)G(s)}{K'} \approx -\frac{1}{\infty^2} - j\frac{1}{-\infty} = -\frac{1}{\infty^2} + j\frac{1}{\infty}$$

From this we see that the magnitudes of both terms are much less than 1 and that the magnitude of the real term is much less than the magnitude of the imaginary term. From the relative magnitudes or by using the angle of $H(s)G(s)/K'$, which is

$$\underline{/H(s)G(s)/K'}\ \bigg|_{\omega \approx -\infty} \approx \tan^{-1}\frac{1/\infty}{-1/\infty^2} = \tan^{-1}(-\infty) = 90°(+)$$

we can deduce that the path starts in the second quadrant near zero, almost on the positive imaginary axis, and its angle of departure is $90°(+)$. Similarly, we shall find that for $\omega \approx +\infty$ the end point is in the third quadrant, near zero, almost on the negative imaginary axis, and the angle of approach is $270°(-)$ or $-90°(+)$.

In Eq. (9-92), it can be noted that the real part equals zero when $\omega^2 = 0.0312$ or $\omega = \pm 0.177$ rad/sec and that the imaginary part equals zero when $\omega^2 = 0.312$ or $\omega = \pm 0.559$ rad/sec.

For $|\omega| \ll 1$

$$\frac{H(s)G(s)}{K'} \approx \frac{0.0312}{\omega^2} + j\,\frac{0.312}{\omega} \tag{9-94}$$

For $\omega = -0(+)$, Eq. (9-94) becomes

$$\frac{H(s)G(s)}{K'} = \frac{0.0312}{[0(+)]^2} - j\,\frac{0.312}{0(+)} \tag{9-95}$$

In Eq. (9-95) the magnitudes of both the real and the imaginary parts become much greater than 1. However, the magnitude becomes larger much faster for the real part than for the imaginary part and the path becomes a horizontal line. For $\omega = +0(+)$, Eq. (9-94) becomes the complex conjugate of Eq. (9-95).

Tabulating the calculations made thus far, we have the accompanying table. The remaining major gaps are between points 4 and 5 and between points 8 and 1. If we

Point	ω	$H(s)G(s)/K'$	$\underline{/H(s)G(s)/K'}$
1	$-\infty$	$j0(+)$	$90°(+)$
2	-0.559	$-1.60 + j0$	
3	-0.177	$0 - j5.65$	
4	$-0(+)$	$\infty^2 - j\infty$	$360°(-)$
5	$0(+)$	$\infty^2 + j\infty$	$0°(+)$
6	0.177	$0 + j5.65$	
7	0.559	$-1.60 + j0$	
8	∞	$-j0(+)$	$270°(-)$

let $s = R_1 e^{j\theta}$ with $R_1 \ll 1$ for the path between 4 and 5 in the s plane, we find from Eq. (9-91)

$$\frac{H(s)G(s)}{K'} \approx \frac{0.0625}{-0.5s^2} = -\frac{0.0625}{0.5R_1^2}\,e^{-j2\theta}$$

$$\approx -\infty e^{-j2\theta} = \infty e^{j(\pi - 2\theta)}$$

Thus, we see that $H(s)G(s)/K'$ moves through twice the angle and in the opposite direction to θ. Therefore, as s moves counterclockwise around the very small radius semicircle from 4 to 5 in the s plane, $H(s)G(s)/K'$ will move clockwise around the very large radius circle from 4 to 5 in the $H(s)G(s)/K'$ plane.

Letting $s = R_2 e^{j\theta}$ with $R_2 \gg 1$ for the path between 8 and 1 in the s plane, we find, from Eq. (9-91),

$$\frac{H(s)G(s)}{K'} \approx \frac{1}{s} = \frac{1}{R_2}\,e^{-j\theta} = [0(+)]e^{-j\theta}$$

Therefore, as s moves clockwise around the very large radius semicircle from 8 to 1 in the s plane, $H(s)G(s)/K'$ will move counterclockwise around a very small radius semicircle from 8 to 1 in the $H(s)G(s)/K'$ plane.

Fig. 9-23 *Critical region for* $H(s)G(s)/K' = (s + 0.25)^2/s^2(s - 0.5)$.

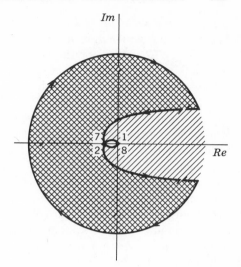

Since the critical region is to the right of the path in the s plane, it also lies to the right of the path in the $H(s)G(s)/K'$ plane. As shown in Fig. 9-23, the critical region in the $H(s)G(s)/K'$ plane includes the entire plane except for the small region bounded by points 1-2-7-8. Thus, the system is conditionally stable.

The real axis within the region not containing roots with positive real parts lies between 0 and -1.60. Therefore, the system will be stable when

$$\frac{1}{K'} < 1.60$$

or

$$K' > \frac{1}{1.60} = 0.625$$

Since $K' = 8K$, from Eq. (9-90), we can conclude that the system will be stable for $8K > 0.625$ or $K > 0.0782$.

9-6. Gain Margin and Phase Margin A major feature of the root-locus method in Secs. 9-3 and 9-4 is that we can deduce the general behavior characteristic to be expected from the system by simply noting (1) the position of the roots in the s plane for a given value of gain K and (2) how the roots move as K is changed. When the system is studied by use of conformal mapping, somewhat less information is available; but we can still deduce more than whether the system is inherently stable, inherently unstable, or conditionally stable, and, if the latter, the values of gain for which it will be stable.

To determine the type of additional information obtainable from conformal mapping, we must first consider the type of information used in mapping the right half of the s plane into the $H(s)G(s)/K'$ plane. Referring to the examples in Sec. 9-5, we see that, except for the circles with $R \ll 1$ and $R \gg 1$, the path of $H(s)G(s)/K'$ is determined solely by the values of s on the $j\omega$ axis, i.e., when $s = 0 + j\omega$. Thus, the path in the finite, nonzero region of the $H(s)G(s)/K'$ plane is a polar plot of $H(j\omega)G(j\omega)/K'$, which for the positive values of ω is a polar plot of the sinusoidal response of the open-loop system.

Consequently, quantitative information obtained from the map will be directly applicable to sinusoidal response only. However, as discussed in Chaps. 4 and 8, a second-order system whose sinusoidal response indicates a large magnification factor (Fig. 4-20) or, the equivalent, a large magnitude of control ratio C/R (Fig. 8-13) can be expected to have a large overshoot in the response to a step input and, in general, an oscillatory transient response. Overshoot and oscillatory transient response can exist only for second-, and higher-, order systems. For such systems, the overshoot and oscillatory nature of the transient response become increasingly pronounced as the system approaches neutral stability.

As shown in Example 9-6, neutral stability of a second-order system will be indicated by two paths passing through, or intersecting at, the critical point $-1/K' + j0$ in the $H(s)G(s)/K'$ plane. Therefore, the closer a path comes to $-1/K' + j0$, the closer the system is to neutral stability, as well as to instability, and the more oscillatory will be the transient response.

As discussed in previous sections, for a system to be unstable,

$$H(s)G(s) = -1 \tag{9-96}$$

which, in view of the discussion above, is the same thing as

$$H(j\omega)G(j\omega) = -1 \tag{9-97}$$

Since $H(j\omega)G(j\omega)$ has both magnitude and phase, we usually find it convenient to consider Eq. (9-97) as

$$|H(j\omega)G(j\omega)| = 1 \tag{9-98}$$

and $\qquad \underline{/H(j\omega)G(j\omega)} = -180° \tag{9-99}$

Therefore, a system will be stable unless the magnitude of the open-loop sinusoidal response is equal to or greater than 1 when the phase angle is $-180°$. Consequently, we now have two ways of describing the *margin*, or difference, between a prescribed operating condition and instability.

The term *gain margin* describes the difference between the amplitude of $H(j\omega)G(j\omega)$ and 1.0 when the angle is $-180°$. In general, it is the factor by which the gain K can be multiplied before the system becomes unstable. For example, in Fig. 9-24a, the magnitude of the open-loop sinusoidal response is 0.56 at the point where the angle is $-180°$. Thus, the gain margin is

$$\text{Gain margin} = \frac{1}{|H(j\omega)G(j\omega)|}\bigg|_{\underline{/-180°}} = \frac{1}{0.56} = 1.79 \tag{9-100}$$

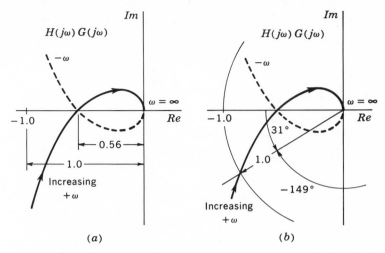

Fig. 9-24   (a) *Gain margin;* (b) *phase margin.*

or, more usefully, in terms of decibels,

$$\text{Gain margin} = 20 \log_{10} 1.79 = 5.07 \text{ db} \qquad (9\text{-}101)$$

It should be noted that *if the gain margin is less than* 1, *or negative decibels, the closed-loop system will be unstable.*

The term *phase margin* describes the difference between the phase angle of $H(j\omega)G(j\omega)$ and $-180°$ when the magnitude equals 1.0. As a matter of convention, the phase margin is the angle from $-180°$ to the vector $H(j\omega)G(j\omega) = 1.0$. For example, in Fig. 9-24b, the phase margin is

$$\gamma = \underline{/H(j\omega)G(j\omega)} - (-180°) = -149° + 180° = 31° \quad (9\text{-}102)$$

It should be noted that *if the phase margin is a negative angle,* the path encircles the critical point $-1 + j0$ and *the closed-loop system will be unstable.*

Although the transient-response characteristics can be related to both the gain and phase margins, the effective damping is more closely related to the phase margin. Gain margins of about 6 db and phase margins of between 45 and 60° are usually considered to be satisfactory.

The concepts of gain and phase margins become difficult to apply, and in some cases have no significance, when a system is conditionally stable in a region not adjacent to zero on the real axis and when the open-loop transfer function has poles in the right half of the s plane. Therefore, we shall limit our consideration of gain and phase margins to systems (1) which are open-loop stable and (2) for which the $H(j\omega)G(j\omega)$ curve crosses the real axis for only one value of $+\omega$.

As shown, the polar plot of the open-loop sinusoidal response, or the path in the $H(s)G(s)$ plane, is ideal for illustrating the concepts of gain and phase margins and for demonstrating the relationship between the sinusoidal response and stability. However, as will be discussed in the next section, these ideas become more useful to the designer when the open-loop response is presented in log-magnitude (decibel) vs. log-frequency and phase-angle vs. log-frequency graphs.

9-7. Attenuation-phase Method In Sec. 8-2 it was shown that the variation of magnitude and phase angle with frequency for the sinusoidal response of closed-loop systems could be presented relatively easily by plotting magnitude and frequency on logarithmic scales. In the case of the magnitude, the use of the decibel was found to be particularly convenient.

The simplicity of plotting the curves is just as useful in relation to the sinusoidal response of the open-loop system, and when combined with the concepts of gain margin and phase margin, the logarithmic plots provide a convenient basis for stability analysis. This approach becomes particularly useful when data from tests with the actual system are available.

As noted in the preceding section, we shall apply the concepts of gain margin and phase margin to only those systems that are open-loop stable. However, this is not a serious limitation, because most real control systems are open-loop stable; and it is not a limiting factor at all when using experimental data, because the fact that open-loop sinusoidal response data could be taken ensures that the system is open-loop stable.

Referring to Fig. 9-24, we can see that the amplitude of $H(j\omega)G(j\omega)$ decreases as the frequency increases, becoming zero at $\omega = +\infty$. The reader should explain to himself why this must be the case for all physically realizable systems. We also can see that the rate of decrease of magnitude relative to the rate of change of the phase angle as the frequency increases determines whether or not the closed-loop system will be stable.

In the terminology of sound analysis and electrical engineering, *attenuation* is used to describe a decrease in magnitude, particularly as a function of frequency. We shall borrow the term for this application to emphasize the importance of the attenuation of the magnitude in relation to the change of the phase angle.

As for the method of conformal mapping, we shall use examples to illustrate the application of the gain-margin and phase-margin concepts to stability analysis.

Example 9-8 $H(s)G(s) = K'/(s + 2)$, for which

$$H(j\omega)G(j\omega) = \frac{K'}{2 + j\omega} \qquad (9\text{-}103)$$

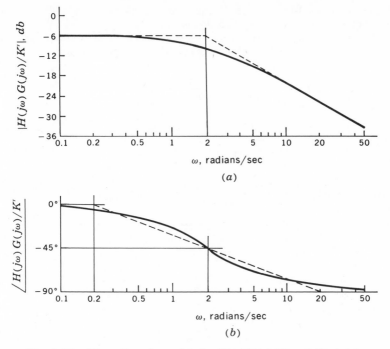

Fig. 9-25 *Attenuation-phase plots for $H(j\omega)G(j\omega)/K' = 1/(2 + j\omega)$.*

However, since K' is a constant, it will be convenient, again, to plot the attenuation-phase diagrams by using only frequency-dependent terms. Then we can determine the maximum value of gain that will provide the desired gain or phase margins. Thus, we shall consider first

$$\frac{H(j\omega)G(j\omega)}{K'} = \frac{1}{2 + j\omega} \tag{9-104}$$

Based on the discussion in Sec. 8-2 and the curves in Fig. 8-7, we can see that the corner frequency for the straight-line asymptotes is 2 rad/sec, that the low-frequency asymptote is horizontal at -6 db, and that the high-frequency asymptote has a slope of -6 db/octave. The resulting curves are shown in Fig. 9-25.

In Fig. 9-25b we can see that the phase angle never becomes more negative than $-90°$ and, therefore, the phase margin cannot be less than 90° for all values of K'. Therefore, we conclude, as in Example 9-3, that this system will be stable for all values of gain. It should also be noted that since the phase angle can never reach $-180°$, the term gain margin has little significance.

Example 9-9 $H(s)G(s) = K'/s^2$, for which

$$\frac{H(j\omega)G(j\omega)}{K'} = \frac{1}{(j\omega)^2} = -\frac{1}{\omega^2} \tag{9-105}$$

From Eq. (9-105), we have

$$\left| \frac{H(j\omega)G(j\omega)}{K'} \right| = \frac{1}{\omega^2}$$

and

$$\underline{/H(j\omega)G(j\omega)} = -180° \tag{9-106}$$

The magnitude will be a straight line passing through 0 db at $\omega = 1$ rad/sec and with a slope of -12 db/octave (40 db/decade).

The phase angle, Eq. (9-106), is always $-180°$; therefore, the phase margin is zero and the system will be unstable (neutrally stable) for all values of gain, as was determined previously in Example 9-6.

Example 9-10 $H(s)G(s) = \dfrac{K(2s + 1)}{s(3s + 1)(s + 1)}$, for which

$$H(s)G(s) = K' \frac{(s + 0.5)}{s(s + 0.333)(s + 1)}$$

and

$$\frac{H(j\omega)G(j\omega)}{K'} = \frac{(0.5 + j\omega)}{j\omega(0.333 + j\omega)(1 + j\omega)} \tag{9-107}$$

where $K' = 0.667K$.

Considering each complex quantity in Eq. (9-107) in turn, we find the following information:

Term	Corner frequency, rad/sec	Slope of asymptotes		ϕ
		Low frequency	High frequency, db/octave	
$0.5 + j\omega$	0.5	0 (at -6 db)	$+6$	0 to $+90°$
$\dfrac{1}{j\omega}$	1	-6 db/oct	-6	$-90°$
$\dfrac{1}{0.333 + j\omega}$	0.333	0 (at $+9.5$ db)	-6	0 to $-90°$
$\dfrac{1}{1 + j\omega}$	1	0 (at 0 db)	-6	0 to $-90°$

The individual curves and the resultant attenuation-phase curves are shown in Fig. 9-26. As can be seen, the phase angle never reaches $-180°$, and, therefore, we conclude that the closed-loop system will be stable for all values of gain.

Projecting downward from the point at which the magnitude = 1, that is, 0 db, we find that the corresponding phase angle is $-142°$ and, therefore, the phase margin is $38°$ when $K' = 1$ or $K = 1.5$.

If, for example, the phase margin is to be $60°$, then projecting upward from the point at which the phase angle is $-120°$, we find that the magnitude is 9 db greater than 0 db. If we designate the new value of K as $K_{60°}$ and the original value as K_0, by definition we must have

$$-9 \text{ db} = 20 \log_{10} \frac{K_{60°}}{K_0}$$

from which

$$\frac{K_{60°}}{K_0} = \text{antilog} \frac{-9}{20}$$

or

$$\frac{K_0}{K_{60°}} = \text{antilog} \frac{9}{20} = 2.82 \tag{9-108}$$

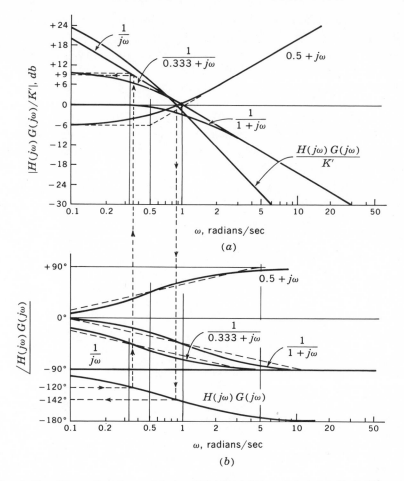

Fig. 9-26 *Attenuation-phase plots for $H(j\omega)G(j\omega)/K' = (0.5 + j\omega)/[j\omega(0.333 + j\omega)$ $(1 + j\omega)]$.*

Substituting $K_0 = 1.5$ into Eq. (9-108), and solving for $K_{60°}$, gives

$$K_{60°} = 0.532$$

which is the maximum value of gain K for a phase margin of 60°.

The relative ease with which the terms can be combined, as in Fig. 9-26, to give the resultant attenuation-phase plots makes this one of the most convenient methods for studying those systems to which it is directly applicable. This becomes particularly important in design when one is trying to improve the performance of the system by modifying components or introducing additional components into the system.

When the open-loop system is a *minimum-phase* system, i.e., one with no open-loop poles and no open-loop zeros in the right half of the s plane, Bode[1] has shown that the attenuation and phase are related to each other in such a manner that if the slope of the attenuation curve is more positive than -12 db/octave (-40 db/decade) at the point it crosses the 0-db line, the closed-loop system will be stable. In most cases, a slope of about -6 db/octave at the 0-db crossover will result in satisfactory performance. In recognition of this contribution to the theory of feedback systems, the term *Bode diagram* has been widely used to describe all plots of log magnitude and phase, and the reader will encounter this usage frequently in the literature.

[1] H. W. Bode, "Network Analysis and Feedback Amplifier Design," chap. 14, D. Van Nostrand Company, Inc., Princeton, N.J., 1945.

chapter 10

NONLINEAR CONTROL
SYSTEMS

The discussion in Chap. 5 applies equally well to control systems as to simple mass-spring-damper systems. As before, there are no truly linear control systems, and we shall again classify systems as (1) those that can be considered linear over a small region about the equilibrium position, (2) those that are piecewise linear, and (3) those that must be considered nonlinear at all times. The major difference is that we are now considering active, rather than passive, systems, and we shall be particularly interested in the effect of nonlinearities on the stability of a closed-loop system.

10-1. Common Sources of Nonlinearities Although in reality every element in the system will be nonlinear, the significant control system nonlinearities are usually associated with (1) the manner in which the elements that control the source of energy sense and react to an error and (2) the manner in which the controlled system reacts to the output of the control elements. In the case of the control elements, the major nonlinearities are due to (1) nonlinear error sensors, such as a thermostat that opens or closes a circuit when the temperature being controlled is above or below, respectively, the reference level, and (2) nonlinear control elements, such as an amplifier that saturates or a fuel valve with only two positions—closed or wide open. The major nonlinearities encountered in controlled mechanical systems are nonlinear damping (Coulomb friction) and/or gear trains and linkages with backlash.

In general, we shall find that functional relationships for the common nonlinearities can be approximated closely by segments of straight lines and that the system then becomes piecewise linear. For very simple cases we can obtain solutions or useful information by use of classical

347

differential equations, Sec. 5-3, or the phase plane, Sec. 5-4. Unfortunately, most real problems will involve systems sufficiently complex that these approaches are no longer practical. Under certain conditions the method of describing functions[1] can be used effectively, but in most cases the designer will be advised to use an analog computer or numerical methods and a digital computer. The relative simplicity with which the most frequently encountered nonlinearities can be simulated and the rapidity with which the effects of changes in parameters can be investigated make the analog computer particularly useful.

It is important to understand that nonlinearity is not necessarily synonymous with undesirability. In fact, in special cases the contrary is more likely to be true. The reasons for spending most of our time with linear systems are simply that we can do more mathematically and, since superposition applies, the conclusions reached and the understanding developed will be generally useful. Although a detailed treatment of nonlinear systems is beyond the scope of this text, some insight into the advantages and limitations of systems in which there are nonlinear elements is worthwhile and can be gained by considering several of the simpler types that are widely used in practice.

10-2. Two-position Control Although often lacking in glamor in comparison with guidance systems for missiles and other complex control problems, the greatest number of systems in use are related to process operations in which the controlled variable is a temperature, the level of a liquid, etc. In many situations the reference level remains essentially static for long periods of time, or it changes so slowly that frequency response has little significance. Under these conditions, the cost of control elements that would provide proportional control is often out of proportion to the benefits, and we commonly find control elements used that have an output that is not proportional to the error but is limited to being one of two constant values. In most situations involving control of temperature or liquid level, the controlled variable changes naturally in one direction or the other, and the control element will normally act to provide, or remove, energy or liquid at either a maximum or zero rate. The descriptive term *on-off control* is commonly used for this case. The concept of a transfer function does not apply, and we must present the functional characteristics of the control element graphically or as a set of piecewise linear equations.

[1] J. E. Gibson, "Nonlinear Automatic Control," chap. 9, McGraw-Hill Book Company, New York, 1963.

J. J. D'Azzo and C. H. Houpis, "Feedback Control System Analysis and Synthesis," 2d ed., chap. 18, McGraw-Hill Book Company, New York, 1966.

J. G. Truxal, "Automatic Feedback Control System Synthesis," pp. 566–612, McGraw-Hill Book Company, New York, 1955.

G. J. Thaler and R. G. Brown, "Analysis and Design of Feedback Control Systems," 2d ed., chaps. 13 and 14, McGraw-Hill Book Company, New York, 1960.

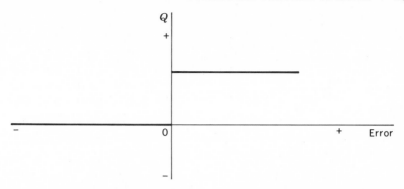

Fig. 10-1 *Input-output characteristic of an on-off controller for a heating system.*

For example, let us consider a heating system, such as discussed in Sec. 7-1, where the control element is an electrically operated valve that is closed as long as the temperature is above the reference level, and the contact points in the thermostat are open, and is open as long as the temperature is below the reference level, and the contact points are closed. If we assume that the thermostat contacts open and close the instant the temperature rises above and falls below, respectively, the reference level, and, if we further assume that the valve operates, the fuel flow changes, etc., instantaneously, the relationship between the heat output Q of the furnace and the error will be as shown in Fig. 10-1.

However, we know from experience that the real situation is not this simple, because nothing happens instantaneously and there is always a finite difference between the temperatures at which the thermostat contact points open and close. The time required for opening and closing the valve is generally so short in comparison with the other time constants in a heating system that it can be neglected without appreciable error. In most cases, it is desirable to make contact points open and close with a snap action to eliminate chatter (making and breaking of contact due to vibration) and to minimize arcing. Snap action requires a toggle effect,[1] which, in turn, requires a finite displacement. The effects of this *delay* or *lag* in switching are important and will be considered in Sec. 10-3. However, we shall first consider the performance of systems with instantaneous switching, as illustrated by the characteristic in Fig. 10-1.

First-order controlled system As discussed in Sec. 8-2, a linear first-order controlled system will have a transfer function of the form

$$G = \frac{1}{aD} \qquad (10\text{-}1)$$

[1] R. M. Phelan, "Fundamentals of Mechanical Design," 2d ed., pp. 12–14, McGraw-Hill Book Company, New York, 1962.

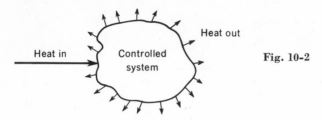

Heat out

Heat in

Fig. 10-2

There is no inertia (or equivalent), and the response to a constant-magnitude input is a constant rate of change (velocity) output.   If we apply this to a heating system, as illustrated in Fig. 10-2, where the rate of heat input is greater than the rate of heat loss and the temperature differences are small enough that we can consider both rates to be constant quantities, the coefficient a in Eq. (10-1) is the thermal capacity of the controlled system.   The change in temperature within the system when heat is being supplied will be a linear function of the difference between the rate of heat input and the rate of heat loss.

For these conditions, the block diagram becomes that in Fig. 10-3 and the response to a step change in reference level will be as shown in Fig. 10-4.   This is not a practical system, if for no other reason than, to maintain the temperature at exactly $r_{0(+)}$, the contact points of the instantaneously switching thermostat would have to open and close at an infinite rate.   The simple change required to make this system practical will be considered in Sec. 10-3.

The response of a first-order control system was shown in Sec. 8-2 to be an exponential curve.   Assuming that the initial rate of heat input is the same for a proportional control system as for the on-off control system, we would expect the response to be similar to that shown by the dashed line in Fig. 10-4.   As can be seen, the on-off system has the faster response, because the rate at which energy is supplied to the system remains constant as long as the error is positive.   This feature, in addition to the relative simplicity and economy of the components, is a major reason for the popularity of two-position control.

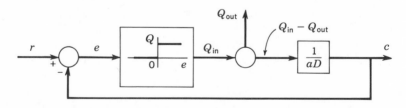

Fig. 10-3   *Block diagram of a heating system with on-off control.*

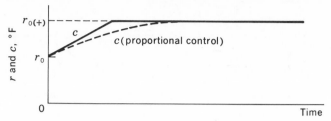

Fig. 10-4   *Comparison of the responses of heating systems with on-off and proportional control to a step change in reference level.*

Second-order controlled system   As discussed in Sec. 8-2, the transfer function of a second-order controlled system will be of the form

$$G = \frac{1}{ID^2 + aD} \tag{10-2}$$

where I is inertia, or its equivalent, and a is viscous damping, or its equivalent.

Although some second-order controlled systems are subjected to forces or other inputs tending to drive the controlled variable in one direction and an on-off controller would be applicable, most require the input to drive in both directions.   For example, if the controlled variable is the position of a mass, we often use reversible control elements, such as electric or hydraulic motors, translation screws, and hydraulic cylinders. Under these conditions the output of the control elements with instantaneous switching can be represented as in Fig. 10-5.

If the controlled variable is the position of a rigid mass, the output of the control element will be a constant-magnitude force F_0, and the system

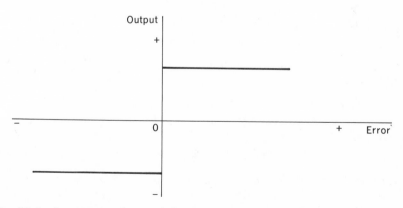

Fig. 10-5   *Input-output characteristic of a two-position controller for use with second-order controlled systems.*

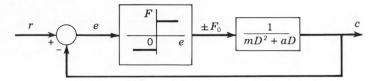

Fig. 10-6 *Block diagram for a system containing a two-position controller and a second-order controlled system.*

becomes that in Fig. 10-6. The piecewise linear equations of motion for the system in Fig. 10-6 are:

For a positive error

$$m\,D^2c + a\,Dc = +F_0 \tag{10-3}$$

For a negative error

$$m\,D^2c + a\,Dc = -F_0 \tag{10-4}$$

In general, this type of system is most likely to be considered for use when the reference input changes in steps. Therefore, we shall limit our discussion to this case, and we shall find it convenient to rewrite Eqs. (10-3) and (10-4) in terms of the error e. To do this, we write, by definition,

$$e = r - c \tag{10-5}$$

from which

$$De = Dr - Dc \tag{10-6}$$

and

$$D^2e = D^2r - D^2c \tag{10-7}$$

If both r and $c = 0$ at $t = 0$ and $r = r_0$ at $t = 0(+)$, Eqs. (10-5) through (10-7) become, respectively, for $t \geq 0(+)$,

$$e = r_0 - c \tag{10-8}$$

$$De = -Dc \tag{10-9}$$

and

$$D^2e = -D^2c \tag{10-10}$$

Substituting from Eqs. (10-9) and (10-10) into Eqs. (10-3) and (10-4) results in:

For e positive

$$m\,D^2e + a\,De = -F_0 \tag{10-11}$$

For e negative

$$m\,D^2e + a\,De = +F_0 \tag{10-12}$$

Although Eqs. (10-11) and (10-12) can be solved by use of classical methods of differential equations, we shall use the phase plane in considering the general behavior characteristics both when damping is negligible and when it cannot be neglected.

When damping is zero, Eqs. (10-11) and (10-12) become, respectively:

For e positive

$$m D^2 e = -F_0 \qquad (10\text{-}13)$$

For e negative

$$m D^2 e = +F_0 \qquad (10\text{-}14)$$

Following the discussion in Sec. 5-4, we shall plot $v = De$ as a function of e in the phase plane. For Eqs. (10-13) and (10-14), we find, respectively:

For e positive

$$v = De = \frac{de}{dt} \qquad (10\text{-}15)$$

$$\frac{dv}{dt} = D^2 e = -\frac{F_0}{m} \qquad (10\text{-}16)$$

and

$$\frac{dv}{de} = \frac{dv/dt}{de/dt} = -\frac{F_0}{mv} \qquad (10\text{-}17)$$

For e negative

$$v = De = \frac{de}{dt} \qquad (10\text{-}18)$$

$$\frac{dv}{dt} = D^2 e = +\frac{F_0}{m} \qquad (10\text{-}19)$$

and

$$\frac{dv}{de} = \frac{dv/dt}{de/dt} = +\frac{F_0}{mv} \qquad (10\text{-}20)$$

Equations (10-17) and (10-20) can be solved readily for v as functions of e by use of the integral calculus, but noting that the slope of the trajectory is positive in the second and fourth quadrants and negative in the first and third quadrants and that the magnitude of the slope is a function of v only (for given values of F_0 and m) leads to the conclusion that we can draw the trajectory in the phase plane with a minimum of effort.

A line that is the loci of points at which the slope of the trajectories is a constant is called an *isocline*. In general, isoclines are curved lines, but they will be straight lines if the system is linear. In the present case, the system is piecewise linear and the isoclines are horizontal straight lines, as shown in Fig. 10-7.

If we assume that the reference level has been changed by a positive step r_0, the trajectory for the two-position control system without damping will be as shown in Fig. 10-7. As can be seen, the trajectory is a closed path and the system will oscillate indefinitely. Therefore, we can conclude that two-position control will not be satisfactory if the controlled system is a second-order system without damping.

It should also be noted that the trajectory for e positive is not tangent to that for e negative at $e = 0$. Thus, we can also conclude that the

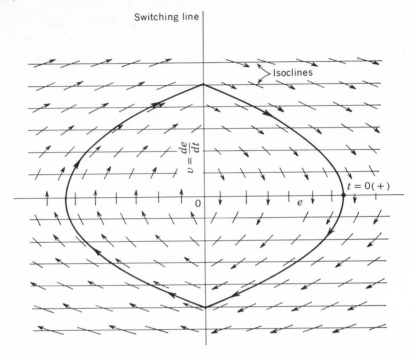

Fig. 10-7 *Phase-plane plot for a system containing a two-position controller and an undamped second-order controlled system.*

motion is not simple-harmonic.   For simple harmonic motion not only would the trajectories have to be tangent but the closed path would have to be an ellipse.

Actually, in this simple case, we could have deduced the behavior characteristics directly without any computations by observing that a constant force acting on a mass with zero friction results in uniformly accelerated motion.   Consequently, we shall expect to have a parabolic relationship between velocity and displacement, and the change in kinetic energy during the interval the force acts in one direction must equal the change during the interval in which the force acts in the other direction.

When viscous damping is present, we find:

For e positive

$$v = \frac{de}{dt} \tag{10-21}$$

$$\frac{dv}{dt} = D^2 e = \frac{-av - F_0}{m} \tag{10-22}$$

and

$$\frac{dv}{de} = \frac{-av - F_0}{mv} = -\frac{a}{m} - \frac{F_0}{mv} \tag{10-23}$$

For e negative

$$v = \frac{de}{dt} \tag{10-24}$$

$$\frac{dv}{dt} = D^2 e = \frac{-av + F_0}{m} \tag{10-25}$$

and

$$\frac{dv}{de} = \frac{-av + F_0}{mv} = -\frac{a}{m} + \frac{F_0}{mv} \tag{10-26}$$

Again, Eqs. (10-23) and (10-26) can be solved directly for v as functions of e by use of the integral calculus, but we can quickly plot the trajectory in the phase plane by noting that the magnitudes of the slope are functions of v only for given values of a, m, and F_0.

Figure 10-8 is the phase-plane plot of Eqs. (10-23) and (10-26) for the same values of m, F_0, and r_0 used in plotting the trajectory in Fig. 10-7. From the plot in Fig. 10-8, it appears that two-position control with instantaneous switching will be satisfactory if some damping is present. It is true that the trajectory in Fig. 10-8 is to be preferred to that in Fig. 10-7, but the performance still leaves much to be desired. The main problem is that the effect of viscous damping decreases as the velocity

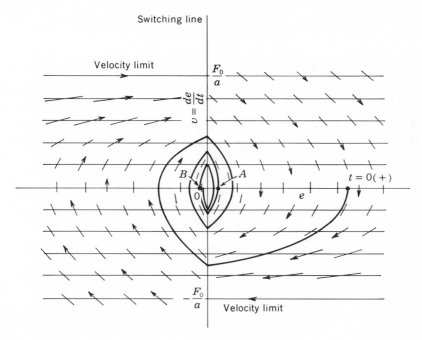

Fig. 10-8 *Phase-plane plot for a system containing a two-position controller and a second-order controlled system with viscous damping.*

decreases, and as $v \to 0$ Eqs. (10-23) and (10-26) approach, respectively,

$$\frac{dv}{de} \approx - \frac{F_0}{mv} \tag{10-27}$$

and
$$\frac{dv}{de} \approx + \frac{F_0}{mv} \tag{10-28}$$

Equations (10-27) and (10-28) are identical with the equations for the undamped case, Eqs. (10-17) and (10-20), respectively, and we can expect the system behavior to approach that in Fig. 10-7 as the equilibrium point is approached. This conclusion agrees with the behavior exhibited in Fig. 10-8. For example, in the first half cycle the amplitude decreases to 35 percent of the initial amplitude. If this were a linear proportional control system, the decrease would indicate a damping ratio $\zeta \approx 0.31$. If we consider the half cycle between A and B, we find the amplitude at A has decreased to only 81 percent of the amplitude at B and the "effective damping ratio" has decreased to $\zeta \approx 0.067$.

10-3. Two-position Control with Hysteresis In the discussion of the heating control system in Sec. 10-2 it was concluded that with instantaneous switching a specified temperature could be maintained with zero error, *provided* the switching frequency was infinite. Similar behavior would be found for all such systems where the control element must supply a constant level of output, whether force, energy, or whatever, to maintain the controlled variable at a steady value.

Since an infinite rate of switching is physically impossible, as is instantaneous switching, we must consider alternatives and their effects on system behavior.

The approach that is commonly used in temperature-control systems is to design the thermostat so that there is a finite difference between the temperature at which the contacts close and that at which the contacts

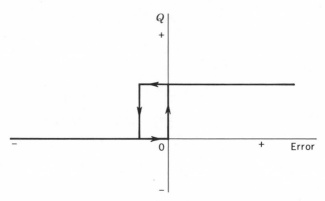

Fig. 10-9 *Input-output characteristic for an on-off controller with hysteresis for use in a heating system.*

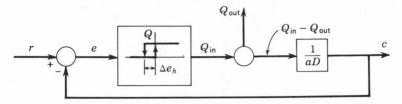

Fig. 10-10 *Block diagram of a heating system with on-off control with hysteresis.*

open. If we assume that the contacts close at the instant the temperature drops below the reference level (the error becomes positive) and that the contacts remain closed until the temperature has risen several degrees above the reference level (the error has become negative by a prescribed amount), the relationship between the output of the control element and the error becomes that shown in Fig. 10-9, rather than that in Fig. 10-1. The loop is characteristic of the phenomenon called *hysteresis*. A heating system with this type of control element will be classed as an on-off control system with hysteresis and can be represented as in Fig. 10-10.

If we assume that the rate of heat input is four times that of heat dissipation, the response to a step change in reference level will be approximately as shown in Fig. 10-11. As can be seen, the temperature will vary between $r_{0(+)}$ and $r_{0(+)} + \Delta e_h$, where Δe_h is the temperature difference (hysteresis effect) between closing and opening of the contact points in the thermostat. The frequency of the cyclic variation in temperature will be affected by changes in Δe_h, the rate of heat input, and the rate of heat dissipation; but in all cases, the system will be stable.

When hysteresis is introduced in a two-position control system where the controlled element is a second-order system, the relationship of the output of the control element to the error signal is usually shown as in Fig. 10-12, where the zero position is located halfway between the values of error at which switching occurs.

If we use the relationship in Fig. 10-12 in the control of the position of a mass, as in Sec. 10-2, we find that Eqs. (10-17), (10-20), (10-23), and

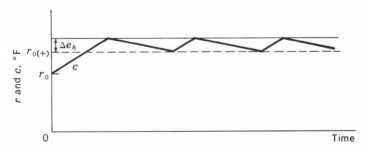

Fig. 10-11 *Response of a heating system with on-off control with hysteresis to a step change in reference level.*

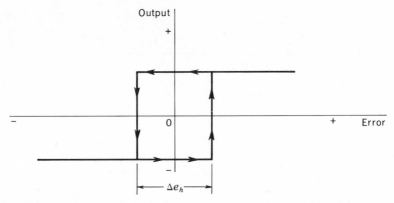

Fig. 10-12 *Input-output characteristic for a two-position controller with hysteresis for use with second-order controlled systems.*

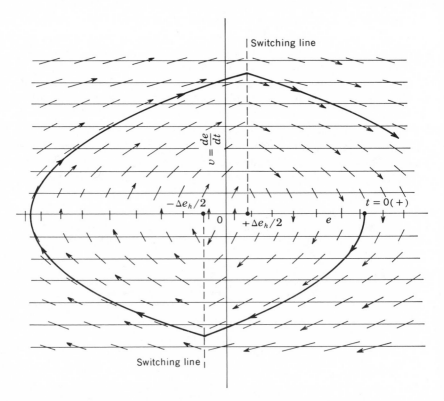

Fig. 10-13 *Phase-plane plot for a system containing a two-position controller with hysteresis and an undamped second-order controlled system.*

(10-26) still apply; but the switching line is shifted to the right $(+\Delta e_h/2)$ in the first quadrant and to the left $(-\Delta e_h/2)$ in the third quadrant.

If we consider the case with zero damping, as in Fig. 10-7, the phase-plane trajectory for the response to a step change in reference level becomes that in Fig. 10-13. Comparing Fig. 10-13 with Fig. 10-7 shows that the addition of hysteresis to a two-position control system without damping results in an ever-increasing amplitude of oscillation and, there-fore, an unstable system. This could have been anticipated by observing that the hysteresis effect is actually a delay in reversing the output of the control element until after the system has passed through the equi-librium point. The result is that the kinetic energy increases at the end of each half cycle, because the force acts through a greater distance when increasing the velocity of the mass than it does when bringing it to rest.

If we consider the case with appreciable damping, as in Fig. 10-8, the phase-plane trajectory becomes that in Fig. 10-14. Although the delay in switching does not result in an unstable system when there is appreci-able damping, the trajectory starting at A in Fig. 10-14 is quite different from that for the equivalent system without hysteresis in Fig. 10-8.

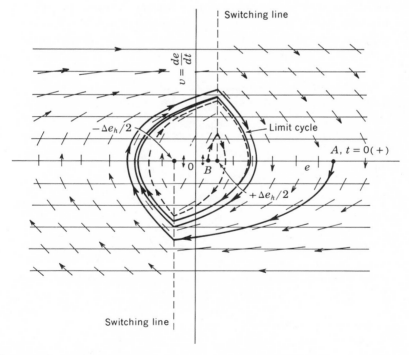

Fig. 10-14 *Phase-plane plot for a system containing a two-position controller with hysteresis and a second-order controlled system with viscous damping.*

In Fig. 10-14, the trajectory "spirals" inward and then continues to follow the same closed path indefinitely. This closed path is called a *limit cycle* and can exist only for a nonlinear system. It should be noted that if the initial point is inside the limit cycle, for example at *B*, the trajectory "spirals" outward until the limit cycle is reached. The continual oscillation associated with a limit cycle is called *hunting*.

On the basis of the discussion in this and the preceding section, we can conclude that the two-position control of a second-order system has limitations under the best of conditions, because the damping becomes relatively ineffective near the equilibrium position in comparison with the constant full-power output of the control element. One of the simplest methods for eliminating hunting will be discussed in the next section.

10-4. Three-position Control—Dead Zone   When a static error can be tolerated in the response to a step change in input, the response of a second-order system in which the output of the control element is a constant magnitude can be improved by modifying the control element so that its output will be zero for small magnitudes of error. Since the system will not respond to an error within this region about zero error, the region is called a *dead zone*.

Dead zones are also encountered in vibrating systems (Sec. 5-3) and proportional control systems involving Coulomb damping. However, in this case the control elements are designed to provide a dead zone, and we now have a three-position control system: (1) full power in one direction, (2) zero, and (3) full power in the other direction. Since there are other possibilities for three-position control, we shall add dead zone in parentheses when discussing this particular type of system. The relationship between the output of the control element and the error signal for a three-position (dead zone) controller will be as shown in Fig. 10-15, where Δe_d is the dead zone.

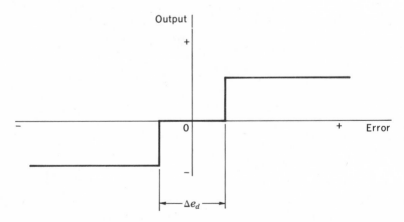

Fig. 10-15   *Input-output characteristic for a three-position (dead-zone) controller.*

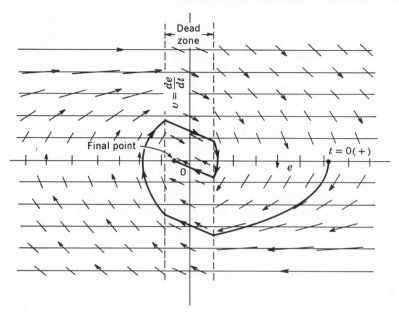

Fig. 10-16 *Phase-plane plot for a system containing a three-position (dead-zone) controller and a second-order controlled system with viscous damping.*

If we consider the second-order controlled system discussed in Secs. 10-2 and 10-3, in which the controlled variable is the position of a mass, we find that Eqs. (10-23) and (10-26) still apply, except within the dead zone where $F_0 = 0$.   In the dead zone we have, from Eq. (10-23) or Eq. (10-26),

$$\frac{dv}{de} = -\frac{a}{m} \tag{10-29}$$

which is constant throughout the dead zone.

If we assume that m, a, F_0, and the step input are the same as for the systems considered in Figs. 10-8 and 10-14, the trajectory in the phase plane will be as shown in Fig. 10-16.   It should be noted that, as for the case of the vibrating system with Coulomb damping in Fig. 5-8, the system remains at rest once the velocity reaches zero within the dead zone.

10-5. Two-position Control with Backlash   In Secs. 10-3 and 10-4 we were concerned primarily with nonlinearities that could be considered to be associated with the error-sensing element and with the response of the power-control element to the error.   The major nonlinearities associated with the controlled element are the presence of Coulomb damping (or its equivalent) and backlash somewhere in the system between the control element and the controlled variable.

In the case of a proportional control system, the effect of Coulomb damping is identical with that discussed in Secs. 5-3 and 5-4 in relation

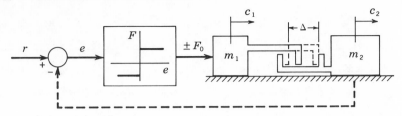

Fig. 10-17 *Schematic diagram of a system containing a two-position controller and a controlled system with backlash.*

to a vibrating system.   The response to a step input will be characterized by (1) a linearly decreasing amplitude of oscillation, and (2) the presence of a dead zone, within which the controlled system will remain at rest because the output from the control elements will not be great enough to overcome the constant magnitude by which the controlled system resists change.

Backlash is a term that we shall apply to any situation in which, upon reversing direction, the input must act for a finite length of time or through a finite distance before it has any effect on the controlled variable. It is most frequently encountered in mechanical systems involving gears and linkages.   The result is a delay or dead region, and, in many respects, its effect is similar to that of hysteresis and dead zone.   However, the relationship between the action of the output of the control element on the controlled system and the error is considerably more complicated, because one consequence of backlash is the impact between bodies.   The result is that not only is the error important, but we must consider the transfer of momentum during impact.

If we consider the case of a two-position control system in which the controlled variable is the position of a mass, we can represent the system schematically as in Fig. 10-17, where Δ is the backlash and $m_1 + m_2$ is the total mass of the controlled system.   As can be seen, the output force from the control element acts directly on m_1, and m_2 is unaffected until m_1 has moved through the backlash distance.   There will be two equations of motion for each direction: one for F_0 acting on m_1 only in the backlash interval and one for F_0 acting on the sum of the masses after m_1 has made contact with m_2.   If there is no damping, the principle of conservation of momentum applies during the impact of m_1 on m_2. If we assume further that the bodies are absolutely rigid and that the impact is inelastic, conservation of momentum requires, for the system in Fig. 10-17,

$$m_1\, Dc_{1,0} + m_2\, Dc_{2,0} = (m_1 + m_2)\, Dc_{2,0(+)} \qquad (10\text{-}30)$$

where the velocities in the left-hand side of the equation are those at the instant of impact and that in the right-hand side is for the two masses moving as one body an instant after impact.

When damping is present, momentum is not conserved during impact, but, in view of other assumptions, the error resulting from ignoring the effect of damping during impact will be negligible for all practical purposes. Therefore, we shall assume Eq. (10-30) is applicable at all times.

If we consider the case where damping is negligible at all times and the bodies are absolutely rigid, the equations of motion become:

(*a*) For *e* positive and $Dc_1 \neq Dc_2$

$$m_1 D^2 c_1 = F_0 \qquad (10\text{-}31)$$
and
$$Dc_2 = \text{const} \qquad (10\text{-}32)$$

(*b*) At impact, from Eq. (10-30)

$$Dc_{2,0(+)} = Dc_{1,0(+)} = \frac{m_1 Dc_{1,0} + m_2 Dc_{2,0}}{m_1 + m_2} \qquad (10\text{-}33)$$

(*c*) For *e* positive and $Dc_1 = Dc_2$

$$(m_1 + m_2) D^2 c_2 = (m_1 + m_2) D^2 c_1 = F_0 \qquad (10\text{-}34)$$

(*d*) For *e* negative and $Dc_1 \neq Dc_2$

$$m_1 D^2 c_1 = -F_0 \qquad (10\text{-}35)$$
and
$$Dc_2 = \text{const} \qquad (10\text{-}36)$$

(*e*) For *e* negative and $Dc_1 = Dc_2$

$$(m_1 + m_2) D^2 c_2 = (m_1 + m_2) D^2 c_1 = -F_0 \qquad (10\text{-}37)$$

The phase plane cannot readily be used to obtain a solution in this case, because the effect of switching cannot be related simply to velocity and error. Since the time required for m_1 to move through the backlash interval relative to m_2 must be considered, the solution can most easily be carried out in the time domain. In this case, the system is piecewise linear, and for negligible damping we need consider only uniformly accelerated and constant-velocity motions and we can arrive at a solution without undue effort without recourse to a computer.

If we consider the system in Fig. 10-17 to be initially at rest in the center of the backlash interval, we shall find the response to a step input

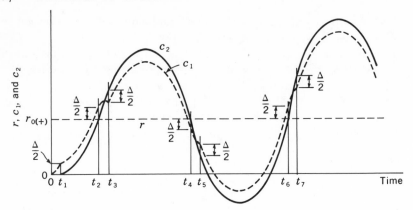

Fig. 10-18 *Response of a system containing a two-position controller and an undamped second-order controlled system with backlash to a step change in reference level.*

will be as shown in Fig. 10-18. As can be seen, the effect of backlash is to introduce a delay in the action of the control element on the controlled system (m_2), and the result is an ever-increasing amplitude of oscillation.

As noted above, this delay is similar to the effect of hysteresis, and the general response characteristics of a system with backlash will be quite similar to, although not identical with, those for a similar system with hysteresis. Therefore, we can expect an ever-increasing amplitude of oscillation, as in Figs. 10-13 and 10-18, if there is no damping, and we can expect a limit cycle, as in Fig. 10-14, if there is damping.

It is important to note that, no matter how small it may be, backlash results in an unstable system if there is zero damping. In a practical case there is always some damping and the amplitude of oscillation will not build up to infinity. However, the amount of damping can be very small and the amplitude of the resulting limit cycle may be too large. Consequently, gears and linkages should have as nearly zero backlash as possible. Methods for eliminating, or at least minimizing, backlash range from requiring extreme precision in all parts down to using spring forces to maintain parts in contact.[1]

---

[1] For discussion related to control of backlash in gears, see J. E. Gibson and F. B. Tuteur, "Control System Components," pp. 328–332, McGraw-Hill Book Company, New York, 1958, and J. G. Truxal (ed.), "Control Engineers' Handbook," pp. 13-27–13-32, McGraw-Hill Book Company, New York, 1958.

THE ANALOG COMPUTER

The basic building block for an analog computer is the *operational amplifier*, consisting of a high-gain dc amplifier, an input impedance Z_i, and a feedback impedance Z_f, as shown in Fig. A-1. The input, grid, and output voltages, e_i, e_g, and e_o, respectively, are measured with respect to ground (not shown). The gain of the amplifier is $K/180°$ or, as indicated in Fig. A-1, simply $-K$. The output voltage will be

$$e_o = -Ke_g \tag{A-1}$$

Summing the voltages, we find

$$e_o = e_i + e_{g/i} + e_{o/g} \tag{A-2}$$

where $e_{g/i}$ is the voltage at the grid relative to the input and $e_{o/g}$ is the voltage at the output relative to the grid. Equation (A-2) can be written as

$$e_o = e_i - i_i Z_i - i_f Z_f \tag{A-3}$$

Summing the currents, we find

$$i_i = i_f + i_g \tag{A-4}$$

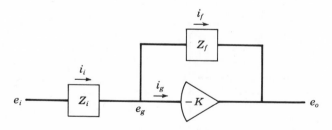

Fig. A-1 *Basic operational amplifier.*

From Fig. A-1, we see that

$$i_f = -\frac{e_{o/g}}{Z_f} = -\frac{e_o - e_g}{Z_f} \tag{A-5}$$

which, in terms of Eq. (A-1), becomes

$$i_f = -\frac{e_o(1 + 1/K)}{Z_f} \tag{A-6}$$

The gain K usually has a magnitude in the order of 10^5 to 10^8, and Eq. (A-6) becomes[1]

$$i_f = -\frac{e_o}{Z_f} \tag{A-7}$$

The grid current is normally $< 10^{-10}$ amp; the impedances usually have magnitudes in the order of 10^5 to 10^7 ohms; and the currents i_i and i_f are in the order of 10^{-3} amp. Therefore, $i_g \ll i_f$, and for practical purposes we can write Eq. (A-4), in terms of Eq. (A-7), as

$$i_i = i_f = i = -\frac{e_o}{Z_f} \tag{A-8}$$

Substituting from Eq. (A-8) into Eq. (A-3) and rearranging leads to

$$e_o = -\frac{Z_f}{Z_i} e_i \tag{A-9}$$

or
$$\frac{e_o}{e_i} = -\frac{Z_f}{Z_i} \tag{A-10}$$

When several inputs are applied through their *individual* input impedances to the grid of the amplifier, as in Fig. A-2, the current through the

[1] Although not too important for present purposes, it should be noted that $e_g \approx 0$.

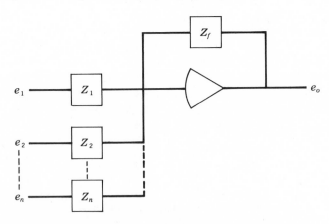

Fig. A-2 *Operational amplifier with multiple inputs.*

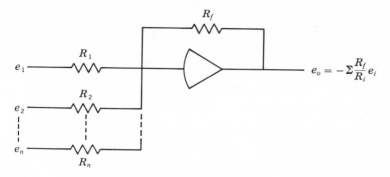

$$e_o = -\Sigma \frac{R_f}{R_i} e_i$$

Fig. A-3 *Summing amplifier.*

feedback impedance is the sum of the currents through the input impedances, and Eq. (A-9) becomes

$$e_o = -\frac{Z_f}{Z_1} e_1 - \frac{Z_f}{Z_2} e_2 - \cdots - \frac{Z_f}{Z_n} e_n = -\sum \frac{Z_f}{Z_i} e_i \qquad \text{(A-11)}$$

Equation (A-10) is the ratio of the output to the input and is, therefore, the transfer function[1] of the operational amplifier.   As such, it will be particularly useful in studying the behavior of control systems in terms of block-by-block (or element-by-element) simulation of the system on the computer.   This approach will be discussed below; but we shall first consider the implications of Eq. (A-11), because it is more generally useful in solving the differential equations associated with vibrations.

When Z_f and Z_i are both resistors, Fig. A-2 becomes Fig. A-3, and Eq. (A-11) becomes

$$e_o = -\sum \frac{R_f}{R_i} e_i \qquad \text{(A-12)}$$

The functions of this operation are (1) multiplication by a constant, (2) addition, and (3) inversion of the signs of the inputs.   To ensure that i_f and i_i are very small but still much greater than i_g, R_i and R_f are specified in megohms; for example, $R = 1$ means $R = 1$ megohm.

When Z_f is a capacitor and Z_i is a resistor, Fig. A-2 becomes Fig. A-4 and Eq. (A-11) becomes

$$e_o = -\sum \int \frac{1}{R_i C_f} e_i \, dt + e_o(0) \qquad \text{(A-13)}$$

The validity of Eq. (A-13) can be shown most readily by considering the case where e_i varies sinusoidally.   In this case, using complex numbers,

[1] See Sec. 7-2 for further discussion of transfer functions.

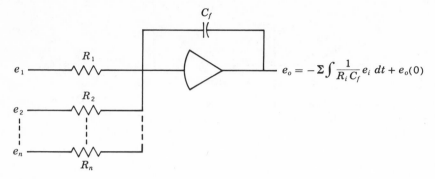

Fig. A-4 *Summing-integrating amplifier.*

we can write

$$e_i = E_i e^{j\omega t} \qquad (A\text{-}14)$$

and

$$Z_f = \frac{1}{j\omega C} \qquad (A\text{-}15)$$

In terms of Eqs. (A-14) and (A-15), Eq. (A-11) becomes, for a single input,

$$e_o = -\frac{1}{j\omega R_i C_f} E_i e^{j\omega t} \qquad (A\text{-}16)$$

which in turn can be shown to be

$$e_o = -\frac{1}{R_i C_f} \int E_i e^{j\omega t} \, dt \qquad (A\text{-}17)$$

In general, the magnitude of $R \times C$ will be about one. Thus, for R in megohms, it is convenient to use C in microfarads; for example, $C = 1$ means $C = 1 \, \mu\text{f}$.

Integration always requires consideration of the initial conditions. Since the grid voltage is practically zero, or grounded, the initial value of e_o will be the voltage across the capacitor at $t = 0$. The circuit in Fig. A-5 shows the components required when the operation is to be integra-

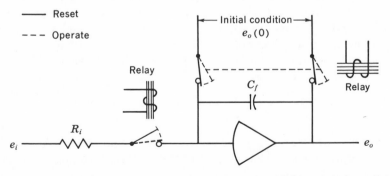

Fig. A-5 *Integrating amplifier showing method for establishing initial conditions.*

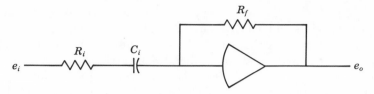

Fig. A-6 *Operational amplifier to provide approximate differentiation.*

tion.   At $t = 0$, the relay-operated switches are in the reset position (solid lines in the figure), the initial-condition voltage $e_o(0)$ is impressed across the capacitor, and the input signal is disconnected from the amplifier.   Then, when the operate button is pushed, the relay-operated switches move to the dashed positions, simultaneously disconnecting the initial-condition voltage and connecting the input signal to the amplifier so that computation begins.

When Z_f is a resistor and Z_i is a capacitor, we find that Eq. (A-11) becomes

$$e_o = -\sum R_f C_i \frac{de_i}{dt} \tag{A-18}$$

and the operation is differentiation.

In general, differentiation is to be avoided whenever possible, because of the problems introduced by noise in the signal and by the overloading of amplifiers when the derivative becomes too great, as for a step function. When differentiation cannot be avoided, the circuit in Fig. A-6 can be used to advantage.   In this case, for a sinusoidal input,

$$Z_i = R_i + \frac{1}{j\omega C_i} = \frac{j\omega R_i C_i + 1}{j\omega C_i} \tag{A-19}$$

and Eq. (A-11) becomes, for a single input,

$$e_o = -\frac{j\omega R_f C_i}{j\omega R_i C_i + 1} e_i \tag{A-20}$$

If we make R_i much less than R_f, for example, $R_i \approx 0.05 R_f$, we can write Eq. (A-20) for low frequencies as

$$e_o \approx -j\omega R_f C_i e_i \tag{A-21}$$

which is the same as

$$e_o \approx -R_f C_i \frac{de_i}{dt} \tag{A-22}$$

For high frequencies, Eq. (A-20) becomes

$$e_o \approx -\frac{R_f}{R_i} e_i \tag{A-23}$$

From Eqs. (A-22) and (A-23) we see that the operational amplifier in Fig. A-6 provides approximate differentiation at low frequencies and avoids the problems associated with rapid changes in the input signal by providing simple multiplication at high frequencies.[1]

Most linear systems can be studied in great detail by using only the operations of summation, multiplication by a constant, and integration. Nonlinear systems require additional equipment, and the reader is referred to the literature for further information.[2]

Additional practical considerations, such as magnitude and time scaling and the use of potentiometers in the input circuit to provide multiplication by values not readily available in terms of resistors and capacitors with fixed values, become important when working with real problems and a real computer. They will be discussed below. However, for the present, let us consider how the basic ideas given above can be applied in deriving computer programs (circuits) for solving the differential equations of a dynamic system and in simulating a control system in terms of transfer functions of the blocks in the block diagram for the system.

A-1. Solving Linear Ordinary Differential Equations   The simplest approach to programming the analog computer is, first, to write the differential equation so that the highest-order derivative equals the sum of all other terms and, then, to select resistors and capacitors to set up the operational amplifiers to provide the sum of the terms that solve the equation.

Example A-1   We are asked to derive a computer circuit that can be used in studying the response of a single-degree-of-freedom vibration system to an arbitrary forcing function $F(t)$.

The equation of motion, Eq. (4-14), is

$$m\,D^2x = -c\,Dx - kx + F(t) \qquad (A\text{-}24)$$

which is rewritten as

$$D^2x = -\frac{c}{m}\,Dx - \frac{k}{m}\,x + \frac{F(t)}{m} \qquad (A\text{-}25)$$

If we assume that D^2x is available, we need only to integrate once to obtain Dx and a second time to obtain x. However, we must keep in mind that the output of an operational amplifier *always* has a sign opposite to that of the input. Consequently,

---

[1] For discussion of a simplified method for determining the frequency response of functions such as in Eq. (A-20), see Sec. 8-2, and for discussion of other methods of approximate differentiation, see A. S. Jackson, "Analog Computation," pp. 145–148, McGraw-Hill Book Company, New York, 1960.

[2] *Ibid.*

C. L. Johnson, "Analog Computer Techniques," 2d ed., McGraw-Hill Book Company, New York, 1963.

G. A. Korn and T. M. Korn, "Electronic Analog and Hybrid Computers," McGraw-Hill Book Company, New York, 1964.

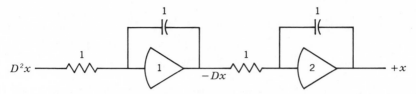

Fig. A-7   *Computer circuit to produce x from D^2x.*

if two integrators are used in series, as in Fig. A-7, the output of the first will be $-Dx$ and the output of the second will be x.   These quantities can be used with the proper values of resistance and capacitance, in combination with the correct number of amplifiers to provide the proper signs, to construct the sum indicated in Eq. (A-25) and shown in Fig. A-8.

At this point, we have D^2x and, as shown by the dashed line, we need only to connect the output of amplifier 3 to the input of amplifier 1 to complete the circuit.   It should be noted that positive values of x, Dx, and D^2x exist at the outputs of amplifiers 2, 4, and 3, respectively.

There are a number of ways of arriving at different programs that will be quite satisfactory.   In many cases, each variation will possess some particular advantage over the circuit in Fig. A-8.   Generally speaking, the advantages are a reduced number of amplifiers and/or a decrease in the variation of the level of the voltages found throughout the circuit.

One basic approach to eliminating one or more amplifiers is to multiply the equation by -1.   If we do this to Eq. (A-25), we find

$$-D^2x = \frac{c}{m} Dx + \frac{k}{m} x - \frac{F(t)}{m} \tag{A-26}$$

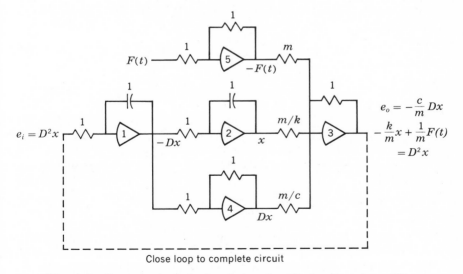

Close loop to complete circuit

Fig. A-8   *Computer circuit for the solution of $m\ D^2x + c\ Dx + kx = F(t)$.*

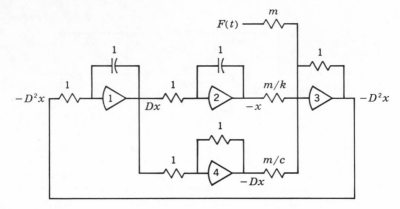

Fig. A-9 *Circuit derived from Fig. A-8 to eliminate one amplifier by using negative quantities for some terms.*

and the computer program becomes that in Fig. A-9. It should be noted that, although Dx is still available (output of amplifier 1) as a positive quantity, we now have only negative values of x and D^2x. Consequently, the user must remember that x and D^2x will actually have signs opposite to those recorded. In this case, it will simplify keeping signs straight if we also record the negative value, rather than the positive value, of Dx. Although eliminating one amplifier in this simple example is not very important, the idea of changing the signs of every term in an equation can be most useful when working with a family of differential equations, as for a system with more than one degree of freedom.

We could have eliminated amplifier 5 in Fig. A-8 by simply letting the forcing function be $-F(t)$ and remembering that the signs of x, Dx, and D^2x would actually be opposite to those recorded.

If we consider simple-harmonic motion, we find that the amplitudes of the velocity

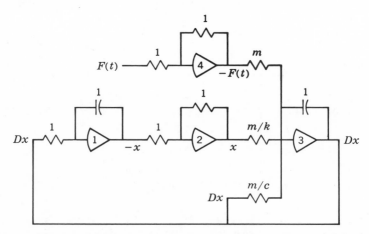

Fig. A-10 *Circuit derived from Fig. A-8 to eliminate D^2x and one amplifier by combining the operations of summing and integration at amplifier 3.*

and acceleration are ω and ω^2, respectively, times the amplitude of the displacement. Consequently, if ω differs greatly from 1 rad/sec, there can be a great difference between the magnitudes of the acceleration and displacement.   In the general case, this must be handled by magnitude scaling, to be discussed in Sec. A-3; but if there is no need for knowing values of acceleration, the problems due to amplitude variations can be minimized by rewriting the differential equation so that the acceleration does not directly appear.   For example, we can rewrite Eq. (A-25) as

$$Dx = \int D^2x \, dt = \int \left[-\frac{c}{m} Dx - \frac{k}{m} x + \frac{F(t)}{m} \right] dt \qquad \text{(A-27)}$$

If we now assume Dx is available, the computer circuit becomes that in Fig. A-10. As above, we can eliminate an additional amplifier by using $-F(t)$ as the forcing function.

A-2. Control System Simulation   As discussed in Secs. 7-2 and 7-5, if the proper impedance mismatch exists between blocks in the system, the ratio of output to input for each block (the transfer function)

Table A-1   Transfer Function Simulation

	Transfer function e_o/e_i	Operational amplifier circuit
1	$-\dfrac{1}{R_iC_fD}$	
2	$-\dfrac{R_f}{R_i} \times \dfrac{1}{1 + R_fC_fD}$	
3	$-\dfrac{1 + R_fC_fD}{R_iC_fD}$	
4	$-\dfrac{R_fC_iD}{1 + R_iC_iD}$	
5	$-\dfrac{C_i}{C_f} \times \dfrac{1 + R_fC_fD}{1 + R_iC_iD}$	

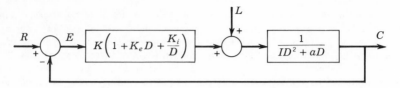

Fig. A-11 *Block diagram for a proportional plus error-rate plus integral control system with a second-order controlled system that is subject to load disturbances.*

is independent of the rest of the system. The simplest way to achieve this degree of isolation is for each block to have a very high input impedance and a very low output impedance. These conditions are adequately met by the operational amplifiers in an analog computer, as shown by the derivation of Eq. (A-10), where the transfer function is simply the negative ratio of the feedback impedance to the input impedance.

The possible combinations of resistors and capacitors are almost unlimited, and only a few of the many circuits available are presented in Table A-1.[1]

Example A-2 We are asked to determine the computer program for simulating the control system whose block diagram is given in Fig. A-11. Since the simple circuits in Table A-1 do not include explicitly any of the transfer functions indicated in the blocks in Fig. A-11, it will be convenient to redraw the diagram with the blocks broken down into sums and products of transfer functions that can be provided, or approximated in the case of the derivative term. Doing this, we find one possibility is the block diagram in Fig. A-12. The transfer functions for the individual blocks

[1] For detailed discussions of this topic and extended tables of amplifier circuits, see:
Jackson, *op. cit.*, chap. 6.
Johnson, *op. cit.*, chap. 4.
G. A. Korn and T. M. Korn, "Electronic Analog and Hybrid Computers," pp. 21-27, McGraw-Hill Book Company, New York, 1964.

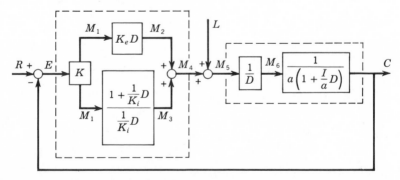

Fig. A-12 *Block diagram in Fig. A-11 broken down into pieces that can be simulated by use of the operational amplifier circuits in Table A-1.*

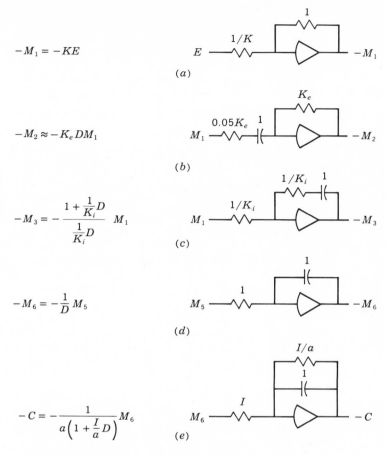

$$-M_1 = -KE$$

(a)

$$-M_2 \approx -K_e D M_1$$

(b)

$$-M_3 = -\frac{1 + \frac{1}{K_i}D}{\frac{1}{K_i}D} M_1$$

(c)

$$-M_6 = -\frac{1}{D} M_5$$

(d)

$$-C = -\frac{1}{a\left(1 + \frac{I}{a}D\right)} M_6$$

(e)

Fig. A-13 *Operational amplifiers to simulate the blocks in Fig. A-12.*

in Fig. A-12 are shown in Fig. A-13, and the final computer program is shown in Fig. A-14.   At the expense of ease in changing parameters and availability for measurement of some terms that could be of interest to the designer, the number of amplifiers required can be reduced significantly by combining operations and/or letting inputs and outputs be negative values, as discussed above.   For example, if one does not need to know the magnitude of the force, M_4, supplied by the control elements or the net force, M_5, on the controlled system, amplifiers 5 and 6 can be eliminated by combining the summing function with the integration at amplifier 7, as shown in Fig. A-15.

A-3. Magnitude Scaling   The object of magnitude scaling is to keep all voltages well within the optimum ranges for computing and recording equipment.   Low levels of voltage are to be avoided because of errors introduced by the ever-present electrical noise.   On the other hand, if

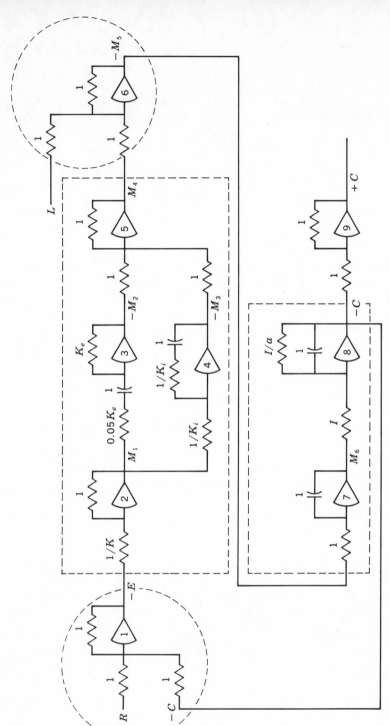

Fig. A-14 *Complete circuit for computer simulation of the system in Fig. A-11.*

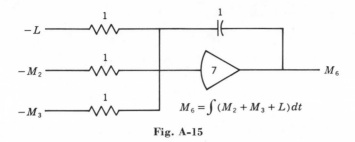

$$M_6 = \int (M_2 + M_3 + L) dt$$

Fig. A-15

the levels become too high, the amplifiers will overload and become non-linear. As a general rule, it is recommended that amplitude scale factors be so chosen that each amplifier will be operating near the middle of its usable range.

For complicated systems it is often most convenient, if not necessary, to determine, experimentally, suitable values of magnitude scale factors by trying different values on the computer until satisfactory operation is obtained. However, in many cases, some knowledge of the physical characteristics of the system and/or study of the equations will permit estimates of magnitudes that will be reasonably close to reality, so that only minor corrections, if any, will need to be made while operating the computer.

Since, for linear systems, amplitude scaling involves only multiplication by a constant, there is little point in setting up a formal procedure—it becomes simply a matter of keeping track of the scale factors on both the computer and the recorder. The procedure will be illustrated in the example below.

Example A-3   A vibration system is described by

$$5 D^2 x + 800 Dx + 18{,}000x = F(t) \tag{A-28}$$

The computer to be used will perform satisfactorily with voltage variations in the range of ± 100 volts and frequencies not exceeding 500 cps. We are asked to program the computer for investigation of the behavior of the system when the amplitude of vibration is to be limited to a maximum value of about 1.0 in.

The first step is to use the discussion above to arrive at a circuit that is at least theoretically possible. One such circuit is shown in Fig. A-16.

To avoid relying completely on trial-and-error operation of the computer, it is worthwhile to study the system to see what information can be deduced relative to expected maximum values of voltages. Since we expect maximum amplitudes near the natural frequency, or frequencies, of the system, the first step is to determine, or estimate, these frequencies. For systems with several degrees of freedom the lowest natural frequencies will usually be most important, and the discussion in Example 6-1 related to approximating systems can be helpful in estimating the natural frequencies. However, in this simple example, we recognize Eq. (A-28) as a second-order linear ordinary differential equation, and we know we can rewrite it as

$$D^2x + 160\,Dx + 3{,}600x = \tfrac{1}{5}F(t) \tag{A-29}$$

or, more usefully in this case,

$$D^2x + 2\zeta\omega_n\,Dx + \omega_n{}^2x = \tfrac{1}{5}F(t) \tag{A-30}$$

where $\omega_n{}^2 = 3{,}600$ and $2\zeta\omega_n = 160$. Thus,

$$\omega_n = \sqrt{3{,}600} = 60 \text{ rad/sec} \tag{A-31}$$

and we can expect

$$Dx\bigg|_{\text{max}} \approx \omega_n X$$

and

$$D^2x\bigg|_{\text{max}} \approx \omega_n{}^2 X$$

which, for $X = 1.0$ in. and $\omega_n = 60$ rad/sec, become, respectively,

$$Dx\bigg|_{\text{max}} \approx 60 \times 1 = 60 \text{ in./sec}$$

and

$$D^2x\bigg|_{\text{max}} \approx 60^2 \times 1 = 3{,}600 \text{ in./sec}^2$$

If we consider displacement first, we find that, to be close to the middle of the operating range of the computer, the magnitude scale should be such that 50 volts on the computer equals 1 in. of displacement. Thus, we shall specify 50 volts/in. for the displacement scale factor.

Since we are working with a linear system, we find that, if we use a displacement scale of 50 volts/in., the circuit in Fig. A-16 becomes that in Fig. A-17, where all

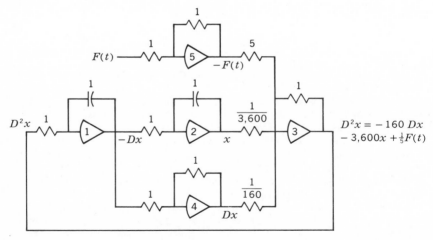

Fig. A-16 *Basic computer circuit for the solution of* $5\,D^2x + 800\,Dx + 18{,}000x = F(t)$.

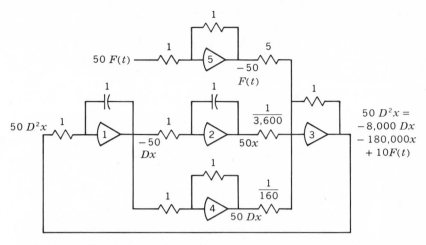

Fig. A-17   *Circuit in Fig. A-16 with a magnitude scale of 50 volts = 1.0 in.*

voltages have magnitudes 50 times those of the quantities to which they are analogous. Showing the scale factors directly on the circuit, as in Fig. A-17, helps to keep things straight.   In terms of the maximum values of x, Dx, and D^2x, we find that, for the circuit in Fig. A-17, we would have to work with the following values of voltage at the amplifier outputs:

Amplifier 2

$$x\Big|_{max} \approx 50 \times 1 = 50 \text{ volts}$$

Amplifier 3

$$D^2x\Big|_{max} \approx 50 \times 3,600 = 180,000 \text{ volts}$$

Amplifiers 1 and 4

$$Dx\Big|_{max} \approx 50 \times 60 = 3,000 \text{ volts}$$

The latter voltages are obviously beyond the 100-volt capacity of the computer, and we must now magnitude-scale the several parts of the circuit.   The output of amplifier 3 is 3,600 times too high, and the outputs of amplifiers 1 and 4 are 60 times too high.   Consequently, we want to select new input and feedback impedances that will provide for division by 3,600 and 60, respectively.   In other words, we want the output of amplifier 3 to be

$$\frac{50}{3,600} D^2x = 0.0139 D^2x$$

and the outputs of amplifiers 1 and 4 to be, respectively,

$$-\tfrac{50}{60} Dx = -0.833 Dx$$

and
$$\tfrac{50}{60} Dx = 0.833 Dx$$

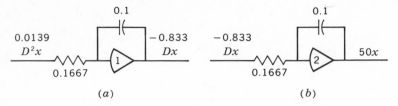

(a)   (b)

Fig. A-18

If we now assume that the value $0.0139\,D^2x$ is available at the input of amplifier 1 and we want to have $-0.833\,Dx$ at the output, we find that the gain of the integrator must be increased from 1 to 60. Consequently, from Eq. (A-13), we want

$$\frac{1}{R_iC_f} = 60$$

or
$$R_iC_f = 0.01667$$

which can be obtained by using $R_i = 0.1667$ and $C_f = 0.1$ to give the operational amplifier in Fig. A-18a.

With $-0.833\,Dx$ as the input to amplifier 2, we find that to have $50x$ as the output, the gain of the integrator must again be 60 and the circuit for operational amplifier 2 becomes that shown in Fig. A-18b. Amplifier 4 will still be a simple inverter and its output will be $0.833\,Dx$.

The output of amplifier 3 is to be

$$0.0139\,D^2x = -2.22\,Dx - 50x + 0.00278F(t) \tag{A-32}$$

The input to amplifier 3 from amplifier 4 is $0.833\,Dx$ and, therefore, the new gain for this part of the circuit must be

$$\frac{2.22}{0.833} = 2.67$$

which, from Eq. (A-12), will be satisfied for $R_f = 1$ if

$$R_i = \frac{1}{2.67} = 0.375$$

The input to amplifier 3 from amplifier 2 is $50x$ and, therefore, the new gain for this part of the circuit must be 1.0, which for $R_f = 1$ requires $R_i = 1$.

If we use the input to amplifier 3 from amplifier 5 as $-50F(t)$, as shown in Fig. A-17, the gain for this part of the circuit will have to be

$$\frac{0.00278}{50} = 0.0000556$$

which is not practical. The problem is that we have not yet considered the possible magnitude of the forcing function. For example, from Eq. (A-28) we see that the spring rate is 18,000 lb/in. Therefore, for a displacement at zero frequency of 1 in., a force of 18,000 lb will be required. A forcing function with an amplitude of $50 \times 18,000$ volts is even more impractical than a gain of 0.0000556. Consequently, another approach is required.

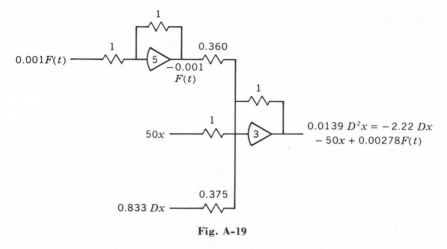

Fig. A-19

One way to handle the forcing function is to designate it as $0.001F(t)$ and then introduce a gain of 2.78 at amplifier 3.   For a gain of 2.78 with $R_f = 1$,

$$R_i = \frac{1}{2.78} = 0.360$$

In terms of the above discussion, the circuit associated with operational amplifiers 5 and 3 becomes that shown in Fig. A-19.

Combining Figs. A-18 and A-19 with the discussion above results in the complete circuit shown in Fig. A-20.   The factors that convert the voltages shown into force- and displacement-time quantities are given in Table A-2.

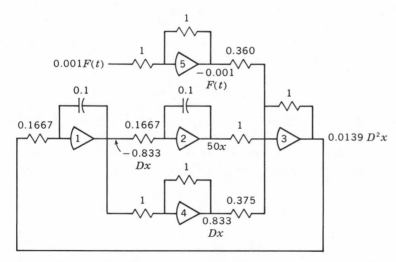

Fig. A-20   *Circuit in Fig. A-16 after magnitude scaling to limit the voltage at all amplifiers to about 50 volts for a maximum displacement of 1.0 in.*

Table A-2

Point in circuit	Function	Conversion factor
Input to amplifier 1	Forcing function	1,000 lb/volt
Output of amplifier 2	Displacement	0.02 in./volt
Output of amplifier 4	Velocity	1.2 in./sec/volt
Output of amplifier 3	Acceleration	72 in./sec²/volt

The remaining practical problems related to frequency limitations of the computing and recording equipment and to avoiding the difficulty of accurately setting up fractional values of resistance will be discussed in the next two sections.

A-4. Time Scaling   When practical, it is simpler and more convenient to use the computer in "real-time" operation, where the times at which events occur on the computer are identical with those for the same events in the real system.   Unfortunately, in many cases the time required to reach a solution in real time may be so great, particularly for process control problems, that it is not economically feasible to tie up the computer for so long, and the errors inherent in integration over a long time interval can result in a solution that has no meaning.   Also, computers and recorders have definite frequency limitations and the frequencies associated with vibrations and control of many systems can be well above the capabilities of the equipment available.   Consequently, we often find it necessary to solve the problem by using a time scale on the computer that is quite different from that of the real system.

There are two approaches for handling time scaling.   One method is to time-scale the problem by rewriting the differential equations with computer time rather than real time as the independent variable, and the other method is to change parameters on the computer so that it operates slower or faster, as the situation may require, than the real system.   Only the latter method will be considered here, because it does not involve changing the original equations and, therefore, (1) the initial conditions and magnitude-scale factors are unaffected, and (2) it can be applied equally well to simulation problems in which we use transfer functions rather than differential equations.

If we use t for real time and τ for computer time, the time-scale factor n is defined by

$$n = \frac{\tau}{t} \tag{A-33}$$

or

$$\tau = nt \tag{A-34}$$

where n can be interpreted as the number of seconds of computer operation corresponding to one second of operation of the actual system in real time.   To see how this can be applied, let us first take a look at the process of integration.

For integration, the operational amplifier is that shown in Fig. A-5 and the relationship of the output to the input is

$$e_o = - \frac{1}{R_i C_f} \int_0^t e_i \, dt + e_o(0) \qquad (A\text{-}35)$$

Substituting from Eq. (A-34) for dt into Eq. (A-35) leads to

$$e_o = - \frac{1}{n R_i C_f} \int_0^\tau e_i \, d\tau + e_o(0) \qquad (A\text{-}36)$$

Comparing Eq. (A-36) with Eq. (A-35) shows that to ensure that e_o at τ is equal to e_o at t we need only to replace the resistor-capacitor combination for the real-time system with one having a product n times as large.

We can arrive at the same conclusion by noting that in Eq. (A-35) the change in output is actually the product of a constant and the integral of the input signal and, therefore, if the value of $R_i C_f$ calculated for the real-time system is doubled, it will take twice as long for the output of the computer to reach a particular level.   In other words, 2 sec of computer time corresponds to 1 sec of real time, and $n = 2$.

For the case of perfect differentiation, the amplifier circuit will be that shown in Fig. A-6 when $R_i = 0$, and from Eq. (A-18),

$$e_o = - R_f C_i \frac{de_i}{dt} \qquad (A\text{-}37)$$

For the values of e_o and e_i to be equal at corresponding values of real time and computer time, Eq. (A-37) must become, in terms of computer time,

$$e_o = - n R_f C_i \frac{de_i}{d\tau} \qquad (A\text{-}38)$$

From a physical viewpoint, we can arrive at Eq. (A-38) by noting that if $n = 2$, the computer is acting half as fast as the real system and thus $de_i/d\tau$ is half as great as de_i/dt and, therefore, to reach the same value of e_o at corresponding values of τ and t, the computer value must be multiplied by 2.   The required multiplication can be accomplished most readily by replacing the real-time values of R_f and C_i with a combination of resistance and capacitance giving a product equal to $n R_f C_i$.

From the above, we conclude that when solving differential equations, the computer can be time-scaled by simply multiplying the real-time values of $R_i C_f$ and $R_f C_i$ by the time-scale factor n.   Time scaling of the computer by multiplying the real-time RC products by the time-scale factor also applies when simulating systems by programming the computer in terms of transfer functions. However, the transfer functions, as in Table A-1, contain different combinations of R and C, and care must be taken to ensure that constant multiplying factors are not changed. This needs to be considered only when values of R are being changed.

Example A-4   The system in Example A-3 is to be studied by use of a computer having an upper frequency limitation of 500 cps. The time-varying signals are to be recorded by use of a direct-writing oscillograph that has a linear response for all frequencies up to 100 cps. We are asked to time-scale the computer to ensure that the maximum frequency of interest is well within the capabilities of the equipment and to make the overall operation as simple as practical.

As in Example A-3, we wish to avoid relying completely on trial-and-error operation of the computer by obtaining as much information as we can from knowledge of the real system. At this time we are most interested in estimating the range of frequencies that must be covered. In general, we shall want to investigate the behavior of the system under forcing functions containing frequencies well above the highest natural frequency of the system.

On the basis of the curves in Figs. 4-20, 6-4, 6-5, 6-12, and 6-13, we can conclude that little additional significant information will be obtained by exciting the system with frequencies greater than four times the highest natural frequency. For the system being considered, we found in Example A-3 that there is one natural frequency, at 60 rad/sec. Therefore,

$$f_n = \frac{\omega_n}{2\pi} = \frac{60}{2\pi} = 9.55 \text{ cps}$$

Consequently, we need not use frequencies above 40 cps, which is well within the capabilities of the computer and recorder. However, when using direct-writing oscillographs, it is desirable to keep the frequencies of the signals being recorded as

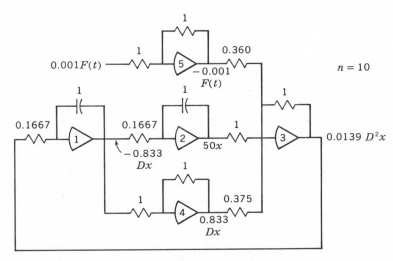

Fig. A-21 *Circuit in Fig. A-20 after time scaling to slow down the solution on the computer so that 10 sec of computer time equals 1 sec of real time.*

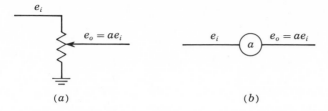

Fig. A-22 *Coefficient potentiometer.* *(a) Schematic circuit; (b) simplified representation.*

low as possible.   A compromise that is reasonable for most analog-computer solutions is to limit the maximum frequency to a few cycles per second.

In this example, slowing the computer down by a factor of 10 will mean that the natural and maximum frequencies will be about 1 and 4 cps, respectively.   These values should be satisfactory, and we shall use, from Eq. (A-33), $n = 10$.

The magnitude-scaled computer program for the investigation in real time is given in Fig. A-20, and the modified circuit after time scaling is given in Fig. A-21.   It should be noted that the only changes are the substitution of 1-μf capacitors for the 0.1-μf capacitors of each integrator, so that the time-scaled values of R_iC_f are 10 times the real-time values.

A-5. Coefficient Potentiometers   Although adjustable resistors can be used to provide fractional values of resistance and circuits derived in the preceding sections can be used, all large—and most small—computers provide facilities that permit programming the computer in a slightly different way that is often more convenient, particularly when solving differential equations where values of resistance and capacitance show up only as simple products or ratios.   The general approach is to use a potentiometer to give a fractional value of a signal, which can then be multiplied as required by combinations of a relatively small number of resistors and capacitors with fixed values.   It should be noted that variable values will still be required to provide the desired complex impedances when using the computer to simulate a system.

The potentiometer in Fig. A-22a is called a coefficient potentiometer, and the accepted symbol is shown in Fig. A-22b.   The ratio of the output to the input is the *coefficient setting a* and is shown in the circle.   The only complication is that $a \leq 1$, and if the multiplication by a number greater than 1 is required, the gain constant of the associated operational amplifier must be specified to provide the overall multiplication required.

The gain constants are normally integers, and the drawing of the computer circuit can be greatly simplified by using the simplified representations of the summing and integrating amplifiers in Fig. A-23a and b, respectively.   The numbers show the gain constants for each input.

Figure A-24 shows the general integrator in both symbolic and component forms. q is a factor by which the coefficient b must be divided to

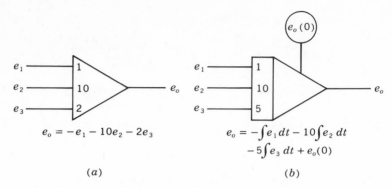

Fig. A-23   *Simplified representation of operational amplifiers.*   *(a) Summing amplifier; (b) summing and integrating amplifier.*

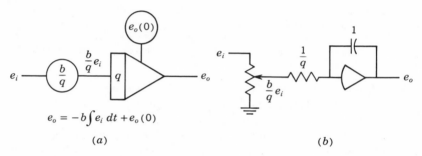

Fig. A-24   *Coefficient potentiometer and operational amplifier for multiplying by a constant greater than 1 and integrating.*   *(a) Simplified representation; (b) computer circuit.*

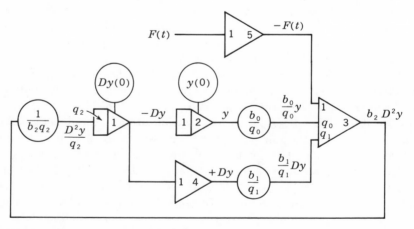

Fig. A-25   *Simplified representation of computer circuit for solving* $b_2 \, D^2y + b_1 \, Dy + b_0y = F(t)$.

make $a \leq 1$. The value of q is not critical, but it is usually convenient to make it a small integer that can be conveniently set as a gain constant on the operational amplifier.

One of the major advantages of using coefficient potentiometers is that each coefficient in the differential equation and the gain constant associated with each block in a system-simulation block diagram can usually be assigned to a single potentiometer, so that the effect of varying any parameter can be investigated simply by changing the setting of the appropriate potentiometer.

The actual procedure for accurately setting the potentiometer varies from computer to computer, and the manufacturer's instruction manual should be consulted for details.

Example A-5   A system is described by

$$b_2 D^2 y + b_1 Dy + b_0 y = F(t) \tag{A-39}$$

We are asked to derive a general computer program for investigating the effect of varying the coefficients b_2, b_1, and b_0 on the response of the system to an arbitrary forcing function.

One of the numerous ways of approaching this problem is to rewrite Eq. (A-39) as

$$b_2 D^2 y = -b_1 Dy - b_0 y + F(t) \tag{A-40}$$

and assume that $b_2 D^2 y$ is available for starting to set up the circuit.   A complete program for this case is shown in Fig. A-25.

appendix **B**

BENDING FORMULAS

L = length of beam, in.
x = distance along beam, in.
y = deflection, in.
M = bending moment, lb-in.
θ = slope of beam, rad
E = modulus of elasticity, psi
I = moment of inertia of section, in.^{4}
w = uniformly distributed load, lb/in.

Beam loading and support	M	y	θ
	$-Px$	$-\dfrac{P}{6EI}(x^3 - 3L^2x + 2L^3)$	$-\dfrac{PL^2}{2EI}$
	$-\dfrac{wx^2}{2}$	$-\dfrac{w}{24EI}(x^4 - 4L^3x + 3L^4)$	$-\dfrac{wL^3}{6EI}$

(diagram)	M	Deflection	Slope
	In length a $+\dfrac{Pbx}{L}$ In length b $+\dfrac{Pa}{L}(L-x)$	In length a $-\dfrac{Pbx}{6EIL}(L^2 - b^2 - x^2)$ In length b $-\dfrac{Pa(L-x)}{6EIL}[2Lb - b^2 - (L-x)^2]$	$\theta_1 = -\dfrac{P}{6EI}\left(bL - \dfrac{b^3}{L}\right)$ $\theta_2 = \dfrac{P}{6EI}\left(2bL + \dfrac{b^3}{L} - 3b^2\right)$
	$\dfrac{wL}{2}\left(x - \dfrac{x^2}{L}\right)$	$-\dfrac{wx}{24EI}(L^3 - 2Lx^2 + x^3)$	$\theta_1 = -\dfrac{wL^3}{24EI}$ $\theta_2 = -\theta_1$
	$M_1 = -\dfrac{Pab^2}{L^2}$ $M_2 = -\dfrac{Pa^2b}{L^2}$	In length a $-\dfrac{Pb^2x^2}{6EIL^3}(3aL - 3ax - bx)$ In length b $-\dfrac{Pa^2(L-x)^2}{6EIL^3}[3bL - (3b + a)(L - x)]$	
	$M_1 = -\dfrac{wL^2}{12}$ $M_2 = M_1$	$y = -\dfrac{wx^2}{24EI}(L^2 + x^2 - 2Lx)$	

appendix C

HELICAL COMPRESSION SPRINGS
UNDER COMBINED AXIAL AND LATERAL LOADS

The spring rate for a helical spring when the load is applied along the spring axis is shown in books on strength of materials and mechanical design to be

$$k = \frac{Gd^4}{8D^3N} \qquad \text{lb/in.} \qquad \text{(C-1)}$$

where G = shear modulus of elasticity, about 11,500,000 psi for steel
  d = wire diameter, in.
  D = mean coil diameter, in.
  N = number of active coils

The relationship between the active number of coils and the total number of coils N_t depends upon the type of end construction of the springs. For most purposes we can assume for springs with (1) plain ends, $N = N_t - 0.5$, (2) plain ends ground, $N = N_t - 1.0$, and (3) squared ends, $N = N_t - 1.75$.

In addition to the parameters in Eq. (C-1), the spring rate for a helical spring under a lateral load, i.e., a load acting perpendicularly to the spring axis, is a function of the manner in which the ends of the spring are held and the magnitude of the axial load present, if any.

To minimize the possibility of instability (buckling of the springs, which would result in collapse of the system), practically all helical compression springs used for vibration isolators will have fixed ends, i.e., the ends will be restrained from rotating. Also, although the lateral spring rate varies with the magnitude of the force acting in the axial direction and, consequently, with the instantaneous magnitude of deflection, useful results will be obtained by considering the lateral spring rate to be corresponding to the equilibrium (static) position of the suspension system.

390

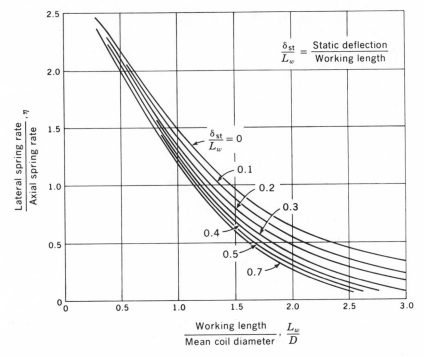

Fig. C-1 *Ratio of the lateral to the axial spring rate as a function of the ratio of the working length to the mean coil diameter for a helical compression spring with fixed ends.* (After C. E. Crede and J. P. Walsh, The Design of Vibration-isolating Bases for Machinery, *Trans. ASME*, vol. 69, pp. A7–14, 1947, and A. M. Wahl, "Mechanical Springs," 2d ed., pp. 70–72, McGraw-Hill Book Company, New York, 1963.)

The equation for calculating the lateral spring rate is considerably more complicated than Eq. (C-1) for the axial spring rate. However, as discussed in Sec. 6-6, we are interested primarily in the ratio η of the lateral to the axial spring rates, and the curves in Fig. C-1 permit its determination with a minimum of calculations.

The effect of the lateral force on the maximum stress in the spring can usually be neglected when designing a helical compression spring for use as a vibration isolator, because the stress due to the axial force is normally much more important, since it includes the effect of the deadweight load. If conditions are such that the effect of the lateral load cannot be neglected, Wahl[1] recommends that it be included by multiplying the stress, calculated in the usual manner for an axial load, by a factor C, which is defined as

[1] A. M. Wahl, "Mechanical Springs," 2d ed., p. 284, McGraw-Hill Book Company, New York, 1963.

$$C = 1 + \frac{\delta_l}{D} + \frac{FL_w}{PD} \tag{C-2}$$

where δ_l = lateral deflection, in.

F = lateral force, lb

L_w = working length of spring, in.

P = axial force, lb

PROBLEMS

Chapter 1.   Rigid Bodies

1-1. The success of many investigations in space is dependent on a satellite maintaining the correct attitude with respect to the sun or another object in space.   The initial orientation after the satellite is put into orbit is usually achieved by use of gas jets.   However, for indefinite operation a more practical source of energy is the sun.   Solar cells are used to convert radiant energy into electric energy, and batteries are used to store the energy for delivery when needed.   The problem then becomes how to utilize electric energy in changing the attitude, or angular orientation, of an object in space.

According to Newton's first law, a particle or a system of particles will continue in motion at a constant speed in a straight line unless acted upon by an external force.   The major consequence in space is that the momentum of the body or system of particles remains constant, and we find that the center of gravity of the system continues to move at the same velocity regardless of the forces that may act *within* the system.

If we apply the above reasoning to angular motion, we again find that in the absence of an external torque the angular momentum of a body or a system of particles will also remain constant.   Consequently, if one member within the system exerts a torque on another member, the reaction torque will cause the first member to accelerate in a direction opposite to the second.

In practice, this effect is utilized in attitude control by using an electric motor driving a flywheel (called a reaction wheel) for each of three orthogonal axes of the satellite.   The details of the control system cannot be considered here, but we can note that one possibility is the use of three-position control (see Sec. 10-4), in which the motor is either off or exerting full torque in either direction of rotation.

A new satellite is being designed.   Preliminary studies indicate that the satellite will weigh about 600 lb and the radii of gyration about the three orthogonal axes x, y, and z will be $\rho_x = 15$ in., $\rho_y = 25$ in., and $\rho_z = 50$ in.   The motor being considered for this application has a maximum torque output of 0.64 oz-in., and the moment of inertia of its rotor is 3.6×10^{-4} lb-in.-sec^2.

For rotation about the $x(y,z)$ axis, and neglecting the effects of minor friction forces, determine (a) the angular acceleration of the satellite in degrees per second per second, (b) the minimum time required for the satellite to rotate from at-rest through an angle of $0.5°$, (c) the weight required for the reaction wheel if its radius of gyration is to be 2.0 in. and its speed is not to change by more than 4,000 rpm during the maneuver described in (b), and (d) the maximum rate at which power must be supplied by the battery if the overall efficiency of the drive is 75 percent and the top operating speed of the motor is 5,000 rpm.

1-2. The use of reaction wheels to control the attitude of space vehicles is discussed in Prob. 1-1.   Additional design parameters of importance to this problem are (1) the

393

maximum motor speed will be limited to 5,000 rpm, (2) the control circuit will permit the motor to deliver full torque at all speeds up to 5,000 rpm in either direction, (3) the mass moment of inertia of the motor rotor plus the reaction wheel will be 1.18×10^{-3} lb-in.-sec^2, and (4) the control circuit will supply full power to drive the reaction wheel in the negative direction, as long as the attitude angle is more negative than $-0.3°$ and the speed is less than 5,000 rpm, and full power to drive the reaction wheel in the positive direction, as long as the attitude angle is more positive than $+0.3°$ and the speed is less than 5,000 rpm. The interval between -0.3 and $+0.3°$ for attitude angle is called a dead zone because there can never be any power supplied to the motor in that region.

Assuming that friction losses are negligible and that the attitude of the satellite is $+3.0°$ about the $x(y,z)$ axis at a time $(t = 0)$ when both the satellite and the reaction wheel have zero angular velocity, determine (a) the maximum angular velocity of the satellite, (b) the time at which the attitude angle first equals zero, (c) the time at which the torque reverses direction, and (d) the maximum value of the attitude angle, in degrees, reached in the negative direction.

1-3. Same as Prob. 1-2 except at $t = 0$ the attitude angle is $+0.2°$, the reaction wheel is rotating at a speed of 3,000 rpm in the negative direction, and the satellite is rotating at a speed of $0.023°/\text{sec}$ in the positive direction.

1-4. A disk cam and a radial roller follower are being designed to provide a maximum rise of 0.500 in. and the following motion characteristics: rise with uniformly accelerated motion in 108° of cam rotation, dwell for 50° of cam rotation, return with uniformly accelerated motion in 100° of cam rotation, and dwell during the remaining 102° of cam rotation during each cycle. Neglecting friction and assuming that the maximum force in the spring is to be 40 percent greater than the minimum, (a) what value of spring rate $(k = P/\delta)$ do you recommend and (b) what should be the initial deflection of the spring if the 1.67-lb follower is to remain in contact with the cam at all times when the cam is rotating at 2,000 rpm?

1-5. Same as Prob. 1-4 except the rise and return motions are to be simple-harmonic.

1-6. Same as Prob. 1-4 except the follower weighs 39.0 lb and the speed of rotation is 300 rpm.

1-7. Same as Prob. 1-4 except the rise is to take place in 90° and the return is to take place in 118° of cam rotation.

1-8. (a) Apply the concept of the equivalent offset inertia force to the link in Fig. P-1 to show that the offset force always intersects the line of centers at the same point H. Will this be true for links that are not rotating about a fixed center? Why?

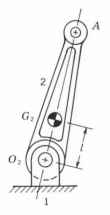

Fig. P-1

(b) Point H is the center of percussion for the link rotating about O_2. Explain how the above statement is consistent, or how it is inconsistent, with the definition encountered in previous studies in physics and mechanics.

1-9. Link 2 in Fig. P-2 is rotating in the clockwise direction at a constant speed of 100 rpm, and $F_A = 90$ lb. Determine the bearing and torque reactions F_{12} and T_{12}, respectively, that are required for equilibrium.

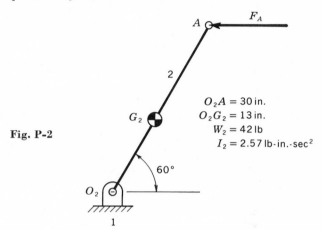

$O_2A = 30$ in.
$O_2G_2 = 13$ in.
$W_2 = 42$ lb
$I_2 = 2.57$ lb-in.-sec^2

Fig. P-2

1-10. Assuming link 2 in Fig. P-2 is keyed to the shaft of a gear reducer at O_2, what is the instantaneous magnitude of the power being transmitted between the gear reducer and the linkage for the conditions in Prob. 1-9? In which direction is the power being transmitted?

1-11. Same as Prob. 1-9 except, also, link 2 is accelerating in the counterclockwise direction at the rate of 650 rpm/sec.

1-12. Same as Prob. 1-10 except apply to Prob. 1-11.

1-13. Same as Prob. 1-9 except, also, link 2 is accelerating in the clockwise direction at the rate of 650 rpm/sec.

1-14. Same as Prob. 1-10 except apply to Prob. 1-13.

1-15. Link 2 in Fig. P-2 is acted on by a force F_A of 125 lb and a torque T_{12} of 2,100 lb-in. in the clockwise direction. If the instantaneous angular velocity is 80 rpm in the counterclockwise direction and we neglect the effects of acceleration on the force and torque, what are the magnitudes and directions of (a) the angular acceleration of link 2 and (b) the bearing reaction F_{21}?

Note: The solution will require several iterations unless the work is simplified by making use of the solution to Prob. 1-8, which for this problem means that the equivalent offset force will always pass through the point H. H will be located between G_2 and A at a distance $G_2H = \rho^2/O_2G_2$, where ρ is the radius of gyration about the center of gravity. Since the component of inertia force due to the normal acceleration is independent of the angular acceleration and acts along the link, we conclude that the component due to the angular acceleration must pass through H and only it will contribute to the inertia torque.

1-16. Same as Prob. 1-15 except that the magnitude of T_{12} is 4,000 lb-in.

1-17. At the instant considered in Example 1-1, a load torque T_{14} is acting on link 4.

If T_{14} is 3,000 lb-in. in the clockwise (counterclockwise) direction, what must be the magnitude and direction of the resultant external torque T_{12} on link 2?

1-18. Same as Prob. 1-17 except determine the resultant bearing reactions at B and O_4.

1-19. Same as Prob. 1-17 except determine the resultant bearing reactions at A and O_2.

1-20. Same as Prob. 1-17 except also determine all of the resultant bearing reactions.

1-21. At the instant considered in Example 1-1, a horizontal force of 150 lb acts to the left (right) at B. Determine the magnitude and direction of the resultant external torque T_{12} required for equilibrium.

1-22. Same as Prob. 1-21 except determine the resultant bearing reactions at O_4 and B.

1-23. Same as Prob. 1-21 except determine the resultant bearing reactions at O_2 and A.

1-24. Same as Prob. 1-21 except also determine all of the resultant bearing reactions.

1-25. Link 4 of the linkage in Example 1-1 has been counterbalanced by the addition of mass diametrically opposite G_4. The center of gravity is now located at the center of rotation O_4 and the mass moment of inertia about the new center of gravity is 1.465 lb-in.-sec^2. Determine the bearing reactions, the shaking force, and the external torque required on link 2 if all other conditions are identical with those in Example 1-1.

1-26. Same as Prob. 1-25 except include a load torque of 4,000 lb-in. acting in the clockwise (counterclockwise) direction on link 4.

1-27. Determine the bearing reactions, the shaking force, and the external torque required on link 2 for the linkage in Example 1-1 when the crank angle has increased to 150° while the angular velocity of link 2 remains constant.

1-28. Same as Prob. 1-27 except the crank angle has increased to 240°.

1-29. How much weight would have to be added at a point 2.00 in. to the left of A on the connecting rod in Fig. 1-12a to make the statically equivalent connecting rod identical with the kinetically equivalent connecting rod?

1-30. The connecting rod in Fig. 1-12 is to be used with a piston weighing 1.312 lb in an engine that has a stroke of 3.00 in. Assuming that the crankshaft is counterbalanced by itself, the engine speed is 4,400 rpm, and the crank angle (measured from head-end dead center) is 30° (45°, 60°, 90°, 180°), determine the shaking force for a single cylinder by using (a) the kinetically equivalent connecting rod in Fig. 1-12b and (b) the statically equivalent connecting rod in Fig. 1-13.

1-31. The connecting rod in Fig. 1-12 is to be used with a piston weighing 1.312 lb in an engine that has a stroke of 3.25 in. Assuming that the crankshaft is counterbalanced to eliminate the rotating inertia force, what will be the magnitudes of the first three harmonics of the shaking force, per cylinder, when the engine is running at 4,000 rpm? What error will be introduced by using the secondary shaking force in place of the second harmonic?

1-32. A punch press is being designed for use in an operation that requires delivery of 65,000 ft-lb of energy. It is proposed that a flywheel be used to supply the major share of the energy by engaging a friction clutch between the flywheel and the toggle linkage of the punch. The flywheel is to be connected at all times to a three-phase, high-slip (Design-D) electric motor through a gear-reduction unit and a belt drive. The motor will have a nominal (synchronous) speed of 1,200 rpm and the torque-speed curve in Fig. P-3. The flywheel is to be rotating at 60 rpm at the time the clutch is engaged. The desired operation frequency is 10 cpm, and the useful delivery of energy is to occur during 45° rotation of the flywheel. It is further desired that the motor speed decrease not more than 10 percent during a cycle.

A rigorous analysis of this problem would require more information and a point-by-point solution. However, as a first approximation, we can assume that all of the punching energy comes from the kinetic energy of the rotating members, the inertia of everything except the flywheel can be neglected, friction losses are negligible, and the motor torque can be considered to be a constant value equal to that corresponding to the speed halfway between the extreme values in a cycle. Using these approximations, and assuming that the motor is running at 95 percent of synchronous speed each time the clutch is engaged, what do you recommend for (a) the moment of inertia of the flywheel and (b) the rated horsepower of the motor?

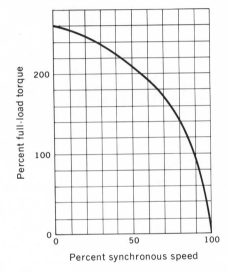

Fig. P-3 *Torque-speed curve for Design-D electric motors.*

1-33. Same as Prob. 1-32 except there are to be 30 cycles of operation each minute.

1-34. The immediate source of energy for many machines is an electric motor. For economic reasons, with 60-cps alternating current, the most common nominal (synchronous) speed is 1,800 rpm—with 1,200 and 3,600 the next most common speeds. Since these speeds, or the somewhat lower actual speeds for most motors, seldom coincide with the speeds of rotation required in the machine, means for reducing or increasing the speed must usually be provided. In most cases, the required machine speed will be lower than the motor speed. Gear units, as in Fig. P-4a, belt drives, or chain drives are commonly used to provide the change in speed.

In many cases where the need for flywheel effect is very great, as for the punch press in Prob. 1-32, the flywheel will be so large that the inertia of the other members can be ignored without introducing serious error. However, in many other cases, the inertia of all parts must be considered, and it is often most convenient to convert the actual system with shafts rotating at different speeds to an equivalent system with all inertias rotating at the same speed.

When torsional vibrations of the system are being considered, as in Example 6-1, the equivalent system must take into account shaft stiffnesses as well as inertia of rotors, and we could use the system in Fig. P-4b, where the equivalent system rotates at the speed of the motor, or the system in Fig. P-4c, where the equivalent system rotates at the speed of the machine. When the relationship between shaft stiffnesses and

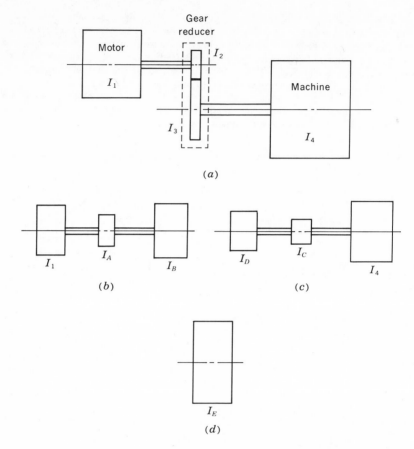

Fig. P-4

inertias is such that vibrations need not be considered, as in many control systems, we can simply lump all of the inertias together, as in Fig. P-4d.

Considering the system in Fig. P-4a and letting r equal the ratio of the output speed to the input speed of the gear reducer, prove that the inertia of a kinetically equivalent system with a single rotor will be (a) at the input-shaft speed (inertia reflected to the input shaft),

$$I_E = I_1 + I_2 + r^2 I_3 + r^2 I_4$$

and (b) at the output-shaft speed (inertia reflected to the driven shaft),

$$I_E = I_4 + I_3 + \frac{1}{r^2} I_2 + \frac{1}{r^2} I_1$$

1-35. The system in Fig. P-5 consists of an internal-combustion engine, a double-reduction gear unit, and a driven machine. For proper operation, the speed of the driven machine must not vary by more than 0.02 percent during a revolution. Measurements made by use of torsional accelerometers and associated equipment (Sec. 4-22) show that the cyclic variation in speed is 0.122 rpm when the machine is running at its nominal speed of 250 rpm.

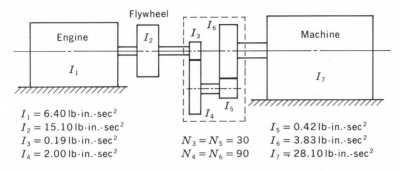

$$I_1 = 6.40\ \text{lb-in.-sec}^2$$
$$I_2 = 15.10\ \text{lb-in.-sec}^2$$
$$I_3 = 0.19\ \text{lb-in.-sec}^2 \qquad N_3 = N_5 = 30 \qquad I_5 = 0.42\ \text{lb-in.-sec}^2$$
$$I_4 = 2.00\ \text{lb-in.-sec}^2 \qquad N_4 = N_6 = 90 \qquad I_6 = 3.83\ \text{lb-in.-sec}^2$$
$$I_7 = 28.10\ \text{lb-in.-sec}^2$$

Fig. P-5

Two solutions are proposed. One is to add a flywheel to the shaft of the driven machine and the other is to increase the size of the engine flywheel. Further study brings forth the information that the engine is a 4-cylinder diesel engine and the load-torque variations on the machine are negligible. The chief engineer then states that the correction will be made by replacing the present flywheel with one having the proper mass moment of inertia. (*a*) Using the results from Prob. 1-34 for calculating equivalent inertias, determine the new value of I_2 required to meet the specifications. (*b*) What are the factors upon which the chief engineer based his decision to replace the present flywheel rather than add a new flywheel to the shaft of the driven machine?

1-36. The motor for the system in Fig. P-6*a* is a 20-hp Design-D (high-slip) motor with a synchronous speed of 1,800 rpm and the torque-speed characteristics shown in Fig. P-3. The torque load variation on the driven machine can be approximated by the torque-angle curve for one revolution in Fig. P-6*b*. Specifications state that the speed of rotation of the driven machine must not vary more than 0.05 percent during a revolution. Preliminary calculations indicate that this cannot be done without adding flywheel effect.

Two solutions are proposed. One is to add a flywheel in the dashed position *A* and the other is to add a flywheel in the dashed position *B*. The chief engineer looks at the situation and says that the flywheel will be added at *B*. (*a*) Using the results from Prob. 1-34 for calculating equivalent inertias, determine the value of mass moment of inertia required for a flywheel at *B*. (*b*) What are the factors upon which the chief engineer based his conclusion? Do you agree with his conclusion? (*c*) Is the specified size of motor appropriate for this application?

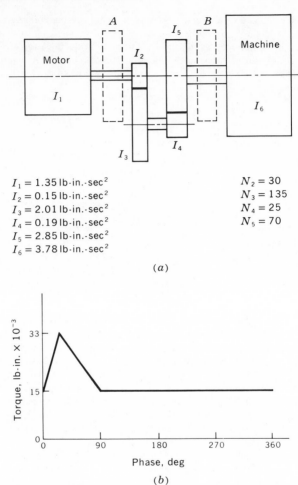

$I_1 = 1.35\,\text{lb-in.-sec}^2$ $N_2 = 30$
$I_2 = 0.15\,\text{lb-in.-sec}^2$ $N_3 = 135$
$I_3 = 2.01\,\text{lb-in.-sec}^2$ $N_4 = 25$
$I_4 = 0.19\,\text{lb-in.-sec}^2$ $N_5 = 70$
$I_5 = 2.85\,\text{lb-in.-sec}^2$
$I_6 = 3.78\,\text{lb-in.-sec}^2$

(*a*)

(*b*)

Fig. P-6

1-37. The torque-crank-angle diagram in Fig. P-7 is a straight-line approximation of the true curve, such as in Fig. 1-19, for a single cylinder of a large diesel engine running at a speed of 1,200 rpm. Considering its possible use as a single-cylinder engine, (*a*) what would be the horsepower rating and (*b*) what total mass moment of inertia (flywheel plus load plus crankshaft plus equivalent inertia of connecting rod, etc.)

would be required to ensure a speed variation of less than 0.10 percent during each revolution?

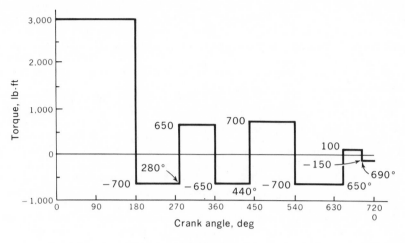

Fig. P-7

1-38. Same as Prob. 1-37 except two cylinders are to be used with equally spaced power strokes.

1-39. Same as Prob. 1-37 except six cylinders are to be used with equally spaced power strokes.

1-40. A 200-hp 1,750-rpm electric motor is installed in the cab of a large power shovel. The motor axis is horizontal, and the rotor inertia is given as $WR^2 = 104.5$ lb-ft^2, where W is the weight and R is the radius of gyration. The rotor bearings are 40.0 in. apart, and the motor shaft is connected to the driven member by use of a flexible coupling. If the maximum slewing speed of the shovel is 6°/sec, what will be the additional bearing loads due to the gyroscopic effect?

1-41. Force transducers are usually devices that convert force into an electrical signal, which can then be measured, recorded, or integrated or become a signal in a control system. Since precession of a gyroscope requires the action of a couple, measurement of the forces making up the couple will provide a direct measurement of the rate of precession. Such a device, called a rate gyro, can be a vital component in fire control and inertial guidance systems.

A rate gyro is to be designed to have a sensitivity such that an angular velocity of 0.005°/sec will provide an electrical signal of 0.010 volt. It is proposed to use a rotor with a mass moment of inertia about its axis of rotation of 0.105 lb-in.-sec^2 running at a speed of 20,000 rpm. The minimum distance between the transducers is 1.50 in. (a) Sketch the arrangement you would recommend for measuring the velocity of rotation about a vertical axis. (b) What is the minimum permissible sensitivity of the transducer and associated circuitry in terms of volts output per pound force on the transducer?

1-42. One of the steam turbines on board a ship has a rotor that weighs 50 tons. The turbine runs at 11,000 rpm and is mounted with its axis of rotation parallel to the hull of the ship. The main bearings are 186.0 in. apart, and the radius of gyration of the rotor is 9.75 in.   Determine the direction and magnitudes of the maximum bearing reactions due to the gyroscopic effect when the ship is pitching through an angle of $\pm 5°$ at a rate of one cycle in 15 sec.   Assume that the pitching is simple-harmonic and that when viewed from the rear the turbine is rotating in the clockwise direction.

1-43. A locomotive is being designed for high-speed freight service.   The maximum speed will be 85 mph, and a 6° curve (radius to track center line of 955.4 ft) will be the sharpest curve it must take at maximum speed.   The drive system will consist of a 5,000-hp gas turbine, a reduction-gear unit, a dc generator, and a motor for each of eight driving axles.

The axial-flow-compressor rotor weighs 3,500 lb, has a radius of gyration of 7.30 in., is supported on bearings that are 55 in. apart, and will have a maximum speed of 13,000 rpm.   The generator rotor weighs 14,500 lb, has a radius of gyration of 12.58 in., is supported on bearings that are 88 in. apart, and will have a maximum speed of 3,000 rpm.   The axes of rotation are to be parallel to the track center line.   (a) What will be the maximum reactions for the compressor and generator bearings resulting from the gyroscopic effect?   (b) The gear-reduction unit can be designed to provide the same or opposite directions of rotation of the turbine and generator rotors. What would be your recommendations in this matter?

1-44. A gas-turbine-powered racing car is being designed for use on tracks, such as the Indianapolis Speedway, where there are only left turns.   Considering only the gyroscopic effect, will there be a preferred and/or a least desirable orientation of the rotor axis and direction of rotation?

1-45. Same as Prob. 1-44 except the car is to be used in road racing over rough terrain.

1-46. (a) Without making another complete analysis, what spring rate would you suggest trying in order to decrease the response time in Example 1-3 to 0.003 sec? (b) What assumptions and approximations were involved in arriving at your answer to (a)?

1-47. The concept of an equivalent link was found in Sec. 1-6 to be useful when determining the inertia forces associated with the slider-crank mechanism.   When studying the response of a system to a specified energy input, as in Sec. 1-9, repetitive calculations can be simplified considerably by finding an equivalent system that consists of a single member that can move only in pure translation or rotation.   Using the data presented in the figures in Example 1-3, determine the equivalent mass at point A as a function of its position.

1-48. Previous calculations have shown that the inertia of the linkage of a certain piece of switchgear can be approximated by a single member with a mass of 0.38 lb-sec²/in. moving in a straight line.   The energy-storage element is a helical compression spring.   Operating specifications call for opening the circuit within $\frac{1}{200}$ sec after the release is tripped.   During this interval, one end of the spring moves 1.00 in.   Assuming that the initial deflection of the spring is 1.00 in., what is your estimate of the spring rate required in this application?

1-49. Same as Prob. 1-48 except the spring is initially compressed 2.00 in.

Chapter 2.   Balancing of Machinery

2-1. The system in Fig. P-8 is to be balanced by adding mass in planes L and R at radii of 3.0 in.   Determine the magnitudes and locations of the required corrections.

2-2. The system in Fig. P-8 is to be balanced by removing mass in planes L and R at radii of 3.5 and 2.5 in., respectively.   Determine the magnitudes and locations of the required corrections.

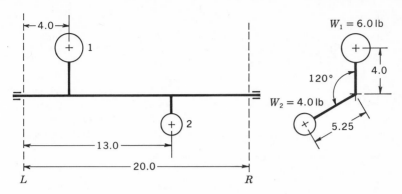

Fig. P-8

2-3. The system in Fig. P-8 is to be balanced by adding mass in the L plane at a radius of 3.0 in. and removing mass in the R plane at a radius of 3.5 in.   Determine the magnitudes and locations of the required corrections.

2-4. Same as Prob. 2-1 except apply to Fig. P-9.
2-5. Same as Prob. 2-2 except apply to Fig. P-9.
2-6. Same as Prob. 2-3 except apply to Fig. P-9.

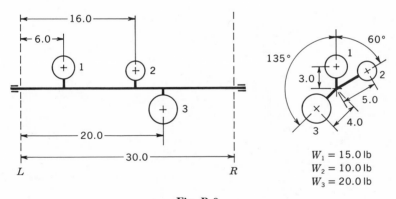

Fig. P-9

2-7. Same as Prob. 2-1 except apply to Fig. P-10.
2-8. Same as Prob. 2-2 except apply to Fig. P-10.
2-9. Same as Prob. 2-3 except apply to Fig. P-10.

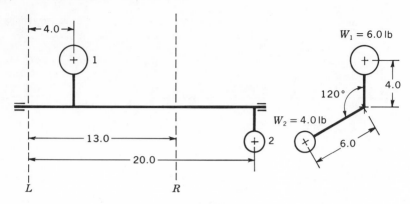

Fig. P-10

2-10. The gear in Fig. P-11 has been balanced in the shop by adding weights, as shown, at identical radii of 45 in. The gears are cast steel, and specifications state that balancing corrections can be made only by boring holes in the $1\frac{1}{2}$-in.-thick side plates. Assuming that the holes in the side plates are to be bored on a radius of 70 in., (a) determine the locations and diameters of the holes required to balance the gear (density of steel is 0.283 lb/in.³) and (b) recommend the tolerance band for the

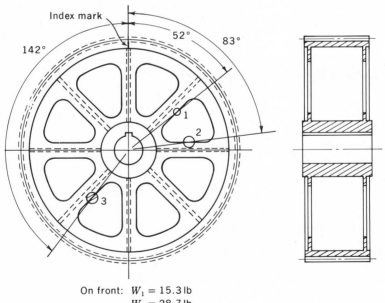

On front: $W_1 = 15.3$ lb
$W_2 = 28.7$ lb
On back: $W_3 = 22.5$ lb

Fig. P-11

weight of material to be removed if the shaking force of all similar gears is not to exceed 50 lb when running at a speed of 100 rpm.

2-11. The need has developed for a force generator that will provide a sinusoidally varying force with an amplitude of 500 lb at a frequency of 600 cpm. It is proposed to attach pieces of steel plate to the sides of two available gears with 15-in. pitch diameters to construct a device similar to that illustrated in Fig. 2-15. If the weights are to be contained within a diameter of 14 in., what is the minimum thickness of plate that could be used?

2-12. Figure P-12 is a schematic drawing of a proposed high-speed, lightweight two-stage double-acting air compressor. The compressor has a stroke of 1.000 in.

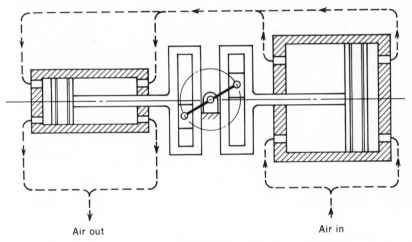

Air out   Air in

Fig. P-12

and is to be driven at 4,800 rpm by a 2-cycle gasoline engine. The cranks are offset 2.25 in. axially, and the rotating inertia forces have been balanced completely. The total weight of the reciprocating parts associated with the low-pressure cylinder is 0.535 lb and with the high-pressure cylinder 0.364 lb.

It is proposed to balance the resulting reciprocating inertia force and couple by using a pair of contrarotating weights, as in Fig. 2-15. (*a*) Where should the weights be located? (*b*) What dimensions do you recommend for the counterweights if they are made from brass (density of 534 lb/ft³) and each must remain within a diameter of 1.25 in.? (*c*) To what accuracy in ounce-inches must final balance be obtained for the maximum value of shaking force to be less than 0.10 lb?

2-13. A new line of 2-cylinder 4-cycle engines is being designed. Six possible crank arrangements are shown in Fig. P-13. Compare *a* with *b* on the basis of turning effort and balance characteristics. Which would you recommend for use in applications such as lawn mowers, portable electric generators, and snow blowers?

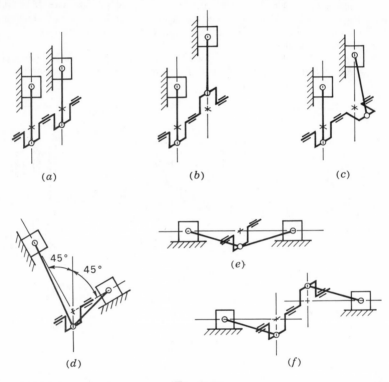

(a)   (b)   (c)

(d)   (f)

(e)

Fig. P-13

2-14. Same as Prob. 2-13 except compare Fig. P-13*b* with *c*.

2-15. Same as Prob. 2-13 except compare Fig. P-13*e* with *f*.

2-16. Same as Prob. 2-13 except compare Fig. P-13*b* with *d*.

2-17. Same as Prob. 2-13 except the engine is to be a 2-cycle engine.

2-18. Same as Prob. 2-13 except the engine is to be a 2-cycle engine and compare Fig. P-13*b* with *c*.

2-19. Same as Prob. 2-13 except the engine is to be a 2-cycle engine and compare Fig. P-13*e* with *f*.

2-20. Same as Prob. 2-13 except the engine is to be a 2-cycle engine and compare Fig. P-13*b* with *d*.

2-21. (*a*) What should be the general relationship between the angles of the throws on a crankshaft for a 6-cylinder in-line 4-cycle engine to provide the smoothest flow of power? (*b*) What crank arrangement and spacing of the cylinders should be used for a 6-cylinder in-line engine to ensure optimum operation with respect to the balance of the primary and secondary forces and couples?

2-22. Two possible crank arrangements for an 8-cylinder in-line 4-cycle engine are shown in Fig. P-14. Which one would you recommend?

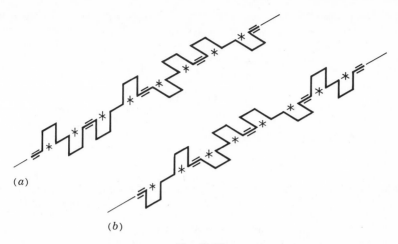

(a)

(b)

Fig. P-14

2-23. A large, slow-speed, 3-cylinder in-line 4-cycle diesel engine is being designed. Considering both turning effort and balance, what do you recommend for (a) the arrangement of throws and locations of counterweights on the crankshaft and (b) the firing order of the cylinders?

2-24. A V-6 engine is being designed. It is to be a high-speed, 4-cycle engine, and the crankshaft is to have three throws. The cylinder block is to be machined on the line used for V-8 engine blocks, and the angle between the banks will, therefore, have to be 90°. Considering both turning effort and balance, what do you recommend for (a) the arrangement of throws and the locations of counterweights on the crankshaft and (b) the firing order of the cylinders?

2-25. Same as Prob. 2-24 except the bank angle is no longer restricted to 90° and your recommendations must also include a value for the bank angle.

Chapter 3.   Vibrations

3-1. Derive an equation for the resultant spring rate $k = F/x$ when three springs with spring rates k_1, k_2, and k_3 are used (a) in parallel, as in Fig. P-15a, and (b) in series, as in Fig. P-15b.

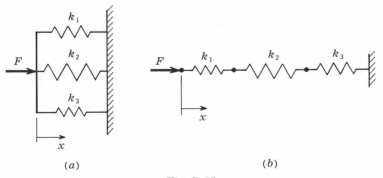

(a)

(b)

Fig. P-15

3-2. Determine the spring rate $k = F/x$ for the arrangement of springs in Fig. P-16.

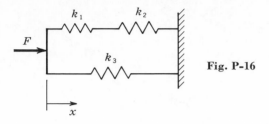

Fig. P-16

3-3. What is the resulting spring rate if a compression spring with a spring rate k is cut into two equal parts and the parts are then used in parallel? Assume all coils are active.

3-4. From elementary strength of materials or Appendix B we find that, for small deflections, the deflection of the end of a uniform cantilever beam is

$$\delta = \frac{PL^3}{3EI} \tag{P-1}$$

where P = force
L = length
E = modulus of elasticity
I = moment of inertia of the section

Assuming the beam is made out of steel ($E = 30 \times 10^6$ psi), derive equations for the spring rate $k = P/\delta$ in terms of the dimensions of the beam when the beam has (a) a rectangular section with a width of w in. and a thickness of t in., (b) a circular section with a diameter of d in.

3-5. In general, for small deflections, the deflection at any point on the simply supported uniform beam in Fig. P-17 can be found by use of the following two equations, from Appendix B:

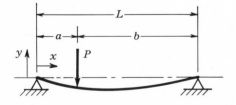

Fig. P-17

In length a

$$y = -\frac{Pbx}{6EIL} (L^2 - b^2 - x^2) \tag{P-2}$$

In length b

$$y = -\frac{Pa(L - x)}{6EIL} [2Lb - b^2 - (L - x)^2] \tag{P-3}$$

where E is the modulus of elasticity and I is the moment of inertia of the section.

Assuming the beam is made of steel ($E = 30 \times 10^6$ psi), derive equations for determining the spring rate $k = P/y$ in terms of the dimensions of the beam when the beam has (a) a rectangular section with a width of w in. and a thickness of t in. and (b) a circular section with a diameter d in.

3-6. Figure P-18 shows schematically a dashpot or damper. The chamber is filled completely with liquid, and the only way for liquid to move from one side of the piston

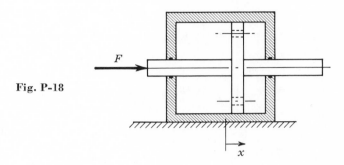

Fig. P-18

to the other is through a series of orifices (holes in the piston) having a total area of A_o in.^{2} The relationship between pressure and flow for a constant-density fluid through an orifice is shown in elementary texts on fluid mechanics to be

$$Q = C_d A_o \sqrt{2gh} \tag{P-4}$$

where Q = flow rate through the orifice, in.^{3}/sec
$\quad C_d$ = orifice coefficient
$\quad A_o$ = orifice area, in.^{2}
$\quad g$ = 386 in./sec^2
$\quad h$ = head of liquid, in.

The coefficient C_d varies with conditions, but for many purposes it can be assumed to be 0.60. The head h is a measure of the pressure drop across the orifice and equals $\Delta p/\gamma$, where Δp is the pressure drop in pounds per square inch and γ is the density of the liquid in pounds per cubic inch. Using Eq. (P-4) and designating the net area of the piston, i.e., the total area minus the area of the piston rod, as A_p, derive an equation that relates the force F to the velocity of the piston Dx. Does this device provide viscous damping?

3-7. The shear stress τ in a fluid is related to the viscosity and rate of shear by

$$\tau = \mu \frac{dU}{dh} \tag{P-5}$$

where μ is the absolute viscosity and dU/dh is the velocity gradient. If the surfaces remain parallel and one is stationary, the shear stress becomes

$$\tau = \mu \frac{U}{h} \tag{P-6}$$

where U is the velocity of the moving surface and h is the thickness of the fluid film between the surfaces.

Figure P-19 shows schematically a torsional vibration damper.   The clearance h between the rotor and the case is uniform everywhere and is much smaller than l and

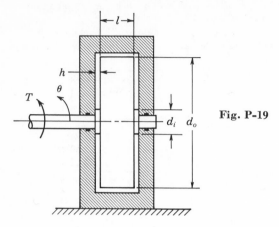

Fig. P-19

d_o.   The clearance space is completely filled with fluid.   Using Eq. (P-6) and the dimensions in Fig. P-19, derive an equation that shows the relationship between the torque T and the angular velocity $D\theta$.   Does this device provide viscous damping?

3-8. Using the results from the solution of Prob. 3-7, determine the dimensions required for the damper in Fig. P-19 if the damping torque is to be 15 lb-in./rad/sec when $h = 0.005$ in., $\mu = 6.0 \times 10^{-6}$ lb-sec/in.^{2} (about SAE 10 oil at 100°F), $d_i = 1.50$ in., and $l = 0.2d_o$.

Chapter 4.   Linear Single-degree-of-freedom Systems

4-1. The equation for displacement of a point as a function of time is $x = 3 \cos \omega t + 5 \sin \omega t$.   Determine the constants and angles required to rewrite the equation in the forms (a) $x = Xe^{j(\omega t + \theta)}$ and (b) $x = X \cos (\omega t + \theta)$.

4-2. The equation for the displacement of a point as a function of time is $x = 4e^{j(\omega t - 60°)}$.   Determine the constants and angles required to rewrite the equation in the forms (a) $x = X \cos (\omega t + \theta)$ and (b) $x = C_1 \cos \omega t + C_2 \sin \omega t$.

4-3. Same as Prob. 4-1 except for (b) write $x = X \sin (\omega t + \theta)$.

4-4. Same as Prob. 4-2 except for (a) write $x = X \sin (\omega t + \theta)$.

4-5. Determine the natural frequency of the system in Fig. P-20.

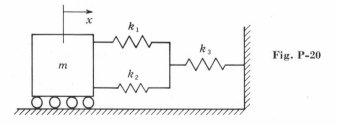

Fig. P-20

4-6. What will be the natural frequency in cycles per second for the system in Fig. P-20 if $m = 10$ lb-sec^2/in., $k_1 = 50$ lb/in., $k_2 = 80$ lb/in., and $k_3 = 70$ lb/in.?

4-7. Determine the natural frequency of the system in Fig. P-21*a*.

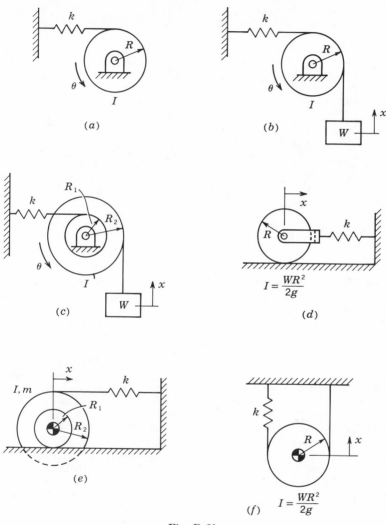

Fig. P-21

4-8. Determine the natural frequency of the system in Fig. P-21*b*.

4-9. For the system in Fig. P-21*b*, derive an equivalent system in which all values are related to rectilinear motion.

4-10. Determine the natural frequency of the system in Fig. P-21*c*.

4-11. Determine the natural frequency of the system in Fig. P-21*d*.

4-12. Determine the natural frequency of the system in Fig. P-21*e*.

4-13. Determine the natural frequency of the system in Fig. P-21*f*.

4-14. A 20-lb lead weight is attached to one end of a white-pine log that is 6.0 ft long and has a diameter of 6.0 in. If the density of white pine is 27 lb/ft³, what will be the frequency in cycles per second of the vertical oscillations of the log when placed in water that has a density of 62.4 lb/ft³?

4-15. Determine the natural frequency of oscillation of the water column in the tube in Fig. P-22*a*. Assume that the diameter of the tube is small relative to the other dimensions.

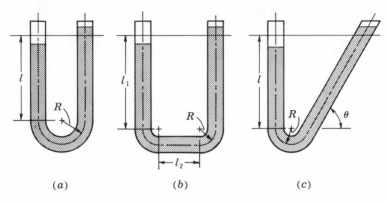

(*a*)   (*b*)   (*c*)

Fig. P-22

4-16. Determine the natural frequency of oscillation of the water column in the tube in Fig. P-22*b*. Assume that the diameter of the tube is small relative to the other dimensions.

4-17. Determine the natural frequency of oscillation of the water column in the tube in Fig. P-22*c*. Assume that the diameter of the tube is small relative to the other dimensions.

4-18. The load carried by one wheel of an automobile is 1,150 lb and the suspension system (spring plus shock absorber and neglecting the tires) is designed to have an undamped natural frequency of 1.2 cps and a damping ratio of 0.80. Determine (*a*) the spring rate in pounds per inch, (*b*) the static deflection in inches, (*c*) the damping coefficient in lb/in./sec, and (*d*) the damped natural frequency in cycles per second.

4-19. The load carried by one wheel of an automobile is 830 lb. It is desired that the suspension system (springs plus shock absorber and neglecting the tires) be designed to have a damped natural frequency of 0.85 cps and a damping ratio of 0.80. What values should be specified for (*a*) the spring rate and (*b*) the damping coefficient *c*?

4-20. Hundreds of millions of dollars of damage result each year from shock occurring during the coupling of railroad freight cars. The schematic drawing in Fig. P-23

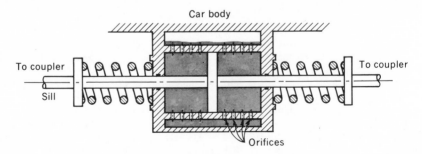

Fig. P-23

shows the major features of one method that can be used to minimize the magnitude of forces transmitted to the car body.

If we assume that the empty car body weighs 40,000 lb, that the load capacity is 70 tons, that the orifices are so arranged that the damping force closely approximates viscous damping, that the maximum acceleration of the loaded car body is not to exceed $4g$ when impacting at 10 mph a long string of stationary conventional cars (with relatively little shock-absorbing capability), and that the damping ratio is to be 0.70, what values would you specify for (a) the spring rate and (b) the damping coefficient? (c) For what magnitude of spring deflection must the system be designed? (d) What will be the maximum acceleration if the car is practically empty?

4-21. The recoil system of a new artillery piece must absorb and dissipate 150,000 ft-lb of energy. The moving parts weigh 3,800 lb, and the travel during recoil is to be 36 in. It is proposed that four identical sets of helical springs and dampers be used and that the dampers be equipped with one-way valves (check valves) so that they are ineffective during recoil and operate only when the barrel is being returned to its initial position. During the return stroke, critical damping is desired. Viscous (linear) damping will be provided by using a shaped rod (metering pin) that passes through a circular hole in an orifice plate to give the correct orifice area as a function of the position of the barrel relative to the carriage.

What values would you specify for (a) the spring rate and (b) the damping coefficient for each of the four shock absorbers? (c) How long will it take for the barrel to return from its position of maximum deflection to within 1 in. of its initial position?

4-22. The spring buffer system is being designed for an elevator. The maximum load, passengers and freight, will be 4,000 lb, and the empty car will weigh 5,000 lb. The buffer system is to stop the loaded car with a maximum acceleration of $2.0g$ if it fails to stop at the bottom landing when traveling at its maximum service speed of 200 fpm. Neglecting friction, (a) what do you recommend for the spring rate of each of the 10 springs, (b) what will be the maximum deflection of the springs, (c) how long will the car be in contact with the springs, and (d) what will be the speed of the car when it first breaks contact with the springs?

4-23. Same as Prob. 4-22 except linear dampers are to be added. The dampers will be designed to provide zero damping while the springs are being compressed and to provide a damping ratio of 0.80 while the elevator is rebounding. Also, determine the value of damping coefficient required for each of the 10 dampers.

4-24. An elevator is supported by four $\frac{5}{8}$-in. 6 × 19 traction-steel wire ropes. The total weight of the car and passengers is 8,000 lb. From previous calculations, we know that the static deflection of the ropes when the elevator is at the basement landing is 6.14 in. and that the friction force on the car can be approximated by viscous damping with a coefficient of 400 lb/ft/sec. The traction motors are located at the top of the elevator shaft and the cable speed at the top can be assumed to increase at a linear rate from zero to 300 fpm in 2.0 sec. (a) Derive the equation for the position of the elevator as a function of time while accelerating, with $t = 0$ corresponding to the instant the elevator motor is energized. (b) What will be the frequency of the transient vibrations of the loaded car? (c) Assuming the car weighs 5,500 lb, what will be the frequency of the transient vibrations of the empty car?

4-25. An undamped single-degree-of-freedom system is subjected to a forcing function $F(t) = F_1 \cos \omega_1 t + F_2 \cos \omega_2 t + F_3 \cos \omega_3 t$. Write the general equation for the steady-state displacement response x.

4-26. Same as Prob. 4-25 except the system includes viscous damping.

4-27. A system has a natural frequency of 15.0 cps, a damping ratio of 0.25, and a spring rate of 750 lb/in. What will be the maximum possible amplitude of the steady-state response to a forcing function that has components of 900 lb at 4.0 cps, 400 lb at 12 cps, and 2,000 lb at 50 cps?

4-28. The Fourier series expansion for the triangular forcing function in Fig. P-24 is

$$F = \frac{8F_0}{\pi^2} \left(\sin \omega t - \frac{1}{3^2} \sin 3\omega t + \frac{1}{5^2} \sin 5\omega t - \cdots \right) \qquad \text{(P-7)}$$

Write the general equation for the steady-state displacement response x of an undamped single-degree-of-freedom system to a triangular forcing function. Consider only the first three harmonics of the forcing function.

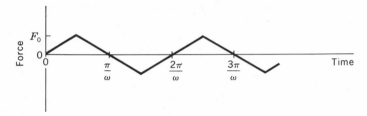

Fig. P-24

4-29. An undamped system with a natural frequency of 80 cps and a spring rate of 220 lb/in. is to be subjected to a triangular forcing function with an amplitude of 100 lb and a frequency of 20 (30, 50, 70, 100) cps. Use the Fourier series expansion given in Prob. 4-28 and sketch the time history of the first three harmonics and their sum for (a) the forcing function and (b) the steady-state displacement response x.

4-30. Same as Prob. 4-29 except the system has a damping ratio of 0.40.

4-31. The Fourier series expansion for the square-wave forcing function in Fig. P-25 is

$$F = \frac{4F_0}{\pi} \left(\sin \omega t + \tfrac{1}{3} \sin 3\omega t + \tfrac{1}{5} \sin 5\omega t + \cdots \right) \qquad \text{(P-8)}$$

Write the general equation for the steady-state displacement response x of an undamped single-degree-of-freedom system to a square-wave forcing function. Consider only the first three harmonics of the forcing function.

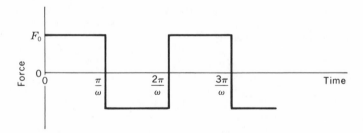

Fig. P-25

4-32. Same as Prob. 4-29 except use the square-wave forcing function in Prob. 4-31.

4-33. The ac-dc motor-generator set for an elevator drive in an apartment building will rotate at 1,750 rpm. The rotors will be accurately balanced, but it is desired that the effect of the remaining small unbalance be minimized as much as practical.   The total weight of the motor-generator set and base will be 7,200 lb.   Assuming this to be a system with a single degree of freedom, what do you recommend for (a) the spring rate of each of four helical springs and (b) the degree to which damping should be added?

4-34. The work done on or by a system equals the change in energy in a given interval of displacement.   Thus,

$$\Delta E = \int_{x_1}^{x_2} F \, dx \tag{P-9}$$

or, more usefully in vibrations, where time is the independent variable,

$$\Delta E = \int_{t_1}^{t_2} F \frac{dx}{dt} \, dt \tag{P-10}$$

(a) Using Eq. (P-10), derive the equation for the change in energy during one cycle when $F = F_0 \sin \omega t$ and $x = X \sin (\omega t + \phi)$.   (b) At what value of phase angle is the energy dissipated a maximum?   (c) Explain how the answer to (b) agrees or disagrees with your own experiences, such as pushing someone in a swing or working with viscous damping.

4-35. Using Eq. (P-10) or the solution to Prob. 4-34, (a) derive the equation for the energy dissipated per cycle when a single-degree-of-freedom system with a linear spring and viscous damping is vibrating at the undamped natural frequency of the system.   (b) What will be the amplitude of vibration at $\omega/\omega_n = 1.0$ for a system with a natural frequency of 21.0 cps and a damping coefficient of 0.050 lb/in./sec when the average rate of energy dissipation is 2.0 watts?

4-36. Using Eq. (P-10) or the solution to Prob. 4-34, determine the amplitude of vibration at a frequency of 17.0 cps of a linear system with a natural frequency of 21.0 cps, a damping ratio of 0.35, and a damping coefficient of 0.050 lb/in./sec when the average rate of energy dissipation is 2.0 watts.

4-37. A gasoline-engine-driven motor-generator set is to be installed in a van as a source of power for demonstrating electronic instruments.   The single-cylinder 4-cycle 10-hp engine will run at 2,800 rpm, and the 3.5-kw generator will run at 3,600 rpm.   The engine has a stroke of $3\frac{1}{4}$ in., the connecting-rod length is 5.75 in., and the total weight of the reciprocating parts is 0.547 lb.   The remaining rotating unbalance of the engine is negligible.   The balance specifications for the generator rotor call for it to be balanced within 0.078 oz-in. at each end of the rotor.   The total weight of the set and frame is 315 lb, and previous experience has indicated that in this application the noise level in the van will be satisfactory as long as the amplitude of the force transmitted does not exceed 2 lb.

Assuming that this system can be studied as a single-degree-of-freedom system, (a) what do you recommend for the spring rate of each of four isolators?   (b) To what degree should damping be provided?   (c) What will be the static deflection of the system?   (d) What will be the maximum amplitude of vibration due to the primary shaking force?   (e) What will be the maximum amplitude of vibration due to the secondary shaking force?

4-38. A single-phase 60-cycle electric motor is subjected to a 120-cps variation in torque because the magnetic field varies from zero to a maximum each half cycle of voltage variation.   Since many of these motors are used in household appliances where noise and vibration are particularly undesirable, it is customary to use a mounting, such as illustrated in Fig. P-26, in which the frame of the motor is isolated from the mounting bracket by rubber bushings at each end. The rubber bushings are bonded to the rings on the inside and outside diameters, and relative motion requires deformation of the rubber. Rubber has a Poisson's ratio of about one-half and is practically incompressible. Consequently, when used, as in this case, in a thin layer bonded to rigid surfaces, it is very stiff with respect to normal loads. However, the same restraints do not apply to torsional or shear deformation, and the mounting in Fig. P-26 can provide both high radial stiffness to resist belt forces and other radial loads and low torsional stiffness to provide reasonable isolation of the 120-cps torque variation.

Assuming that the motor frame (case) weighs 19.5 lb and has a radius of gyration of 2.28 in., what value of torsional spring constant should be specified for the rubber bushings at each end if not more than 10 percent of the torque variation is to be transmitted to the bracket?

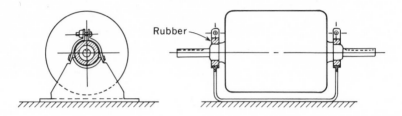

Fig. P-26 *Rubber-mounted single-phase ac motor.*

4-39. Coal and other granular materials are commonly shipped in hopper-bottom railroad cars. Inducing the material to flow out of the bottom of the car through the hoppers can be a considerable problem. Devices called car shakers are often used to break the material loose and to keep it flowing steadily. Most methods in use involve impact on the car body. Someone has proposed eliminating the impact by using the increased transmissibility near resonance to magnify the effect of the inertia force of a rotating unbalanced body. It is further proposed that the shaking force have an amplitude of 500 lb at a frequency of 600 cpm and that the transmissibility not exceed 10. Assuming, as a first approximation, that this system will be linear and have a single degree of freedom, (*a*) what values of natural frequency could possibly be used? (*b*) Which of the values found in (*a*) would you recommend? Why? (*c*) What magnitude of unbalance would be required? (*d*) Will you expect the actual shaking force transmitted to the car and its contents to be greater or less than the desired 500 lb? Why?

4-40. A 95-lb electronic instrument is to be attached to a bulkhead of a ship. The installation specifications call for a suspension system that will limit the amplitude of vibration of the instrument case to not more than 0.00015 in. Vibration measurements of the motion of the bulkhead have indicated that the worst conditions are an amplitude of 0.0007 in. at 20 cps, an amplitude of 0.0012 in. at 35 cps, and an amplitude of 0.0018 in. at 53 cps. What spring rate do you recommend for each of the four isolators?

4-41. If we assume the stiffness of the tires is much greater than that of the springs and if we assume that part of the weight is carried independently by each wheel, we can use Fig. P-27 as an approximation of an automobile on a washboard road. Assuming, further, that the undamped natural frequency of the system is 1.20 cps and that

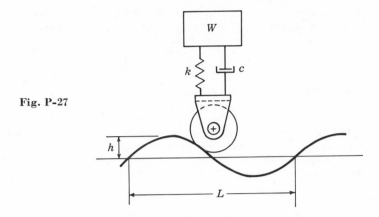

Fig. P-27

the damping ratio is 0.85, what will be the amplitude of the motion of the car body when the car is moving at 20 mph and the profile of the road varies sinusoidally with $h = 2.0$ in. and $L = 20$ ft?

4-42. Same as Prob. 4-41 except find the maximum speed permissible if the acceleration of the car body is not to exceed $0.5g$.

4-43. For the system in Fig. P-28a, (a) write the equation of motion. (b) What is the undamped natural frequency? (c) Derive the equations for the amplitude ratio and phase angle of the response when $y = Y \sin \omega t$.

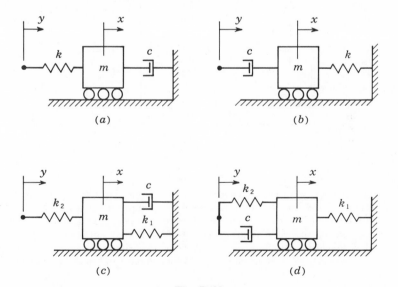

Fig. P-28

4-44. Someone suggests applying a step function of displacement ($t \leq 0$, $y = 0$ and $t > 0$, $y = 1$) to the system in Fig. P-28a and asks for your opinion about the conditions under which this can be done. Considering only the physical character-istics of the elements making up the system, what is your answer?

4-45. What happens to the system in Fig. P-28*a* if a forcing function $F(t)$, rather than $y(t)$, is applied to the left end of the spring?

4-46. Same as Prob. 4-43 except apply to the system in Fig. P-28*b*. Leave the equation for the phase angle in terms of the sum or difference of angles.

4-47. Same as Prob. 4-44 except apply to the system in Fig. P-28*b*.

4-48. Same as Prob. 4-45 except apply to the system in Fig. P-28*b*.

4-49. Same as Prob. 4-43 except apply to the system in Fig. P-28*c*.

4-50. Same as Prob. 4-43 except apply to the system in Fig. P-28*d*. Leave the equation for the phase angle in terms of the sum or difference of angles.

4-51. Figure P-29 shows a simple setup that is used in a shop for balancing large-diameter narrow rotors. In practice, the rotor is brought up to a speed of 10 times

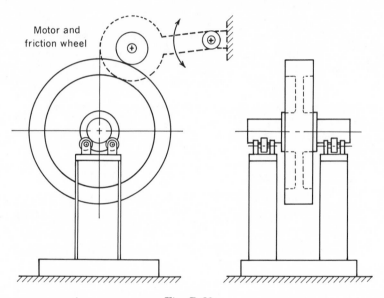

Motor and friction wheel

Fig. P-29

the natural frequency for lateral vibrations of the supporting frame. The motor is then raised to let the system operate as free of external restraints as possible. While coasting, the amplitude of the frame vibration is measured and the angular position of the rotor, with reference to an arbitrary index line, at the point of maximum ampli-tude is determined by using a displacement transducer and a triggered stroboscopic light (see Sec. 4-22). A brake (not shown) is then applied to bring the rotor rapidly to rest.

Assuming the rotor weighs 500 lb and that the equivalent mass of the supporting frame for horizontal vibrations is 150 lb, (*a*) what percent error will be introduced by assuming $\omega/\omega_n = \infty$ rather than 10? (*b*) What must be the sensitivity (smallest read-able value of displacement) if the rotor is to be balanced within 0.10 oz-in.? (*c*) What percent error in magnitude will be introduced if the corrections specified are based upon a nominal rotor speed of 10 times the natural frequency and the speed has decreased to 8 times the natural frequency by the time measurements are made? (*d*) What is the maximum permissible value of damping ratio if using a phase angle of −180° is not to introduce an error of more than 1° in the angular position of the correction? (*e*) For the damping ratio determined in (*d*) how many cycles will be required before the amplitude of free vibrations of the system (rotor not rotating) will decrease to 80 percent of its initial value?

4-52. A 750-lb steel rotor is being balanced in the setup described in Prob. 4-51. The amplitude of horizontal motion is measured as 0.008 in. and the index line makes an angle of 35° ccw with respect to the vertical center line when the rotor is in its extreme position to the right. The correction is to be made by drilling identical ½-in.-diam holes in both faces of the rotor at a radius of 19.00 in. Neglecting the effect of the point of the drill, what would you specify for the location and depth of holes if (*a*) there are no restrictions on the depth and (*b*) if no hole is to be deeper than ¾ in.?

4-53. The rotor in Fig. P-30 is to be balanced in the field. Preliminary measurements have determined that the natural frequency of the entire assembly for horizontal

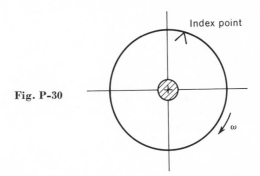

Fig. P-30

vibrations is 173 cps. When running at its normal speed of 5,000 rpm, the maximum displacement to the right has been observed to occur when a radial line to an index point on the rotor makes an angle of 145° ccw from the vertical. If the trial correction is to be made by adding weight, in what general location on the rotor should it be placed for maximum effectiveness?

4-54. Same as Prob. 4-53 except the normal speed of the rotor is 15,000 rpm.

4-55. A large, high-speed, double-suction, centrifugal compressor handles a corrosive gas and, as a result, requires periodic balancing in the field. This problem was anticipated, and 72 evenly spaced, drilled and tapped radial holes, numbered 1 to 72, have been provided in the periphery of the 36-in.-diam impeller. The holes are 2.0 in. deep and are tapped for 1/2-12UNC threads. Self-locking balancing screws (slugs) are available with lengths in ¹⁄₁₆-in. increments from ¼ in. to 2.0 in. The slugs weigh 0.0390 lb/in. of length. An access panel is provided at the top of the casing.

During a periodic overhaul, the impeller is being rebalanced. Vibration measurements (see Sec. 4-22) show that the compressor assembly vibrates horizontally with an amplitude of 8.65×10^{-5} in. It is further noted that when looking at the compressor from the end corresponding to the hole numbers increasing in the counterclockwise direction, the maximum displacement to the right occurs when hole 43 is at the top point. The compressor speed is well below the critical speed of the entire system, and a trial correction of a ½-in.-long slug is screwed into hole 61 far enough to locate its center of gravity at a radius of 17.50 in. Vibration measurements now show that a maximum displacement of 6.23 × 10⁻⁵ in. to the right occurs when hole 48 is at the top point. (a) What are the magnitude and position of the initial unbalance? (b) What do you recommend for the numbers of the holes, the lengths of slugs, and the depths at which the slugs should be located?

4-56. The six-bladed axial-flow fan in Fig. P-31 has been overhauled in the field and the vibration was found to be excessive when the fan was put back into service. The overall diameter is 50.0 in., and the normal operating speed is 1,150 rpm. Measurements made while running show that as reinstalled the amplitude of horizontal

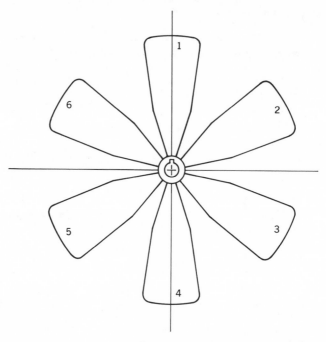

Fig. P-31

vibration of the frame at the fan center line is 0.00225 in. and the center line of blade 1 makes an angle of 70° measured counterclockwise from the vertical when the frame is at its maximum displacement to the right. A trial weight of 0.05 lb is clamped to the tip on the center line of blade 6, and new measurements show a maximum displacement to the right of 0.00228 in. when the center line of blade 1 makes an angle of 95° measured counterclockwise from the vertical. (a) What are the magnitude and location relative to blade 1 of the shaking force of the unbalanced fan? (b) Assuming that the blades are made of steel, that corrections can be made only by drilling ¼-in.-diam holes radially into the ends of the blades, and that no hole is to be over ⅜ in. deep and neglecting the effect of the point of the drill, what do you recommend for the locations and depths of holes to balance the fan?   (c) What is the direction of rotation of the fan?

4-57. Using an energy method, determine the natural frequency of the system in Fig. P-20.

4-58. Using an energy method, determine the natural frequency of the system in Fig. P-21a.

4-59. Using an energy method, determine the natural frequency of the system in Fig. P-21b.

4-60. Using an energy method, determine the natural frequency of the system in Fig. P-21c.

4-61. Using an energy method, determine the natural frequency of the system in Fig. P-21d.

4-62. Using an energy method, determine the natural frequency of the system in Fig. P-21e.

4-63. Using an energy method, determine the natural frequency of the system in Fig. P-21f.

4-64. Using an energy method, determine the natural frequency of the system in Fig. P-22a.

4-65. Using an energy method, determine the natural frequency of the system in Fig. P-22b.

4-66. Using an energy method, determine the natural frequency of the system in Fig. P-22c.

4-67. Approximately, what percent error is introduced by neglecting the mass of the shaft in Example 4-5 and considering the shaft to act as a simple spring?

4-68. Determine the lowest natural frequency for lateral vibrations of the shaft and rotor system in Example 4-5 for the case when the rotor is located 10 in. from the left bearing by using the sine-curve approximation for the deflection curve.

4-69. Determine the lowest natural frequency for lateral vibrations of the shaft and rotor system in Example 4-5 for the case when the rotor is located 10 in. from the left bearing by using the static-deflection curve resulting from the weight of the rotor alone as the deflection curve.

4-70. Approximately, what percent error will be introduced in determining the lowest natural frequency of lateral vibrations for the shaft and rotor system in Example 4-5 for the case when the rotor is located 10 in. from the left bearing, if the shaft is considered to be massless and to act as a simple spring?   Use the answer to either Prob. 4-68 or Prob. 4-69 as the basis for comparison.

4-71. The journal bearings on the shaft in Example 4-5 are being replaced with tapered roller bearings, and the system closely approaches a beam with fixed ends. (a) What will be your calculated estimate of the lowest natural frequency for lateral vibrations with the new bearings?   (b) Do you expect the true value to differ appreciably from your calculated value?   Will it be higher or lower?

4-72. A solid steel shaft with a uniform diameter of 3.000 in. is supported by tapered roller bearings that are located 50.00 in. apart.   What is your calculated estimate of the lowest natural frequency for lateral vibrations of this system?

4-73. Same as Prob. 4-72 except a rotor weighing 200 lb is located midway between the bearings.

4-74. In the cases of sinusoidal force and motion disturbance, it was noted that, when it is not possible to operate well below the natural frequency, one should try to operate as far above it as practical.   For example, in Sec. 4-16, it was recommended that ω/ω_n should at least be greater than 3.5.   However, in the case of shaft whirl (Sec. 4-20) it is recommended that $\omega/\omega_n > 1.5$ should be satisfactory.   (a) What is the basis for the lower value in the case of shaft whirl?   (*b*) Is there much to be gained in the case of shaft whirl by striving to make $\omega/\omega_n > 3.5$?   (*c*) How does the above relate to the case of inertia-force excitation?

4-75. Figure P-32 shows the proposed drive system for a small wind tunnel.   The 1,150-rpm motor shaft is connected to the propeller shaft through a flexible coupling that will not transmit a bending moment.   The shaft is supported by a cylindrical roller bearing at A and by a matched pair of preloaded angular contact bearings at B. What minimum diameter of solid steel shaft would you recommend if the critical whirling frequency is to be at least 40 percent greater than the running speed?

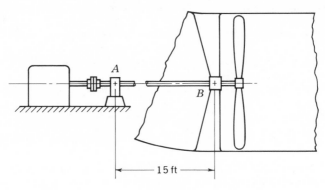

Fig. P-32

4-76. A 20-hp steam turbine is being designed for operation at 12,000 rpm.   The turbine wheel weighs 11.2 lb and is to be located midway between journal bearings that are 6.00 in. apart.   The output of the turbine will be through a flexible coupling that will not transmit a bending moment.   Strength considerations require the diameter of the solid steel shaft to be at least $\frac{3}{8}$ in.

Assuming that operation will be satisfactory if the running speed is below $0.7\omega_n$ or above $1.5\omega_n$, what shaft diameter would you recommend for this application? Consider both smoothness of operation and power loss in the journal bearings.

4-77. Many large mixers use a vertical drive in which a motor is mounted at the top over a tank and a shaft extends down into the tank so that the propeller is located near the bottom of the tank.   As a first approximation, we can consider such a system to be a vertical cantilever beam with a concentrated mass on the end of the shaft. (*a*) If the steel shaft has a diameter of 1.50 in. and a length of 43 in. and the propeller weighs 16 lb, what is the maximum speed of rotation you would recommend to ensure that the system operates at less than one-half of the lowest critical frequency for shaft whirl?

4-78. A vertical mixer, as discussed in Prob. 4-77, is being designed.   It is proposed to use a standard 850-rpm electric motor driving a 20-lb propeller at the end of a 6-ft-long solid steel shaft.   What diameter should the shaft be to ensure that the system operates below one-half the lowest critical speed for shaft whirl?

Chapter 5.   Nonlinear Single-degree-of-freedom Systems

5-1. Determine the natural frequency for small oscillations of the system in Fig. P-33*a*.

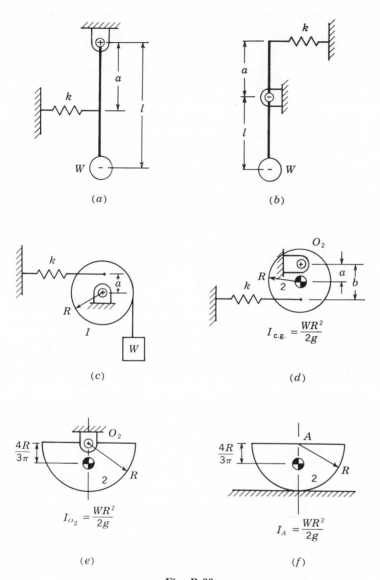

Fig. P-33

5-2. Determine the natural frequency for small oscillations of the system in Fig. P-33*b*.

5-3. Determine the natural frequency for small oscillations of the system in Fig. P-33*c*.

5-4. Determine the natural frequency for small oscillations of the system in Fig. P-33*d*.

5-5. Determine the natural frequency for small oscillations of the system in Fig. P-33*e*.

5-6. Determine the natural frequency for small oscillations of the system in Fig. P-33*f*.

5-7. In Fig. P-34*a* the spring k_1 is attached to the mass with initial tension F_0. Determine the natural frequency for small-amplitude vibrations of the system.

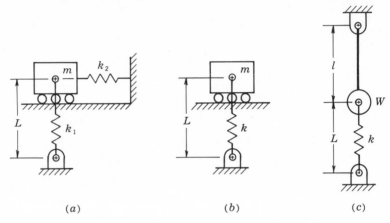

(*a*)   (*b*)   (*c*)

Fig. P-34

5-8. In Fig. P-34*b* the spring is attached to the mass with initial tension F_0. Determine the natural frequency for small-amplitude vibrations of the system.

5-9. In Fig. P-34*c* the spring is attached to the concentrated mass of the simple pendulum. The initial tension in the spring is F_0. Determine the natural frequency for small-amplitude oscillations of the system.

5-10. Same as Prob. 5-9 except the system oscillates in a horizontal plane (about a vertical axis).

5-11. In Sec. 4-22 it was noted that a seismic system used to measure displacement should have as low a natural frequency as possible. If a linear system is being used, the required static deflection may be too large to be practical when trying to measure displacements at frequencies of the order of 1 cps or slower. One scheme that has been useful is the "principle of instability," which is concerned with nonlinear systems in which the effective restoring force is partially counterbalanced by another force that acts to disturb the system. The latter force appears as a negative spring rate, and if the magnitude is sufficiently large, the system will be unstable. Several possible configurations are shown in Fig. P-35. (*a*) Determine the natural frequency for the system in Fig. P-35*a*. (*b*) How would you determine where to put stops to prevent the system from collapsing if the amplitude of oscillation tries to become too large?

5-12. Same as Prob. 5-11 except apply to Fig. P-35*b*. Spring k_1 is installed with initial tension F_0.

5-13. Same as Prob. 5-11 except apply to Fig. P-35*c*.

5-14. Figure P-35*d* shows the basic mechanism of a metronome which uses the principle of instability, discussed in Prob. 5-11, to achieve a low natural frequency. What value of torsional spring rate will be required for small amplitudes of oscillation if W is 0.0325 lb and a frequency of $\frac{1}{2}$ cps is desired when $l = 4$ in.?

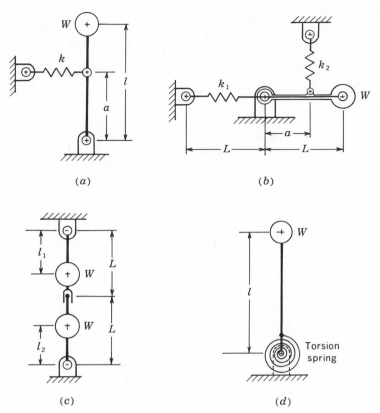

Fig. P-35

5-15. The compound pendulum in Fig. P-36 is often used in determining the moment of inertia of large objects, such as automobiles and airplanes, that cannot be conveniently mounted on the platform of a gravity torsional pendulum, as in Fig. 5-3, with each of its principal axes in turn in the vertical direction. In such cases it is also difficult to determine accurately the location of the center of gravity in the direction of one of the principal axes by using a knife edge or finding the vertical reactions at three or four points by using scales or load cells. Swinging the compound pendulum with two different values of a provides enough information to permit determining both the vertical distance of the center of gravity with respect to an arbitrary reference plane and the moment of inertia of the body about the center of gravity. Derive

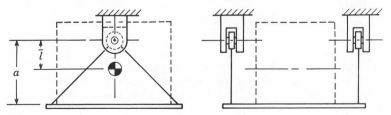

Fig. P-36

the necessary equations and set up the procedure to be followed when both the height of the center of gravity above the base of the body and the moment of inertia are to be determined by use of the compound pendulum. Assume the weight and inertia of the platform are negligible in comparison with the values for the body.

5-16. Same as Prob. 5-15 except the weight and inertia of the platform cannot be neglected.

5-17. Some commercial vibration isolators are so designed that when supporting a weight within the rated capacity of the isolator, the natural frequency for small amplitudes will be independent of the actual value of the load. This feature simplifies the selection of an isolator from a limited number of stock items, but the main reason for using the nonlinear spring is that, for loads within the rated range, the static position of the mass will be somewhere near the middle of the distance between the stops or snubbers in the isolator. What must be the relationship between the force and deflection for an isolator if the natural frequency of the system is to be independent of the weight supported by the isolator?

5-18. The helical-spring suspension system designed for the motor-generator set in Prob. 4-33 has been installed, and the occupants of nearby apartments are still complaining about the noise and vibration. It is proposed to decrease the transmissibility to 0.001 by changing over to an air-spring suspension system. Service air is available at 120 psi. What are your specifications for the dimensions of each of the four air springs and the volume of the reservoir?

5-19. A communications van is being designed for off-the-road use. The natural frequency of the suspension system is to be between 55 and 60 cpm for all loads within the range of an empty weight of 18,000 lb and a fully loaded weight of 30,000 lb. The weight distribution will be 45 percent on the front axles when empty and 35 percent on the front axles when loaded. It is proposed to use an air-spring suspension in which each spring will have its own reservoir and a level-control system. Air will be available at 75 psi. Assuming that each spring and its share of the load can be treated as a single-degree-of-freedom system, what are your specifications for the dimensions of the spring and volume of the reservoir for each of the front axle springs?

5-20. Same as Prob. 5-19 except do for the rear axle springs.

5-21. Figure P-37 shows a plank weighing W lb resting on two identical rollers. The rollers have parallel axes of rotation and are connected to electric motors. They can be driven independently in either direction at any desired speed up to 1,800 rpm.

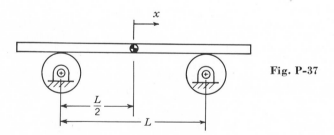

Fig. P-37

This system is actually a control system, because the time variations within the system control the supply of energy to and removal of energy from the system, but it can be treated directly as a problem in rigid-body dynamics. As with many control systems, under certain conditions it will provide steady-state oscillations and is, therefore, often considered as a problem in vibrations. Assuming a coefficient of friction μ

between the board and the rollers, consider the following conditions: (1) both rollers rotating in the same direction, (2) the left roller rotating counterclockwise and the right roller rotating clockwise, and (3) the left roller rotating clockwise and the right roller rotating counterclockwise. Specify (a) for which condition(s) the system will be stable and (b) what the frequency of oscillation will be, if any. (c) Will the degree of alignment of the longitudinal axis of the board relative to the roller axes have any effect on the behavior of the system? (d) Will the speed of rotation of the rollers have any effect on the behavior of the system?

5-22. If the coefficient of friction between the mass and the frame in Fig. P-38 is 0.30, how far will the mass travel before it comes to rest if it is pushed to the right to compress the spring 1.0 in. and then released?

Fig. P-38

5-23. For the system in Fig. P-39, $W = 100$ lb, $k_1 = k_2 = 200$ lb/in., and $a + b = 2.0$ in. What will be the frequency of vibration of the system if damping is negligible and the mass is moved to the right to compress the spring 2.0 in. and then released?

5-24. Same as Prob. 5-23 except $a + b = 0$.

5-25. For the system in Fig. P-39, $W = 100$ lb, $k_1 = 100$ lb/in., $k_2 = 200$ lb/in., and $a + b = 2.0$ in. What will be the frequency of vibration of the system if damping is negligible and the mass is moved to the right to compress the spring 2.0 in. and then released?

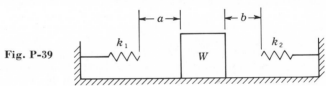

Fig. P-39

5-26. Same as Prob. 5-25 except $a + b = 0$.

5-27. For the system in Fig. P-39, $W = 100$ lb, $k_1 = k_2 = 200$ lb/in., $a + b = 2.0$ in., and the coefficient of friction $= 0.20$ between the mass and the frame. Use the phase-plane method to determine (a) the number of cycles of oscillation and (b) the final position of the mass if it is moved to the right to compress the spring 2.0 in. and then released.

5-28. Same as Prob. 5-27 except the coefficient of friction is 0.50.

5-29. Same as Prob. 5-27 except $a + b = 0$.

5-30. Same as Prob. 5-27 except $a + b = 0$ and the coefficient of friction is 0.50.

5-31. Same as Prob. 5-27 except also determine the time required for the mass to move to the extreme left position.

5-32. Same as Prob. 5-27 except also determine the time required for the mass to come to rest.

5-33. For the system in Fig. P-39, $W = 100$ lb, $k_1 = 100$ lb/in., $k_2 = 200$ lb/in., $a + b = 2.0$ in., and the coefficient of friction $= 0.20$ between the mass and the frame.

Use the phase-plane method to determine (*a*) the number of cycles of oscillation and (*b*) the final position of the mass if it is moved to the right to compress the spring 1.0 in. and then released.

5-34. Same as Prob. 5-33 except the coefficient of friction is 0.50.

5-35. Same as Prob. 5-33 except $a + b = 0$.

5-36. Same as Prob. 5-33 except $a + b = 0$ and the coefficient of friction is 0.50.

5-37. Same as Prob. 5-33 except also determine the time required for the mass to move to the extreme left position.

5-38. Same as Prob. 5-33 except determine the time required for the mass to come to rest.

5-39. For the system in Fig. P-40, $W = 100$ lb, $k_1 = k_2 = 100$ lb/in., $a + b = 2.0$ in., and $c = 4.50$ lb/in./sec. Use the phase-plane method to determine the

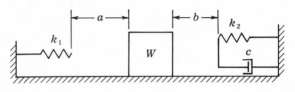

Fig. P-40

number of cycles of oscillation before the amplitude of the velocity has decreased to 10 percent of its maximum value if the mass is moved to the right to compress the spring 2.0 in. and then released.

5-40. For the system in Fig. P-40, $W = 100$ lb, $k_1 = k_2 = 200$ lb/in., $a + b = 2.0$ in., $c = 4.50$ lb/in./sec., and the coefficient of friction = 0.20 between the mass and the frame. Use the phase-plane method to determine (*a*) the number of cycles of oscillation and (*b*) the final position of the mass if it is moved to the right to compress the spring 1.0 in. and then released.

5-41. For the system in Fig. P-41, $W = 500$ lb, $k_1 = k_2 = 200$ lb/in., $k_3 = 300$ lb/in., $a + b = 2.0$ in., and $F_\mu = 100$ lb. Use the phase-plane method to determine (*a*) the number of cycles of oscillation and (*b*) the final position of the mass if it is moved to the right to compress the spring 2.0 in. and then released.

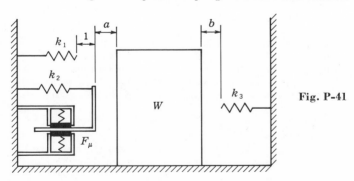

Fig. P-41

5-42. Problem 4-20 is concerned with design parameters for a railroad car shock-absorbing device utilizing viscous damping. One of the alternatives that have been proposed is to substitute Coulomb friction for the viscous damping by having the sill slide between spring-loaded brake shoes.   The maximum travel from the position of zero spring deflection is to be 12 in. in either direction when the loaded car impacts a string of conventional cars (with relatively little shock-absorbing capability) at a speed of 10 mph and 50 percent of the kinetic energy is to be dissipated during the first period of compression of the spring.   (a) What should be the value of the friction force?   (b) What should be the value of the spring rate?   (c) What will be the maximum acceleration in g's experienced by the loaded car?   (d) What will be the maximum acceleration in g's experienced by the unloaded car?   (e) What will be the extent of the dead zone?   (f) What changes, if any, would you recommend to improve the operating characteristics of the system?

5-43. Same as Prob. 5-42 except the maximum acceleration of the loaded car is to be $4g$ and, also, determine the spring deflection required.

5-44. In Example 5-1 it was shown that for small angles of oscillation the simple pendulum can be treated as a linear system.   (a) Use the phase-plane method to determine the natural frequency for oscillation of the pendulum in Fig. P-42 when

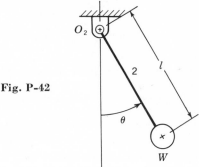

Fig. P-42

$W = 50$ lb, $l = 20$ in., and the pendulum is released from an initial position of $\theta = 120°$.   (b) Compare the answer to (a) with the value calculated for small-amplitude oscillations.

5-45. Same as Prob. 5-44 except the initial position is $\theta = 135°$.

5-46. Same as Prob. 5-44 except the initial position is $\theta = 170°$.

5-47. Using the concepts related to the phase-plane method, discuss the behavior of the pendulum in Fig. P-42 if it is released from (a) an initial position of $\theta = $ exactly $180°$ and (b) θ is arbitrarily close, but not equal, to $180°$.

5-48. A viscous damper for oscillatory motion has been installed at O_2 on the pendulum in Fig. P-42.   If $W = 50$ lb, $l = 20$ in., and the damping coefficient is 150 lb-in./rad/sec, (a) what is the maximum angular velocity and (b) how many cycles of oscillation are required for the amplitude to decrease to $15°$ when the pendulum is released from an initial angle of $135°$?

5-49. Same as Prob. 5-48 except the initial angle is $170°$.

5-50. A Coulomb damper (disk brake) has been installed at O_2 on the pendulum in Fig. P-42.   Assuming that $W = 50$ lb, $l = 20$ in., the magnitude of the friction torque is 200 lb-in., and the pendulum is released from an initial position $\theta = 135°$, what will be (a) the number of cycles of oscillation before the pendulum comes to rest, (b) the final position of the pendulum, (c) the dead zone, and (d) the time required for the pendulum to first reach $\theta = 0°$?

5-51. Same as Prob. 5-50 except the initial angle is $170°$.

5-52. The dashed box in Fig. P-43 contains an overrunning clutch and a disk brake. In operation there is negligible damping when the rotor is turning in the positive

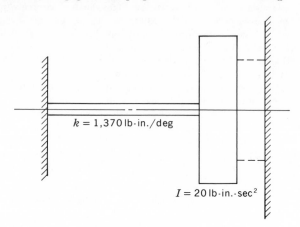

$k = 1,370 \, \text{lb-in./deg}$

$I = 20 \, \text{lb-in.-sec}^2$

Fig. P-43

direction and there is a constant friction torque $T\mu$ with a magnitude of 4,000 lb-in. when the rotor is turning in the negative direction. How many oscillations will the rotor make and what will be its final position if it is rotated to an angle of $+8°$ and released?

5-53. Same as Prob. 5-52 except the rotor is released from an initial deflection of $-8°$.

5-54. Same as Prob. 5-52 except the initial conditions are $\theta \big|_0 = +8°$ and $D\theta \big|_0 = +200°/\text{sec}$.

Chapter 6.   **Multi-degrees-of-freedom Systems**

6-1. Figure P-44 shows schematically the drive train of a new machine to be used in making continuous sheets of plastic. The motor is a ¾-hp single-phase 60-cycle

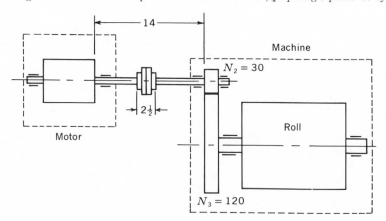

Fig. P-44

ac motor with a rated speed of 1,750 rpm.   The motor shaft and the machine input shaft are made of steel and are ¾ in. in diameter.   The shaft coupling has negligible inertia and negligible torsional flexibility.   The inertia of the motor armature is 0.056 lb.-in.-sec², and the equivalent rotating inertia of the machine, at the input shaft, is 0.28 lb-in.-sec².   The gear teeth and shafts in the machine are much stiffer than the net shaft length between the armature and gear 2.   The 120-cps variation in magnetic field results in a periodic torque variation with a fundamental frequency of 120 cps. (*a*) What amplitude of steady-state oscillation of the roll will be expected owing to the first harmonic of the torque variation on the armature if it is assumed to have an amplitude equal to 35 percent of the rated torque of the motor?   (*b*) Relatively speaking, what will be the effect of the higher-order harmonics with their smaller amplitudes?

6-2. The machine discussed in Prob. 6-1 has been built and tested.   The effect of the 120-cps variation in torque is noticeable in the finished product, and it is proposed to replace the torsionally rigid coupling with a commercially available coupling that operates with a deflection of 15° when transmitting rated torque.   Assuming the new coupling is a linear device and that the motor is delivering its rated torque, and considering only the first harmonic of the torque variation, what reduction (percent) in amplitude of oscillation of the roll will we expect to find after installing the new coupling?

6-3. The machine discussed in Prob. 6-1 has been built and tested.   The effect of the 120-cps variation in torque is noticeable in the finished product, and it is proposed to replace the rigid coupling by one that has a high degree of torsional flexibility. What spring rate should be specified for the new coupling if the amplitude of oscillation of the roll is to be decreased to 10 percent of its value with the original coupling?

6-4. Figure P-45 shows two rotors located a distance L apart on a constant-diameter shaft with a torsional stiffness k.   Damping is negligible, and a forcing function $T_1 = T_0 \sin \omega t$ acts on rotor 1.   Derive the general equation for the steady-state amplitude of the shaft twist per inch of length, $|\theta_1 - \theta_2|/L$.

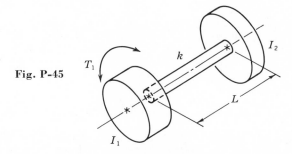

Fig. P-45

6-5. Using the results from Prob. 6-4, (*a*) what diameter of steel shaft do you recommend for the system in Fig. P-45 if $|\theta_1 - \theta_2|$ is not to exceed 0.0010° at a frequency of 700 cpm when $T_0 = 50$ lb-in., $L = 20$ in., $I_1 = 0.085$ lb-in.-sec², and $I_2 = 0.125$ lb.-in.-sec²?   (*b*) Neglecting stress concentration, what will be the maximum shear stress in the shaft for the conditions in (*a*)?

6-6. Using the results from Prob. 6-4, what diameter do you recommend for a steel shaft for the system in Fig. P-45 if the maximum shear stress, neglecting stress concentration, is not to exceed 8,000 psi when $L = 20$ in., $I_1 = 0.085$ lb-in.-sec², $I_2 = 0.125$ lb-in.-sec², and the forcing function has an amplitude of 50 lb-in. and a frequency of 700 cpm?

6-7. Two rotors are attached to a shaft, as shown in Fig. P-45. The amplitude of oscillation of rotor 2 is greater than can be tolerated, and it is proposed to alleviate the situation by attaching directly to one of the rotors an inertia disk I_3. What general statements can be made about the conditions under which I_3 should be attached to one rotor in preference to the other?

6-8. For the system in Fig. P-45, $I_1 = 1.79$ lb-in.-sec², $I_2 = 5.96$ lb-in.-sec², and $k = 900$ lb-in./deg. The amplitude of oscillation of rotor 2 is excessively high, and it is proposed to attach a disk with an inertia of 3.00 lb-in.-sec² directly to one of the rotors. If the frequency of the sinusoidal forcing function is 27 cps, (*a*) to which rotor should the inertia disk be attached for maximum effectiveness and (*b*) what will be the percentage reduction in amplitude of the oscillation of rotor 2?

6-9. Same as Prob. 6-8 except the frequency of the forcing function is 29 cps.

6-10. Same as Prob. 6-8 except the frequency of the forcing function is 32 cps.

6-11. Same as Prob. 6-8 except the frequency of the forcing function is 38 cps.

6-12. Figure P-46 shows a torsional system in which the disturbance is an oscillation $\theta_1(t)$ at one end of a shaft. Derive the equations for the amplitude and phase angle of the response of rotor 2 when $\theta_1(t) = \Theta_1 \sin \omega t$.

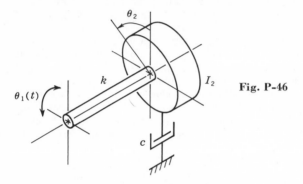

Fig. P-46

6-13. In Fig. P-47, link 4 oscillates through an angle of 45° with approximately simple harmonic motion. The equivalent inertia of the major internal parts of the

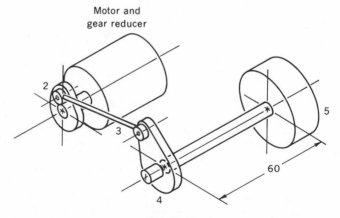

Fig. P-47

machine is represented by rotor 5 with a moment of inertia of 17.5 lb-in.-sec². The frequency of oscillation is to be 3.0 cps. Assuming the shaft connecting link 4 and rotor 5 is steel with a uniform diameter and neglecting stress concentration, (a) what diameter is required if we consider all parts (including the shaft) rigid and the maximum shear stress in the shaft to be 15,000 psi? (b) What will be the maximum stress for the shaft in (a) if all parts are considered rigid except the 60-in. length of shaft between link 4 and rotor 5?

6-14. Same as Prob. 6-13 except determine the diameter required for the 60-in. length of shaft if the maximum shear stress, neglecting stress concentration, is to be 15,000 psi when all parts are considered rigid except the shaft itself.

6-15. Same as Prob. 6-13 except in addition the effect of the working load can be approximated by adding a viscous damper with a damping coefficient of 5.0 lb-in./deg/sec between rotor 5 and the frame.

6-16. Same as Prob. 6-13 except, in addition, the effect of the working load can be approximated by adding a viscous damper with a damping coefficient of 5.0 lb-in./deg/sec between rotor 5 and the frame, and determine the diameter required for the 60-in. length of shaft if the maximum shear stress, neglecting stress concentration, is to be 15,000 psi when all parts are considered rigid except the shaft itself.

6-17. As discussed in Sec. 6-1 and shown in Fig. 6-3, the node for the free vibration of the simple torsional system in Fig. P-45 will be located at a point between the rotors that is determined by the ratio I_1/I_2. When a sinusoidal forcing function acts on one of the rotors, the situation changes and the location of the node depends also upon the frequency. Derive the necessary equations and discuss the change in location of the node for the system in Fig. P-45 as the frequency of the forcing function $T_1 = T_0 \sin \omega t$ increases from zero to infinity.

6-18. Assuming that the pendulums in Fig. P-48 are simple pendulums, all members are rigid except the shaft (with torsional spring rate k) to which the pendulums are

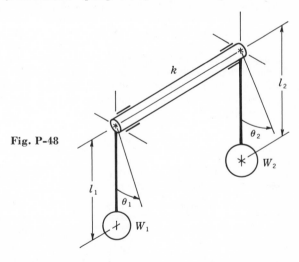

Fig. P-48

attached, $W_1 = W_2 = W$, and $l_1 = l_2 = l$, (a) what are the two natural frequencies of oscillation? (b) Where are the corresponding nodes located? (c) Under what conditions will the phenomenon of beating be observed?

6-19. Same as Prob. 6-18 except $W_1 = W_2 = 10$ lb, $l_1 = l_2 = 20$ in., $k = 0.25$ lb-in./deg, and in place of (c) determine the frequency of beating in cycles per second.

6-20. Same as Prob. 6-18 except $W_1 = 2W_2$.

6-21. Assuming that the pendulums in Fig. P-48 are simple pendulums and all members are rigid except the shaft (with torsional spring rate k) to which the pendulums are attached, what is the general equation for determining the natural frequencies of oscillation?

6-22. Use the answer to Prob. 6-21 to determine the natural frequencies when $W_1 = 20$ lb, $W_2 = 10$ lb, $l_1 = 10$ in., $l_2 = 20$ in., and $k = 2.0$ lb-in./deg.

6-23. Assuming $k_1 = k_2 = k_3 = k$ and $m_1 = m_2 = m$, determine (a) the frequencies and (b) the location of the corresponding nodes for the two principal modes of vibration of the system in Fig. P-49.

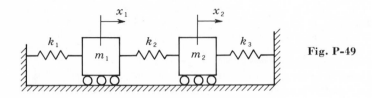

Fig. P-49

6-24. Same as Prob. 6-23 except $k_1 = 3k_2 = 3k_3 = 3k$.

6-25. Same as Prob. 6-23 except $m_1 = 3m_2 = 3m$.

6-26. Same as Prob. 6-23 except $k_1 = 3k_2 = 3k_3 = 3k$ and $m_1 = 3m_2 = 3m$.

6-27. Assuming $k_1 = k_2 = k_3 = k$ and $m_1 = m_2 = m$, derive the equations for the amplitudes of the displacements of the masses in Fig. P-49 when a forcing function $F_1 = F_0 \sin \omega t$ acts in the horizontal direction on m_1.

6-28. Derive the equations for the amplitudes of the displacements of the masses in Fig. P-49 when a forcing function $F_1 = F_0 \sin \omega t$ acts in the horizontal direction on m_1.

6-29. Same as Prob. 6-28 except a viscous damper with a damping coefficient c_3 is installed in parallel with the spring k_3.

6-30. Assuming $k_1 = k_2 = k$ and $m_1 = m_2 = m$, determine (a) the frequencies and (b) the locations of the corresponding nodes for the principal modes of vibration of the system in Fig. P-50.

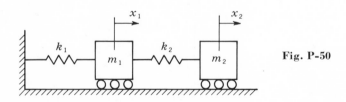

Fig. P-50

6-31. Same as Prob. 6-30 except $m_1 = 3m_2 = 3m$.

6-32. Same as Prob. 6-30 except $k_1 = 3k_2 = 3k$.

6-33. Same as Prob. 6-30 except $k_1 = 3k_2 = 3k$ and $m_1 = 3m_2 = 3m$.

6-34. Assuming $k_1 = k_2 = k$ and $m_1 = m_2 = m$, derive the equations for the amplitudes of the displacements of the masses in Fig. P-50 when a forcing function $F_2 = F_0 \sin \omega t$ acts in the horizontal direction on m_2.

6-35. Derive the equations for the amplitudes of the displacements of the masses in Fig. P-50 when a forcing function $F_2 = F_0 \sin \omega t$ acts in the horizontal direction on m_2.

6-36. Same as Prob. 6-35 except a viscous damper with a damping coefficient c_1 is installed in parallel with spring k_1.

6-37. Derive the equations for the amplitudes of the displacements of the masses in Fig. P-51 for $y(t) = Y \sin \omega t$, $k_1 = k_2 = k$, $m_1 = m_2 = m$, and $c_2 = 0$.

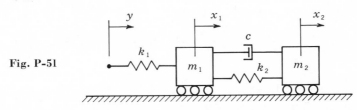

Fig. P-51

6-38. Derive the equations for the amplitudes of the displacements of the masses in Fig. P-51 when $y(t) = Y \sin \omega t$.

6-39. In Prob. 4-41 the flexibility and damping characteristics of the tires were neglected in order to permit considering the automobile suspension system as a single-degree-of-freedom system. A better approximation is that shown in Fig. P-52, where

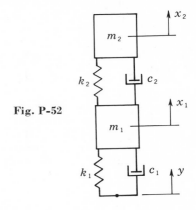

Fig. P-52

m_2 is the part of the mass carried by the spring k_2, c_2 represents the shock absorber, m_1 is the mass of wheel, tire, linkage, etc., not supported by the spring k_2 (the so-called unsprung mass), and k_1 and c_1 are the spring constant and damping coefficient for the tires.

Assuming that $W_2 = 765$ lb, $k_2 = 106$ lb/in., the shock absorber can be approximated by a viscous damper with a damping coefficient of 25 lb/in./sec, the unsprung weight is 135 lb, the tire can be approximated by a linear spring with a spring rate of 1,500 lb/in., the damping in the tire is negligible, and the car is traveling down a road with a sinusoidal profile with peaks spaced 20 ft apart and a peak-to-trough distance of 4 in., (a) what will be the amplitudes of motion of the car body and the center of the wheel if the car speed is 20 mph? (b) How would you determine the speed at which the tires cease to remain continuously in contact with the road surface?

6-40. Derive the equations for the amplitudes of the displacements of the masses in Fig. P-53 when $c_1 = c_2 = c_3 = 0$, $k_1 = k_2 = k_3 = k$, $m_1 = m_2 = m$, and $y(t) = Y \sin \omega t$.

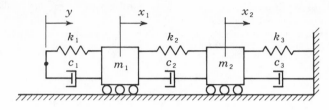

Fig. P-53

6-41. Derive the equations for the amplitudes of the displacements of the masses in Fig. P-53 when $y(t) = Y \sin \omega t$.

6-42. (a) Determine the frequencies and (b) describe the modes of vibration for the principal modes of the system in Fig. P-54a when $I = 0.50$ lb-in.-sec², $k_1 = 40$ lb/in., $k_2 = 20$ lb/in., $W = 50$ lb, and $R = 3.0$ in.

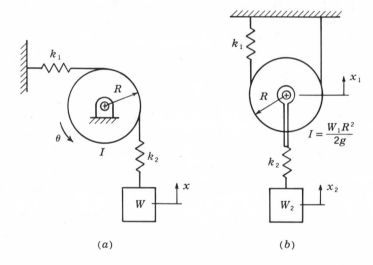

(a)   (b)

Fig. P-54

6-43. Same as Prob. 6-42 except apply to the system in Fig. P-54b.

6-44. The system in Fig. P-55 is in service and the amplitude of vibration is too large to be tolerated. Neither k nor m can be readily changed, and it is not practical

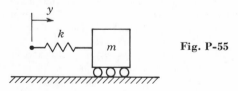

Fig. P-55

to add damping to the system.   The forcing function $y(t)$ has a constant frequency, and someone proposes adding another spring-mass system to absorb the vibration, as was shown in Sec. 6-2 to be practical for the case of a force disturbance.   What are your conclusions about the practicality of the proposal?

6-45. A small hand-held power tool weighs 1.50 lb and contains a part weighing 0.040 lb that reciprocates with simple harmonic motion at the rate of 4,800 cpm.   The amplitude of the reciprocating motion is 0.0625 in.   The resulting vibration is uncomfortable to the user of the tool, and it is proposed to add a vibration absorber in the form of a small mass and spring in line with the reciprocating inertia force.   (a) What do you recommend for the weight of the mass and the spring rate of the absorber? (b) For what maximum deflection must the absorber spring be designed?   *Note:* The natural frequency of the original tool when held by the operator is practically zero.

6-46. Same as Prob. 6-45 except the tool must operate at speeds between 4,000 and 4,800 cpm.   Will it be necessary to add damping to the absorber?   If so, what do you recommend for the damping coefficient?

6-47. A small, lightweight (175-lb) 4-cylinder in-line engine is to be used under conditions where vibrations must be minimized and where the engine suspension system must be relatively stiff.   The engine will accelerate rapidly and will run at a constant speed of 3,600 rpm.   The basic suspension system will have a natural frequency about three times the engine speed.   It is too late to redesign the engine to add rotating counterweights to balance the secondary shaking force (see Sec. 2-8), and it is proposed that a vibration absorber be installed.   Assuming that the mass of the absorber cannot exceed 10 percent of the mass of the engine, the reciprocating weight of each piston and connecting rod is 1.38 lb, the connecting-rod length is 5.50 in., and the stroke of the engine is 3.00 in., what specifications would you set for the design of the elements of the absorber?

6-48. Derive the equation for the amplitude of θ_1 for the system in Fig. P-56 when $T_2 = T_0 \sin \omega t$.

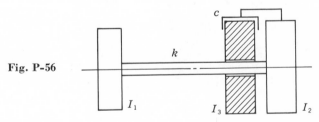

Fig. P-56

6-49. Derive the equation for the amplitude of θ_2 for the system in Fig. P-56 when $T_2 = T_0 \sin \omega t$.

6-50. Derive the equation for the amplitude of $\theta_1 - \theta_2$ for the system in Fig. P-56 when $T_2 = T_0 \sin \omega t$.

6-51. A new line of reciprocating compressors is undergoing field service tests. Each compressor is driven at a speed of 750 rpm by a gear motor.   The equivalent inertia of the gear motor is 20 lb-in.-sec^2, and the equivalent inertia of the compressor is 30 lb-in.-sec^2.   The shaft connecting the two units has a spring rate of 1.21×10^4 lb-in./deg.   Reports from the field have indicated that the shafts are failing prematurely in fatigue.   Subsequent experimental studies, using a torsional accelerometer, have shown that there is a pronounced torsional vibration at a frequency of 37.5 cps superposed upon the steady speed of 750 rpm.   Since the torque variations that are exciting the system undoubtedly originate in the compressor, it is proposed to add a Lanchester damper, using viscous damping, to the system at the compressor end of the shaft. Assuming that the inertia of the damper is equal to that of the compressor, what value of damping coefficient should be used?

6-52. Same as Prob. 6-51 except, also, use the results of the solution to Prob. 6-50 and determine the percent reduction in the amplitude of the stress variation that will result from the installation of the Lanchester absorber.

6-53. The amplitude of the torque variation that is responsible for the torsional vibrations of the compressor drive discussed in Prob. 6-51 has been determined to be 256 lb-in. What specifications would you set up for the detailed design of a centrifugal pendulum absorber for this application if the radius to the point of attachment of the pendulum cannot exceed 4 in., the amplitude of oscillation cannot exceed 5°, and two pendulums (located diametrically opposite to balance the radial forces) are to be used?

6-54. A Lanchester damper using Coulomb friction is shown in Fig. P-57. The coefficient of friction is 0.20, the total force from the springs is 750 lb, and the total

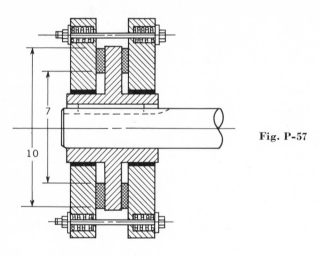

Fig. P-57

moment of inertia of the outer disks is 6.0 lb-in.-sec². (*a*) What value of angular acceleration of the shaft is required before the absorber has any effect? (*b*) In operation, the absorber mass is observed to oscillate through an angle of ±4° about the position corresponding to a uniform speed of rotation. How much energy is dissipated during each cycle?

6-55. Determine (*a*) the frequencies and (*b*) the locations of the nodes for the principal modes of vibration of the system in Fig. P-58*a*. Assume the system is constrained to move in the vertical direction and to rotate in the plane of the paper.

6-56. Same as Prob. 6-55 except apply to the system in Fig. P-58*b*.

6-57. Same as Prob. 6-55 except apply to the system in Fig. P-58*c*.

6-58. Same as Prob. 6-55 except apply to the system in Fig. P-58*d*.

6-59. What should be the spring rate for the right-hand spring, in terms of the value *k* for the left-hand spring, if the vertical and oscillatory modes of vibration of the system in Fig. P-58*a* are to be decoupled?

6-60. Same as Prob. 6-59 except apply to the system in Fig. P-58*d*.

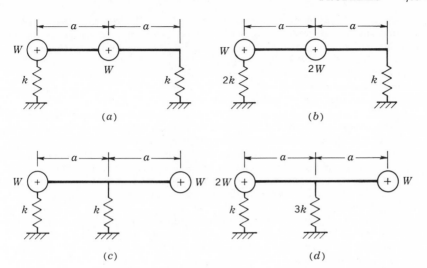

Fig. P-58

6-61. Considering motions of displacement only in the vertical direction and of rotation only in the plane of the paper, determine (a) the natural frequencies and (b) the locations of the nodes for the principal modes of vibration of the system in Fig. P-59a when $I = 0.22$ lb-in.-sec^2, $W = 10$ lb, $k_1 = 10$ lb/in., $k_2 = 20$ lb/in., $a_1 = 7$ in., and $a_2 = 5$ in.

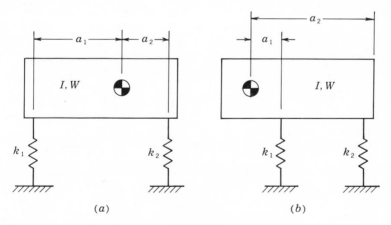

Fig. P-59

6-62. Same as Prob. 6-61 except $I = 0.88$ lb-in.-sec^2, $W = 30$ lb, $k_1 = 20$ lb/in., $k_2 = 20$ lb/in., $a_1 = 10$ in., and $a_2 = 3$ in.

6-63. Same as Prob. 6-61 except $I = 0.73$ lb-in.-sec^2, $W = 20$ lb, $k_1 = 150$ lb/in., $k_2 = 100$ lb/in., $a_1 = 10$ in., and $a_2 = 4$ in.

6-64. Same as Prob. 6-61 except $I = 0.73$ lb-in.-sec^2, $W = 20$ lb, $k_1 = 150$ lb/in., $k_2 = 100$ lb/in., $a_1 = 4$ in., and $a_2 = 10$ in.

6-65. Considering motions of displacement only in the vertical direction and of rotation only in the plane of the paper, determine (a) the natural frequencies and (b) the locations of the nodes for the principal modes of vibration of the system in Fig. P-59b when $I = 0.22$ lb-in.-sec², $W = 10$ lb, $k_1 = 20$ lb/in., $k_2 = 10$ lb/in., $a_1 = 3$ in., and $a_2 = 10$ in.

6-66. Same as Prob. 6-65 except $I = 0.88$ lb-in.-sec², $W = 30$ lb, $k_1 = 450$ lb/in., $k_2 = 100$ lb/in., $a_1 = 4$ in., and $a_2 = 12$ in.

6-67. An automobile weighing 3,900 lb has a wheelbase of 120 in. The center of gravity is located 53 in. behind the front axle, and the radius of gyration for pitching oscillations is 35 in. Assuming the system behaves as a two-degrees-of-freedom system with the coordinates being the vertical position y and the pitch angle θ, (a) where will the nodes be located for the principal modes of vibration if the front and rear springs have the same spring rate? (b) What must be the relationship between the spring rates of the front and rear springs if the coordinate modes are to be decoupled?

6-68. Specify the spring rates for the front and rear springs of the automobile discussed in Prob. 6-67 if the vertical and pitching coordinate modes of vibration are to be decoupled and the natural frequency of the vertical mode is to be 1.2 cps. What will be the natural frequency of the pitching mode vibration?

6-69. Assume that the system in Fig. P-59a is constrained to move in the vertical direction and to rotate in the plane of the paper. (a) Derive the equations for the natural frequencies of the principal modes of vibration and (b) show that when the coordinate modes are coupled, one natural frequency will be lower and one will be higher than the natural frequencies for the case when the coordinate modes are decoupled. *Hint:* In (b) use the binomial series expansion for the square root.

6-70. Determine the natural frequencies for the system in Fig. P-60 for the case when $W = 6,500$ lb, $\rho_{xy} = 23$ in., $\rho_{zy} = 6.0$ in., $\rho_{zz} = 15$ in., $k_y = 4,000$ lb/in., and $k_x/k_y = k_z/k_y = 0.90$.

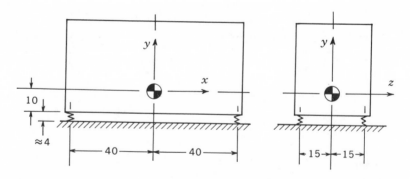

Fig. P-60

6-71. Determine the natural frequencies for the system in Fig. P-60 for the case when $W = 3,000$ lb, $\rho_{xy} = 18$ in., $\rho_{zy} = 10$ in., $\rho_{xz} = 15$ in., $k_y = k_x = k_z = 3,000$ lb/in.

6-72. The system in Fig. P-60 is subjected to inertia forces in the x and y directions. The forces act essentially in the plane containing the x and y axes, and the lowest frequency is 1,140 cpm. What value would you recommend for the spring rates for the helical springs if $k_y = k_x = k_z$, $W = 3,600$ lb, $\rho_{xy} = 23$ in., $\rho_{zy} = 10$ in., $\rho_{xz} = 17$ in., and the minimum ratio ω/ω_n is to be greater than 3.0 for all modes that will be excited?

6-73. Same as Prob. 6-72 except, also, the natural frequencies that are not expected to be excited must not be within 10 percent of the main exciting frequency of 1,150 cpm.

6-74. Springs are to be selected for the system in Fig. P-61, for which $W = 150$ lb, $\rho_{xy} = 4.0$ in., $\rho_{zy} = 2.0$ in., and $\rho_{zx} = 3.5$ in. The forcing function has a frequency of

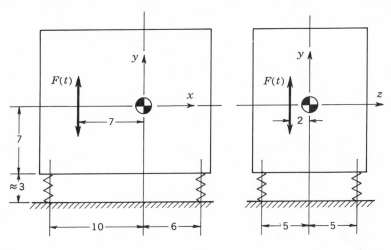

Fig. P-61

2,500 cpm, and it acts in the vertical direction, as shown. What spring rates do you recommend for the springs if $k_y = k_x = k_z$ and it is desired to keep the forcing function frequency at least three times greater than that of any principal mode of vibration it might excite?

6-75. The object shown in Fig. P-62 weighs 80 lb and has the following radii of gyration: $\rho_{xy} = 7.0$ in., $\rho_{yz} = 5.0$ in., and $\rho_{xz} = 6.5$ in. A suspension system is

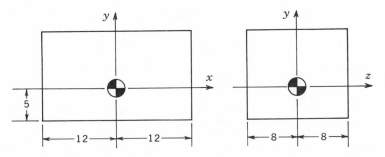

Fig. P-62

required that will be as stiff as possible in all directions without any natural frequency being greater than 15 cps. Assuming that helical compression springs are to be used, what spring rates should they have and where should they be located if $k_y = k_x = k_z$ and their diameters and lengths are about 1.5 in. and 3.0 in., respectively, when the system is at rest?

6-76. Using the step-by-step procedure discussed in Sec. 6-8, determine (*a*) the frequencies and (*b*) the locations of the nodes for the principal modes of vibration of the system in Fig. P-63.

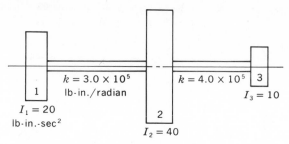

Fig. P-63

6-77. Using the step-by-step procedure discussed in Sec. 6-8, determine the amplitudes of oscillation for the system in Fig. P-63 when a disturbing torque with an amplitude of 500 lb-in. and a frequency of 20 cps acts on rotor 1.

6-78. Same as Prob. 6-77 except the torque acts on rotor 3.

6-79. Same as Prob. 6-77 except the torque acts on rotor 2.

6-80. Using the step-by-step procedure discussed in Sec. 6-8, determine the amplitudes of oscillation for the system in Fig. P-63 when a disturbing torque with an amplitude of 500 lb-in. and a frequency of 30 cps acts on rotor 1.

6-81. Same as Prob. 6-80 except the torque acts on rotor 3.

6-82. Same as Prob. 6-80 except the torque acts on rotor 2.

6-83. Using the step-by-step procedure discussed in Sec. 6-8, determine (*a*) the frequencies and (*b*) the locations of the nodes for the principal modes of vibration of the system in Fig. P-64.

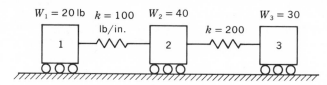

Fig. P-64

6-84. Using the step-by-step procedure discussed in Sec. 6-8, determine the amplitudes of oscillation for the system in Fig. P-64 if a disturbing force with an amplitude of 50 lb and a frequency of 10 cps acts on mass 1.

6-85. Same as Prob. 6-84 except the forcing function acts on mass 3.

6-86. Same as Prob. 6-84 except the forcing function acts on mass 2.

6-87. Using the step-by-step procedure discussed in Sec. 6-8, determine (*a*) the frequencies and (*b*) the locations of the nodes for the principal modes of vibration of the system in Fig. P-65.

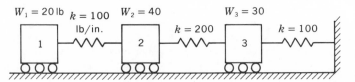

Fig. P-65

6-88. Using the step-by-step procedure discussed in Sec. 6-8, determine the amplitudes of oscillation for the system in Fig. P-65 when a forcing function with an amplitude of 50 lb and a frequency of 10 cps acts on mass 1.

6-89. Same as Prob. 6-88 except the forcing function acts on mass 3.

6-90. Using the step-by-step procedure discussed in Sec. 6-8 and assuming the gears have negligible inertia and are rigid, determine (a) the frequencies and (b) the locations of the nodes for the principal modes of vibration of the system in Fig. P-66.

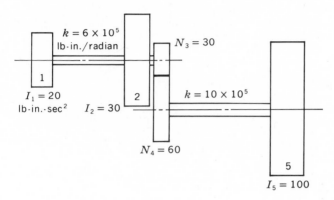

Fig. P-66

6-91. Using the step-by-step procedure discussed in Sec. 6-8 and assuming the gears have negligible inertia and are rigid, determine the amplitudes of oscillation for the system in Fig. P-66 when a forcing function with an amplitude of 200 lb-in. and a frequency of 20 cps acts on rotor 1.

6-92. Same as Prob. 6-91 except the forcing function acts on rotor 5.

6-93. Same as Prob. 6-91 except the forcing function acts on rotor 2.

6-94. Using the step-by-step procedure discussed in Sec. 6-8, determine the critical frequency of the system in Fig. P-67.

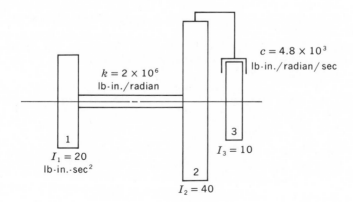

Fig. P-67

6-95. Using the step-by-step procedure discussed in Sec. 6-8, determine the amplitudes of oscillation for the system in Fig. P-67 when a forcing function with an amplitude of 600 lb-in. at a frequency of 60 cps acts on rotor 1.

6-96. Same as Prob. 6-95 except the frequency is 65 cps.

6-97. Same as Prob. 6-95 except the frequency is 70 cps.

6-98. Using the step-by-step procedure discussed in Sec. 6-8, determine the amplitudes of oscillation for the system in Fig. P-67 when a forcing function with an amplitude of 600 lb-in. at a frequency of 60 cps acts on rotor 2.

6-99. Same as Prob. 6-98 except the frequency is 65 cps.

6-100. Determine the (a) frequencies and (b) locations of the nodes for the principal modes of vibration of the system in Fig. P-68.

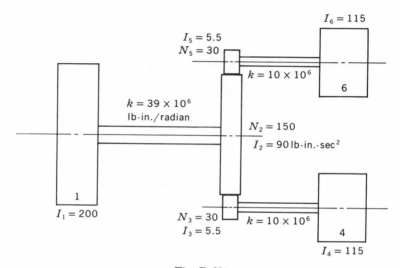

Fig. P-68

6-101. Determine (a) the equation for the natural frequencies and (b) the shape of the deflection curve for the first three modes of free vibration of a uniform cantilever beam.

6-102. Determine (a) the equations for the natural frequencies and (b) the shape of the deflection curve for the first three modes of free vibration of a uniform beam with clamped (fixed) ends.

6-103. Determine (a) the equations for the natural frequencies and (b) the shape of the deflection curve for the first three modes of free vibration of a uniform beam with one end clamped and the other end hinged.

6-104. An automotive-type propeller shaft is 5 ft long and consists of a steel tube with universal joints at each end, as shown in Fig. P-69. The tube has an outside

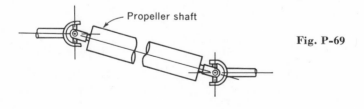

Propeller shaft

Fig. P-69

diameter of 4.00 in. and a wall thickness of 0.0625 in.   What are the first two critical frequencies for lateral vibrations of the shaft?

6-105. As shown in Fig. P-69, automotive-type propeller shafts usually consist of a hollow steel tube with a universal joint at each end.   Assuming the tube has a constant diameter and thickness and using $\pi D^3 t/8$ (D = mean diameter and t = thickness) as an approximation for the rectangular moment of inertia of a thin-walled cylinder about a diameter, (a) show that the critical frequencies for lateral vibrations are independent of the wall thickness and (b) determine the diameter required if the shaft is to be 6 ft long and the lowest critical frequency for whirl is to be 30,000 cpm.

6-106. A torque of 400 lb-ft is to be transmitted between shafts whose axes are to remain parallel while the amount of offset changes continuously.   It is proposed to use an automotive-type propeller shaft (Fig. P-69) with universal joints at each end as the intermediate shaft.   The shaft is to be made of steel, and the distance between the universal joints will be 52 in.   Assuming that the lowest critical frequency for lateral vibrations and shaft whirl is to be 10 times the nominal speed of rotation of 3,600 rpm and that the maximum shear stress due to the rated torque is to be 5,000 psi, determine (a) the outside diameter and wall thickness required if the shaft is a uniform thin-walled tube and (b) the diameter required if the shaft is solid.   (c) What is the ratio of the weight of the major length of the shaft in (b) to that in (a)?

Note: The rectangular moment of inertia of a thin-walled tube about a diameter is approximately $\pi D^3 t/8$, where D = mean diameter and t = thickness.

6-107. Same as Prob. 4-75 except use the equation for the exact solution as a continuous uniform beam instead of Rayleigh's method.

6-108. Same as Prob. 4-75 except the bearing at A is an internally self-aligning ball bearing and use the equation for the exact solution as a continuous uniform beam instead of Rayleigh's method.

6-109. Starting with Eq. (6-352), derive an equation for determining the natural frequencies of longitudinal vibration of a uniform rod when a small rigid mass m is attached to one end and the other end is free.

6-110. The pumping of water or oil from deep wells usually involves the use of a rod, called the sucker rod, to connect the drive unit at the top with the piston at the bottom of the well.   If the rod has a uniform diameter of $1\frac{1}{8}$ in. and the well is 3,800 ft deep, what is the maximum pumping frequency you would recommend to ensure that it will not exceed one-half the lowest natural frequency for longitudinal vibrations of the rod?

6-111. A machine utilizes a $\frac{3}{4}$-in.-diam steel rod to transmit an axial force that varies sinusoidally at a frequency of 40 cps.   Operation has been satisfactory, but the speed has recently been doubled and the rod is now operating close to its lowest natural frequency.   Some of the proposed schemes for eliminating the problem are as follows: (a) replace the present rod with one having a larger diameter, (b) replace the steel rod with one made of aluminum, (c) replace the steel rod with one made of magnesium, and (d) relocate parts as required to permit shortening the present rod. Consider each proposal in turn and select the one you think offers the greatest potential for improvement in this situation.

6-112. One method for using an electrodynamic shaker in vibration testing of heavy objects is illustrated in Fig. P-70, where the shaker is rotated so that its output axis is horizontal and acts on a slip table to which the heavy object is attached.   The amplitude of vibration will be measured, and the signal will be used as the feedback signal in a control system that will maintain the amplitude at the desired level as long as the requirements of force, displacement, energy, frequency, etc., remain within the capabilities of the equipment.   (a) Operation is usually limited to frequencies below the first natural frequency of the major part of the system, i.e., the shaker armature plus the slip table. Why? (*b*) Neglecting the effect of the armature and the connecting links, what is the maximum permissible length of the table if it is to be made from a solid piece of magnesium and the system is to be useful to 2,000 cps with no load on the table? (*c*) For the table length determined in (*b*), what will be the maximum useful frequency if a "rigid" object weighing 200 lb is attached to the slip table at the end away from the connection to the shaker? (*d*) Someone suggests that the natural frequency of the slip table can be raised by boring a large number of non-connecting holes in the bottom of the plate to remove weight. What is your answer to a request for a recommendation as to whether it should or should not be done?

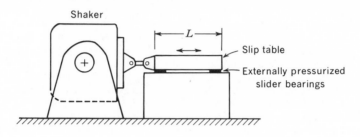

Fig. P-70

6-113. Ultrasonic frequencies are used in diverse applications ranging from locating submarines to speeding up chemical reactions to homogenizing milk. Most methods for generating the high-frequency "sound" waves are based upon the excitation of a solid body at a resonant frequency. The major methods involve the phenomena of piezoelectricity and magnetostriction. In the case of piezoelectricity, the operation is essentially the reverse of that encountered when the phenomenon is used in measuring vibrations, as discussed in Sec. 4-22. In the case of magnetostriction, the dimensional change occurring when ferromagnetic materials are subjected to a change in magnetic field is the basis for the excitation. The coils in Fig. P-71 are supplied with

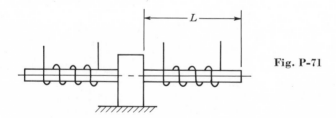

Fig. P-71

energy from an oscillator. The coils are part of the feedback circuit in the oscillator, and the frequency of the magnetic field variation is maintained at a natural frequency for longitudinal vibrations of the rods. (*a*) In most cases, the frequency is the lowest natural frequency. Why? (*b*) If the rods are made of pure nickel with a density of 0.322 lb/in.³ and a tensile modulus of elasticity of 30×10^6 psi, what length L must be used to give a frequency of 25,000 cps?

6-114. In the great majority of vibration problems the error introduced by ignoring the fact that a helical spring is a continuous member with distributed mass, and even neglecting the mass entirely, is insignificant. However, in a few cases, such as in valve mechanisms for automotive engines, where the springs are subjected to a periodic forcing function containing high-frequency harmonics, the possibility of resonance of the spring itself cannot be ignored. A helical spring can be treated as a solid rod made of a fictitious material having the mechanical properties that provide the same density and elastic properties as the spring. The relationship between the deflection and the applied force on a helical spring made from round wire is

$$\delta = \frac{8PD^3N}{Gd^4} \quad \text{in.} \tag{P-11}$$

where P = load, lb
D = mean diameter of the spring, in.
N = number of active coils
G = shear modulus of elasticity, about 11,500,000 psi for steel
d = diameter of the wire, in.

Using Eq. (P-11) to determine the equivalent solid rod, show that when a closely coiled steel spring is used under an initial load sufficiently great to ensure that the ends remain square against the spring seats, the natural frequencies for longitudinal vibrations will be approximately

$$f_n = n \times 14,100 \frac{d}{ND^2} \quad \text{cps} \tag{P-12}$$

where n = 1, 2, 3, ...

6-115. Rotation is imparted to the tool bit in a well-drilling operation through a steel tube. If the tube can be considered uniform with an outside diameter of 3.5 in. and an inside diameter of 2.75 in., what torsional frequencies might we expect to have excited by the cutting action of the tool bit when the bottom of the well is 4,000 ft below the surface of the earth?

6-116. Starting with Eq. (6-385), derive an equation for determining the natural frequencies of torsional vibrations of a uniform rod when one end is fixed and the other end is attached to a small rigid rotor with mass moment of inertia I.

6-117. Starting with Eq. (6-385), derive an equation for determining the natural frequencies of torsional vibrations of a uniform rod when a small rigid rotor with mass moment of inertia I is attached to one end and the other end is free.

6-118. The motor for the wind-tunnel drive in Prob. 4-75 is connected to the propeller shaft through a flexible coupling that is torsionally soft and will effectively isolate the motor from the shaft for torsional vibrations with frequencies greater than 160 cps. The steel shaft has a diameter of $2\frac{1}{2}$ in., and the mass moment of inertia of the eight-bladed propeller is 350 lb-in.-sec². Each time a blade of the propeller passes a strut in the structure supporting the bearing at B, the shaft experiences a torque disturbance. It is proposed to use a propeller with equally spaced blades and bearing support consisting of two struts with an included angle of $67\frac{1}{2}°$. Will the proposed arrangement be satisfactory with respect to torsional vibrations of the shaft as a continuous member?

Chapter 7. Control Systems

7-1. Considering the system in Fig. P-72 when $U = 0$, derive (a) the closed-loop transfer function C/R, (b) the open-loop transfer function B/E_1, and (c) the equation for the error E as a function of the reference input R.

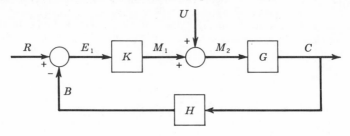

Fig. P-72

7-2. Considering the system in Fig. P-72 when $R = 0$, derive equations for (a) the response C and (b) the error E as functions of the disturbance U.

7-3. Same as Prob. 7-1 except apply to the system in Fig. P-73.

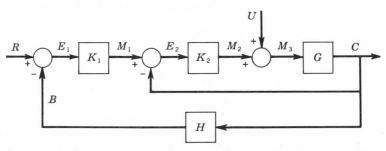

Fig. P-73

7-4. Same as Prob. 7-2 except apply to the system in Fig. P-73.

7-5. Considering the system in Fig. P-74 when $U = 0$ and $L = 0$, derive (a) the closed-loop transfer function C/R, (b) the open-loop transfer function B_1/E_1, and (c) the equation for the error E as a function of the reference input R.

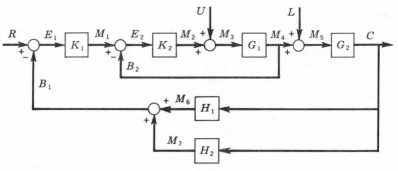

Fig. P-74

7-6. Considering the system in Fig. P-74 when $R = 0$ and $L = 0$, derive equations for (*a*) the response C and (*b*) the error E as functions of the disturbance U.

7-7. Considering the system in Fig. P-74 when $R = 0$ and $U = 0$, derive equations for (*a*) the response C and (*b*) the error E as functions of the load disturbance L.

7-8. Same as Prob. 7-5 except apply to the system in Fig. P-75.

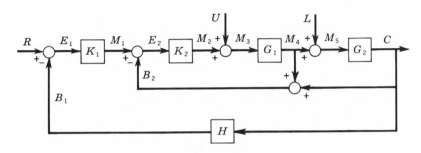

Fig. P-75

7-9. Same as Prob. 7-6 except apply to the system in Fig. P-75.

7-10. Same as Prob. 7-7 except apply to the system in Fig. P-75.

7-11. Derive the differential equation for the system in Fig. P-72 when $U = 0$, $K = 5$, $G = 1/(4D^2 + D)$, and $H = 1$.

7-12. Derive the differential equation for the system in Fig. P-72 when $R = 0$, $K = 5$, $G = 1/(4D^2 + D)$, and $H = 1$.

7-13. Derive the characteristic equation for the system in Fig. P-72 when $K = 5$, $G = 1/(4D^2 + D)$, and $H = 1$.

7-14. Same as Prob. 7-11 except $H = 1 + 1.5D$.

7-15. Same as Prob. 7-12 except $H = 1 + 1.5D$.

7-16. Same as Prob. 7-13 except $H = 1 + 1.5D$.

7-17. Same as Prob. 7-11 except $K = 5 + 1.5D$.

7-18. Same as Prob. 7-12 except $K = 5 + 1.5D$.

7-19. Same as Prob. 7-13 except $K = 5 + 1.5D$.

7-20. Same as Prob. 7-11 except $K = 5 + 0.1/D$.

7-21. Same as Prob. 7-12 except $K = 5 + 0.1/D$.

7-22. Same as Prob. 7-13 except $K = 5 + 0.1/D$.

7-23. Same as Prob. 7-11 except $K = 5 + 1.5D + 0.1/D$.

7-24. Same as Prob. 7-12 except $K = 5 + 1.5D + 0.1/D$.

7-25. Same as Prob. 7-13 except $K = 5 + 1.5D + 0.1/D$.

7-26. Derive the differential equation for the system in Fig. P-73 when $U = 0$, $K_1 = 2$, $K_2 = 5$, $G = 1/(4D^2 + D)$, and $H = 1$.

7-27. Derive the differential equation for the system in Fig. P-73 when $R = 0$, $K_1 = 2$, $K_2 = 5$, $G = 1/(4D^2 + D)$, and $H = 1$.

7-28. Derive the characteristic equation for the system in Fig. P-73 when $K_1 = 2$, $K_2 = 5$, $G = 1/(4D^2 + D)$, and $H = 1$.

7-29. Same as Prob. 7-26 except $K_1 = 2 + 1.5D$.

7-30. Same as Prob. 7-28 except $K_1 = 2 + 1.5D$.

7-31. Same as Prob. 7-26 except $K_1 = 2 + 1.5D + 0.1/D$.

7-32. Same as Prob. 7-28 except $K_1 = 2 + 1.5D + 0.1/D$.

7-33. Same as Prob. 7-27 except $K_1 = 2 + 1.5D + 0.1/D$ and $K_2 = 5 + 0.1/D$.
7-34. Same as Prob. 7-28 except $K_1 = 2 + 1.5D + 0.1/D$ and $K_2 = 5 + 0.1/D$.

Chapter 8.   Response of Basic Linear Control Systems

8-1. If all terms for the system in Fig. P-76 are expressed in units of pounds, inches, and seconds, what value should be specified for K if the response to a step change in reference input is to reach 98.2 percent of the steady-state value in 1.50 sec?

8-2. Same as Prob. 8-1 except the response is to reach 90 percent of the steady-state value in 1.50 sec.

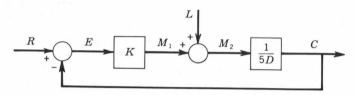

Fig. P-76

8-3. If all terms for the system in Fig. P-76 are expressed in units of pounds, inches, and seconds, what value should be specified for K to ensure that (a) the steady-state error to a ramp-function input of 1.00 in./sec does not exceed 1.00 in. and (b) the response to a step input of 1.00 in. is to reach 0.98 in. in less than 1.5 sec?

8-4. Same as Prob. 8-3 except the steady-state error to the ramp-function input is not to exceed 0.50 in.

8-5. If all terms for the system in Fig. P-76 are expressed in units of pounds, inches, and seconds, what value should be specified for K to ensure that the response to a sinusoidal variation in reference level is down not more than 3 db at 0.5 cps?

8-6. If all terms for the system in Fig. P-76 are expressed in units of pounds, inches, and seconds, what value of K should be specified to ensure that the steady-state error in response to a step change in load of 4.0 lb does not exceed 0.50 in.?

8-7. If all terms for the system in Fig. P-76 are expressed in units of pounds, inches, and seconds, what value of K should be specified to ensure that (a) the response to a step change in reference level reaches 98.2 percent of the steady-state value in 2.0 sec, (b) the steady-state error in response to a ramp change in reference level of 0.50 in./sec is not to exceed 0.30 in., (c) the steady-state error in response to a step change in load of 8.0 lb is not to exceed 0.50 in., and (d) the response to a sinusoidal variation in reference level is to be down not more than 3 db at 0.2 cps?

8-8. For the system in Fig. P-77, (a) what value should be specified for K if the overshoot in the response to a step change in reference level is to be 10 percent of the

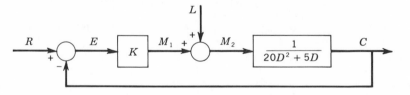

Fig. P-77

steady-state value?   (b) What will be the critical frequency in cycles per unit time when the value of K from (a) is used?

8-9. For the system in Fig. P-77, (a) what value should be specified for K if the magnitude of the control ratio is to be 3 db at the undamped natural frequency of the system?   Assuming the value of K from (a) is used and that all terms are in units of pounds, inches, and seconds, what will be (b) the undamped natural frequency in cycles per second, (c) the damped natural frequency in cycles per second, and (d) the maximum overshoot in inches in the response to a step change in reference level of 0.40 in.?

8-10. If all terms for the system in Fig. P-77 are in units of pounds, inches, and seconds, (a) what value of K would you recommend to provide a minimum settling time for a ± 5 percent tolerance band in the step response of the system?   (b) With the value of K from (a), what will be the value of the settling time in seconds?

8-11. The system in Fig. P-77 has been used with $K = 10$.   The system behavior appears to be satisfactory in all respects except the response is too oscillatory.   It is proposed to correct the situation by adding error-rate to the present proportional control to bring the effective damping ratio up to 0.70.   What value should be specified for the error-rate coefficient K_e?

8-12. In units of pounds, inches, and seconds, the transfer function for a controlled system is $G = 1/(10D^2 + 0.5D)$.   The system specifications call for an undamped natural frequency of 5.0 cps and an effective damping ratio of 0.60.   It is proposed to use proportional plus error-rate control.   What values should be specified for (a) the gain K and (b) the error-rate coefficient K_e?

8-13. For the system in Fig. 8-17, $K = 5$, $K_i = 0.2$, and $G = 1/(10D^2 + 3D)$ and all terms are in units of pounds, inches, and seconds.   Determine (a) the system type number, (b) the type of input function for which the steady-state error will be a constant, (c) the magnitude of the steady-state error for a unit input function of the type specified in (b), and (d) the magnitude in decibels of the control ratio $C(j\omega)/R(j\omega)$ when the frequency is 0.06 cps.

8-14. For the system in Fig. 8-18, $K = 5$, $K_e = 0.6$, $K_i = 0.4$, and $G = 1/(10D^2 + 3D)$ and all terms are in units of pounds, inches, and seconds.   Determine (a) the magnitude of the steady-state error when the reference input remains at zero and the load variation with respect to time is as follows: $L(t) = 5$, $L(t) = 5t$, and $L(t) = 5t^2$, and (b) the magnitude in decibels of the control ratio $C(j\omega)/R(j\omega)$ when the frequency is 0.06 cps.

8-15. The system in Fig. P-78 has proportional plus derivative-feedback plus integral control.   All terms are in units of pounds, inches, and seconds, and $I = 10$,

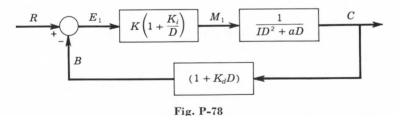

Fig. P-78

$a = 1$, $K = 40$, $K_i = 0.4$, and $K_d = 0.6$.   Determine (a) the steady-state error in the response to the following inputs: $r = 3$, $r = 3t$, and $r = 3t^2$ and (b) the magnitude in decibels of the control ratio $C(j\omega)/R(j\omega)$ when the frequency is 1.0 cps.

8-16. For the system in Fig. P-79, $G_1 = K(D + 0.4)/D$ and $G_2 = 1/(5D^2 + 3D)$. (a) What is the system type number?   (b) For what type of input function, $r(t)$, will

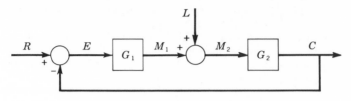

Fig. P-79

the system have a constant steady-state error?   (c) If all terms are expressed in units of pounds, inches, and seconds, what will be the magnitude of the steady-state error for a unit function of the type specified in (b)?

8-17. Same as Prob. 8-16 except $G_1G_2 = 3(4s + 1)^2/s(2s^2 - s)$.

Chapter 9.   Stability Analysis of Linear Systems

9-1. Use the Routh criterion to show that for the system in Fig. 8-17, with proportional plus integral control, when $G = 1/(ID^2 + aD)$, the maximum permissible value of the integral coefficient K_i is independent of the gain K.   Upon what parameter(s) will it depend?

9-2. Use the Routh criterion to determine the range of values for the integral coefficient K_i for stable operation of the system in Fig. 8-18, with proportional plus error-rate plus integral control, when $K = 5$, $K_e = 0.6$, and $G = 1/(10D^2 + 3D)$.

9-3. Use the Routh criterion to determine the range of values of the error-rate coefficient K_e for stable operation of the system in Fig. 8-18, with proportional plus error-rate plus integral control, when $K = 5$, $K_i = 1$, and $G = 1/(10D^2 + D)$.

9-4. Use the Routh criterion to determine the range of values of the integral coefficient K_i for stable operation of the system in Fig. P-78, with proportional plus derivative-feedback plus integral control, when $K = 5$, $I = 10$, $a = 2$, and $K_d = 0.7$.

9-5. Use the Routh criterion to determine the range of values of the derivative feedback coefficient K_d for stable operation of the system in Fig. P-78, with proportional plus derivative-feedback plus integral control, when $K = 4$, $K_i = 0.5$, $I = 8$, and $a = 1$.

9-6. Sketch the root loci for a system with $H(s)G(s) = 3K/(s + 2)$.   What conclusions can you draw from the root-loci plot about the behavior of the system?

9-7. Same as Prob. 9-6 except $H(s)G(s) = 3K(s + 3)/(s + 2)$.

9-8. Same as Prob. 9-6 except $H(s)G(s) = 3K(s + 2)/(s + 3)$.

9-9. Same as Prob. 9-6 except $H(s)G(s) = 3K(s - 2)/(s + 3)$.

9-10. Same as Prob. 9-6 except $H(s)G(s) = 3K(s - 3)/(s + 2)$.

9-11. Sketch the root loci for a system with $H(s)G(s) = 2K/s(s + 1)(s + 2)$.   (a) For what values of K will the system be stable and (b) what value of K should be specified if the transient response is to be somewhat similar to that of a second-order system with $\zeta = 0.70$?

9-12. Same as Prob. 9-11 except $H(s)G(s) = 2K/s(s + 1)^2$.

9-13. Same as Prob. 9-11 except $H(s)G(s) = 2K/s(s^2 + 2s + 2)$.

9-14. Same as Prob. 9-11 except $H(s)G(s) = 2K(s + 2)/s(s + 1)(s + 3)$.

9-15. Same as Prob. 9-11 except $H(s)G(s) = 2K/s(s + 1)(s + 2)(s + 3)$.

9-16. Same as Prob. 9-11 except $H(s)G(s) = 2K(s + 2)/s(s + 1)^2(s + 3)$.

9-17. Same as Prob. 9-11 except $H(s)G(s) = 2K/s(s^2 + 2s + 2)(s + 3)$.

9-18. Same as Prob. 9-11 except $H(s)G(s) = 2K(s + 2)/s(s^2 + 2s + 2)(s + 3)$.

9-19. Same as Prob. 9-11 except $H(s)G(s) = K(4s + 1)^2/s^2(2s - 1)$.

9-20. For a system $H(s)G(s) = 3K(s + 3)/(s + 2)$. Using the method of conformal mapping, determine the range of values of K for stable operation.

9-21. Same as Prob. 9-20 except $H(s)G(s) = 3K(s + 2)/(s + 3)$.

9-22. Same as Prob. 9-20 except $H(s)G(s) = 3K(s - 2)/(s + 3)$.

9-23. Same as Prob. 9-20 except $H(s)G(s) = 3K(s - 3)/(s + 2)$.

9-24. For a system $H(s)G(s) = 2K/s(s + 1)(s + 2)$. Using the method of conformal mapping, determine (*a*) the range of values of K for stable operation and, if appropriate, (*b*) the value of K for a phase margin of 50°.

9-25. Same as Prob. 9-24 except $H(s)G(s) = 2K/s(s + 1)^2$.

9-26. Same as Prob. 9-24 except $H(s)G(s) = 2K/s(s^2 + 2s + 2)$.

9-27. Same as Prob. 9-24 except $H(s)G(s) = 2K(s + 2)/s(s + 1)(s + 3)$.

9-28. Same as Prob. 9-24 except $H(s)G(s) = 2K/s(s + 1)(s + 2)(s + 3)$.

9-29. Same as Prob. 9-24 except $H(s)G(s) = 2K(s + 2)/s(s + 1)^2(s + 3)$.

9-30. Same as Prob. 9-24 except $H(s)G(s) = 2K/s(s^2 + 2s + 2)(s + 3)$.

9-31. Same as Prob. 9-24 except $H(s)G(s) = 2K(s + 2)/s(s^2 + 2s + 2)(s + 3)$.

9-32. Same as Prob. 9-24 except $H(s)G(s) = K(1 - 0.5s)/2s$.

9-33. A control system has recently been modified. It is important that the system not be permitted to become unstable when put back into operation. The transfer functions are not available. The system is open-loop stable, and the following frequency-response data have been taken for the open-loop system with $K = 1$.

Frequency, cps	$|H(j\omega)G(j\omega)|$	ϕ, deg
0	1.0	−0
0.02	0.94	−3
0.05	0.90	−16
0.1	0.83	−40
0.2	0.65	−73
0.4	0.36	−117
0.5	0.27	−135
0.7	0.16	−159
0.8	0.13	−168
0.9	0.10	−179
1.0	0.08	−185
1.1	0.06	−191
1.2	0.05	−196
2.0	0.01	−225
5.0	0.003	−257

Use a polar plot of the frequency-response data to determine (*a*) the values of K for stable operation of the system when the loop is closed and (*b*) the value of K for a phase margin of 60° when the loop is closed.

9-34. Frequency-response data for a system with the feedback loop open at the summing junction are given below.

Frequency, cps	$|H(j\omega)G(j\omega)|$	ϕ, deg
0	15.0	0
0.2	14.2	-3
0.5	13.5	-16
1.0	12.5	-40
2.0	9.7	-73
4.0	5.4	-117
5.0	4.0	-135
7.0	2.5	-159
8.0	2.0	-168
9.0	1.5	-179
10.0	1.2	-185
11.0	1.0	-191
12.0	0.8	-196
20.0	0.2	-225
50.0	0.04	-257

Use a polar plot of the frequency-response data to determine (a) the number of decibels by which the gain K can be increased without the system becoming unstable or the number of decibels by which the gain must be decreased to prevent the system from becoming unstable when the loop is closed and (b) the number of decibels by which the gain K should be changed to provide a phase margin of 45°.

9-35. A control system has recently been modified, and serious complications will result if the system turns out to be unstable when put into operation. Based on previous knowledge, there is reason to believe that the system will be open-loop stable, and it is proposed to obtain open-loop frequency-response data that can be used to predict the behavior of the system when the feedback loop is closed. When open-loop tests were started, it was noticed that the response tended to increase at a constant rate for a dc (zero-frequency) input and that with zero input the system would remain at rest wherever the response happened to be when the input became zero. This behavior did not prevent running frequency-response tests, and the data for $K = 1$ are given below.

Frequency, cps	$|H(j\omega)G(j\omega)|$	ϕ, deg
0.05	19.50	-98
0.1	9.23	-111
0.2	3.94	-131
0.3	2.21	-149
0.4	1.30	-163
0.5	0.87	-174
0.6	0.61	-183
0.7	0.44	-191
1.0	0.21	-207
2.0	0.03	-236

(*a*) What is the significance of the behavior of the open-loop system at zero frequency? Use a polar plot of the frequency-response data to determine (*b*) the range of values of K for stable operation when the feedback loop is closed and (*c*) the value of K that should be specified to give a phase margin of 50°.

9-36. Frequency-response data for a system in operation with the feedback loop open at the summing junction are given below.

Frequency, cps	$|H(j\omega)G(j\omega)|$	ϕ, deg
0.05	54.6	−98
0.1	25.8	−111
0.2	11.03	−131
0.3	6.18	−149
0.4	3.64	−163
0.5	2.44	−174
0.6	1.71	−183
0.7	1.23	−191
1.0	0.58	−207
2.0	0.09	−236

Use a polar plot of the frequency-response data to determine (*a*) the factor by which the gain K can be increased without the system becoming unstable or the factor by which the gain K must be decreased to prevent the system from becoming unstable when the feedback loop is closed and (*b*) the number of decibels by which the gain K should be changed to provide a phase margin of 60°.

9-37. Use the attenuation-phase method to determine the range of values of K for stable operation of a system having $H(s)G(s) = 3K(s + 3)/(s + 2)$.

9-38. Same as Prob. 9-37 except $H(s)G(s) = 3K(s + 2)/(s + 3)$.

9-39. Same as Prob. 9-37 except $H(s)G(s) = 3K(s - 2)/(s + 3)$.

9-40. Same as Prob. 9-37 except $H(s)G(s) = 3K(s - 3)/(s + 2)$.

9-41. Use the attenuation-phase method to determine (*a*) the range of values of K for stable operation and (*b*) the value of K for a phase margin of 50° when $H(s)G(s) = 2K/s(s + 1)(s + 2)$.

9-42. Same as Prob. 9-41 except $H(s)G(s) = 2K/s(s + 1)^2$.

9-43. Same as Prob. 9-41 except $H(s)G(s) = 2K(s + 2)/s(s + 1)(s + 3)$.

9-44. Same as Prob. 9-41 except $H(s)G(s) = 2K/s(s + 1)(s + 2)(s + 3)$.

9-45. Same as Prob. 9-41 except $H(s)G(s) = 2K(s + 2)/s(s + 1)^2(s + 3)$.

9-46. Same as Prob. 9-41 except $H(s)G(s) = 2K/s(s^2 + 2s + 2)$.

9-47. Same as Prob. 9-41 except $H(s)G(s) = 2K/s(s^2 + 2s + 2)(s + 3)$.

9-48. Same as Prob. 9-41 except $H(s)G(s) = 2K(s + 2)/s(s^2 + 2s + 2)(s + 3)$.

9-49. Same as Prob. 9-41 except $H(s)G(s) = K(1 - 0.5s)/2s$.

9-50. Same as Prob. 9-33 except use the attenuation-phase method.

9-51. Same as Prob. 9-34 except use the attenuation-phase method.

9-52. Same as Prob. 9-35 except use the attenuation-phase method.

9-53. Same as Prob. 9-36 except use the attenuation-phase method.

Chapter 10.   Nonlinear Control Systems

10-1. A two-position control system is being designed.   The controlled system has the transfer function $G = 1/(10D^2 + 12D)$, and the controller has the characteristic shown in Fig. P-80a.   All terms are in units of pounds, inches, and seconds, and

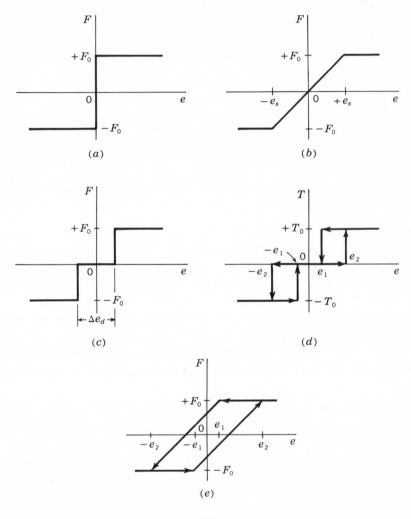

Fig. P-80

$F_0 = 400$ lb.   Show on the phase plane the response of this system to a step change in error of $+10$ in. when the system is initially at rest with zero error.

10-2. Same as Prob. 10-1 except also determine the time required for the response to move from the initial positive error to the maximum negative error.

10-3. Same as Prob 10-1 except the step change in error is $+20$ in.

10-4. Same as Prob. 10-1 except the step change in error is $+20$ in. and also determine the time required for the response to move from the initial positive error to the maximum negative error.

10-5. A controller with the input-output characteristic shown in Fig. P-80b can be quite effective in that it combines proportional control for small errors with maximum output capacity of the controller available when the error is large. This type of controller is sometimes classed as a proportional controller with saturation. Assuming $F_0 = 500$ and $e_s = 0.20$ in., show on the phase plane the response of a controlled system with a transfer function $G = 1/(D^2 + 20D)$ to a step change in error of $+0.5$ in. when the system is initially at rest with zero error. All terms are in units of pounds, inches, and seconds.

10-6. Same as Prob. 10-5 except also determine the time required for the trajectory to cross the line of zero error for the first time.

10-7. Same as Prob. 10-5 except the step change in error is $+1.0$ in.

10-8. The controller described in Prob. 10-5 is to be used to position an object weighing 400 lb when there is both viscous and Coulomb damping. The viscous damping coefficient is 10 lb/in./sec, and the coefficient of friction between the weight and its guides is 0.1. Show on the phase plane the response of this system to a step change in reference input of $+0.5$ in. when the system is initially at rest with zero error.

10-9. Same as Prob. 10-8 except also determine the time required for the trajectory to first cross the zero-error line.

10-10. Same as Prob. 10-8 except also determine the time required for the system to come to rest.

10-11. Same as Prob. 10-8 except the step change in reference input is $+1.0$ in.

10-12. The position of an object weighing 80 lb is to be controlled by using a three-position controller with the input-output characteristics shown in Fig. P-80c. $F_0 = 300$ lb, $\Delta e_d = 0.040$ in., and the viscous damping coefficient is 5.0 lb/in./sec. Show on the phase plane the response of this system to a step change in reference level of -0.15 in. when the system is initially at rest with zero error.

10-13. Same as Prob. 10-12 except also determine the time required for the trajectory to first cross the zero-error line.

10-14. Same as Prob. 10-12 except also determine the time required for the amplitude to become less than 0.030 in.

10-15. Same as Prob. 10-12 except in addition the coefficient of friction between the object and its guides is 0.08.

10-16. Same as Prob. 1-2 except utilize the phase plane to the maximum extent possible.

10-17. Same as Prob. 1-3 except utilize the phase plane to the maximum extent possible.

10-18. The angular position of a platform was to be controlled by use of a three-position controller (on-off, with a dead zone). However, when installed, the controller was found to have the input-output characteristic shown in Fig. P-80d, which indicates hysteresis is also present. $T_0 = 50$ lb-in., $e_1 = 3°$, and $e_2 = 5°$. The platform weighs 1,000 lb and has a radius of gyration of 14 in. The bearing friction can be approximated by a combination of viscous and Coulomb friction, with a viscous damping coefficient of 1.5 lb-in./deg/sec and a Coulomb friction torque of 8 lb-in. Show on the phase plane the response of this system to a step change in reference input of $+15°$ when the system is initially at rest with an error of $-2°$.

10-19. Same as Prob. 10-18 except also determine the time required for the trajectory to first cross the line of zero error.

10-20. The azimuth control system for a radar antenna is based upon a controller that provides a torque that is proportional to the error until its maximum torque capacity is reached. For larger errors the torque output is constant at the maximum value. The input-output characteristic is identical with that in Fig. P-80b except we are now dealing with torsional rather than lineal terms. $T_0 = 30$ lb-in. and $e_s = 5°$. The antenna weighs 300 lb and has a radius of gyration of 13.5 in. The damping is a combination of air resistance, viscous damping, and Coulomb friction, and it can be approximated by a viscous damping coefficient of 1.4 lb-in./deg/sec and a Coulomb friction torque of 5 lb-in. Show on the phase plane the response of this system to a step change in reference level of $+10°$ when the system is initially at rest with zero error.

10-21. Same as Prob. 10-20 except also determine the time required for the trajectory to first cross the zero-error line.

10-22. Same as Prob. 10-20 except also determine the time required for the error to become, and remain, within $\pm1°$.

10-23. Same as Prob. 10-20 except the step change in reference input is $+120°$.

10-24. The system discussed in Prob. 10-20 has been installed and hysteresis has been found to be present. The actual input-output characteristic of the controller is shown in Fig. P-80e, where $e_1 = 4°$ and $e_2 = 6°$. Show on the phase plane the response of the system with hysteresis to a step change in reference input of $+10°$ when the system is initially at rest with zero error.

10-25. A three-position control system contains so little damping that the decay in response to a step input is much too slow for satisfactory operation. The controller input-output characteristic corresponds to that in Fig. P-80c, with $F_0 = 80$ lb and $\Delta e_d = 0.20$ in., and the controlled variable is the position of an object weighing 140 lb. It is proposed to install spring-operated, magnetically released brakes that are applied whenever the error is within the dead zone. Use the phase plane to show the response of the system to a step change in reference level of $+1.00$ in. when the system is initially at rest with zero error and the braking force is 80 lb.

10-26. Same as Prob. 10-25 except the braking force is 140 lb.

ANSWERS TO SELECTED PROBLEMS

1-1. (a) 0.00655 °/sec², (b) 12.4 sec, (c) 0.0791 lb, (d) 3.15 watts.

1-2. (a) 1.766×10^{-3} rad/sec, (b) 37.3 sec, (c) 40.2 sec, (d) 1.08 °.

1-4. (a) 82.4 lb/in., (b) 1.256 in.

1-9. $F_{12} = 134$ lb; $T_{12} = 2{,}340$ lb-in., cw.

1-10. 3.71 hp from gear reducer to linkage.

1-15. (a) 54.8 rad/sec², ccw. (b) 47 lb.

1-21. Sum of load + inertia torques $T_{12} = 118$ lb-in., ccw.

1-22. Sum of load + inertia forces $F_{34} = 111$ lb and $F_{14} = 36$ lb.

1-23. Sum of load + inertia forces $F_{23} = F_{12} = 54$ lb.

1-30. (a) $F_{iW_1} = 1{,}180$ lb; $F_{iW_2} = 1{,}390$ lb; $S = 2{,}490$ lb. (b) $F_{iW_1} = 1{,}140$ lb; $F_{iW_2} = 1{,}440$ lb; $S = 2{,}490$ lb.

1-35. (a) 47.3 lb-in.-sec². (b) (1) Put flywheel next to engine so gears do not have to carry torque variations; (2) inertia on high-speed shaft is 81 times more effective than on low-speed shaft.

1-37. (a) 146.4 hp, (b) 5,630 lb-in.-sec².

1-42. 5,570 lb; ship turns to right when bow is rising.

1-48. 37,500 lb/in.

2-1. $W_R = 4.0$ lb, 80° cw from W_1, and $W_L = 5.6$ lb, 158° cw from W_1.

2-6. $W_R = -6.19$ lb, 137° ccw from W_1, and $W_L = 9.67$ lb, 179° ccw from W_1.

2-11. 0.189 in.

2-13. (a) Evenly spaced power strokes; primary forces and secondary forces add; no couple. (b) Unevenly spaced power strokes; primary forces cancel; secondary forces add; couple due to primary forces. If low-speed, use (a) for smoothest power. If high-speed, use (b) to minimize shaking forces.

2-21. (a) 2 at 0°; 2 at 120°; 2 at 240°. (b) 0-120-240-240-120-0°, symmetrical about midpoint on crankshaft.

3-2. $k_3 + k_1k_2/(k_1 + k_2)$.

3-4. (a) $7.5wt^3 \times 10^6/L^3$, (b) $4.42d^4 \times 10^6/L^3$.

3-7. $T = (\mu\pi/16h)(d_o{}^4 - d_i{}^4 + 4d_o{}^3l)\, D\theta$; linear.

4-1. (a) $x = 5.83e^{i(\omega t - 59.1°)}$, (b) $x = 5.83 \cos(\omega t - 59.1°)$.

4-6. 0.34 cps.

4-8. $\omega_n = [k/(I/R^2 + W/g)]^{1/2}$.

4-11. $\omega_n = (2kg/3W)^{1/2}$.

4-13. $\omega_n = (8kg/3W)^{1/2}$.

4-16. $\omega_n = [2g/(2l_1 + l_2 + \pi R)]^{1/2}$.

4-19. (a) 170 lb/in., (b) 30.5 lb/in./sec.

4-24. (a) $x = e^{-0.806t}(0.458 \cos 7.89t + 0.1441 \sin 7.89t) - 0.458 - 0.768t + 15.0t^2$, (b) 1.26 cps, (c) 1.51 cps.

4-27. 2.53 in.

4-34. (a) $\Delta E/\text{cycle} = -\pi F_0 X \sin \phi$, (b) $+90°$.

4-40. 99 lb/in.

4-43. (a) $m\,D^2x + c\,Dx + kx = ky$, (b) $\omega_n = \sqrt{k/m}$, (c) $X/Y = 1/\{[1 - (\omega/\omega_n)^2]^2 + (2\zeta\omega/\omega_n)^2\}^{1/2}$; $\phi = -\tan^{-1}\{(2\zeta\omega/\omega_n)/[1 - (\omega/\omega_n)^2]\}$.

4-51. (a) 1.01% too high, (b) 9.7×10^{-6} in., (c) 0.57% too high, (d) 0.087, (e) 0.41.

4-56. (a) 108 lb; 159° ccw from 1, (b) blade 5, nine holes $\frac{3}{8}$ in. deep and one hole $\frac{5}{64}$ in. deep; blade 4, fifteen holes $\frac{3}{8}$ in. deep and one hole $\frac{19}{64}$ in. deep. (c) cw.

4-57. $\omega_n = [(k_1 + k_2)k_3/m(k_1 + k_2 + k_3)]^{1/2}$.

4-59. $\omega_n = [k/(I/R^2 + W/g)]^{1/2}$.

4-61. $\omega_n = (2kg/3W)^{1/2}$.

4-63. $\omega_n = (8kg/3W)^{1/2}$.

4-65. $\omega_n = [2g/(2l_1 + l_2 + \pi R)]^{1/2}$.

4-68. 112 cps.

4-70. In comparison with Prob. 4-68, 24.5% too high, and in comparison with Prob. 4-69, 25.5% too high.

4-78. $5\frac{7}{16}$ in.

5-2. $\omega_n = [(ka^2 + Wl)g/Wl^2]^{1/2}$.

5-6. $\omega_n = 0.808(g/R)^{1/2}$.

5-8. $\omega_n = (F_0/mL)^{1/2}$.

5-12. $\omega_n = [(2ka^2 - F_0L)g/2WL^2]^{1/2}$.

5-17. $F = e^{Cx}$.

5-21. $\omega_n = (2\mu g/L)^{1/2}$.

5-23. 3.36 cps.

5-28. (a) One, (b) $+0.09$ in.

5-39. Two.

5-45. 0.45 cps as compared with 0.70 cps for small amplitudes.

5-48. (a) 4.65 rad/sec, cw. (b) $< \frac{3}{4}$.

6-1. 0.00505°.

6-4. $|\theta_1 - \theta_2|/L = I_2T_0/L|k(I_1 + I_2) - I_1I_2\omega^2|$.

6-8. (a) 1, (b) 65.6%.

6-12. $\Theta_2 = k\Theta_1/[(k - I_2\omega^2)^2 + (c\omega)^2]^{1/2}$; $\phi = -\tan^{-1}[c\omega/(k - I_2\omega^2)]$.

6-19. (a) $f_1 = 0.700$ cps; $f_2 = 0.748$ cps. (b) f_1, at ∞; f_2, at midpoint between pendulums. (c) 0.048 cps.

6-23. (a) $\omega_1 = (k/m)^{1/2}$; $\omega_2 = (3k/m)^{1/2}$. (b) ω_1, at ∞; ω_2, at midpoint of spring 2.

6-27. $X_1 = |(2k - m\omega^2)F_0/(m^2\omega^4 - 4km\omega^2 + 3k^2)|$; $X_2 = kF_0/|m^2\omega^4 - 4km\omega^2 + 3k^2|$.

6-34. $X_1 = kF_0/|m^2\omega^4 - 3km\omega^2 + k^2|$; $X_2 = |(2k - m\omega^2)F_0/(m^2\omega^4 - 3km\omega^2 + k^2)|$.

6-37. $X_1 = |(k - m\omega^2)kY/(m^2\omega^4 - 3km\omega^2 + k^2)|$; $X_2 = k^2Y/|m^2\omega^4 - 3km\omega^2 + k^2|$.

6-41. $X_1 = \{[k_1(-m_2\omega^2 + k_2 + k_3) - c_1\omega^2(c_2 + c_3)]^2 + \omega^2[k_1(c_2 + c_3) + c_1(-m_2\omega^2 + k_2 + k_3)]^2\}^{1/2}Y/\{[(-m_1\omega^2 + k_1 + k_2)(-m_2\omega^2 + k_2 + k_3) - \omega^2(c_1c_2 + c_1c_3 + c_2c_3) - k_2^2]^2 + \omega^2[(-m_1\omega^2 + k_1 + k_2)(c_2 + c_3) + (-m_2\omega^2 + k_2 + k_3)(c_1 + c_2) - 2c_2k_2]^2\}^{1/2}$.

6-45. (a) $W = 0.15$ lb; $k = 98.3$ lb/in.   (b) 0.0167 in.

6-49. $\Theta_2 = |I_1\omega^2 - k|(I_3{}^2\omega^2 + c^2)^{1/2}T_0/\omega^2\{I_3{}^2\omega^2[I_1I_2\omega^2 - k(I_1 + I_2)]^2 + c^2[I_1(I_2 + I_3)\omega^2 - k(I_1 + I_2 + I_3)]^2\}^{1/2}$.

6-53. $R = 4.00$ in.; $r = 0.444$ in.; $W = 5.18$ lb; pins must carry $F_c = 368$ lb.

6-55. (a) $\omega_1 = 0.872(kg/W)^{1/2}$; $\omega_2 = 2.69(kg/W)^{1/2}$.   (b) ω_1, node is $2.12a$ to right of center of gravity; ω_2, node is $1.12a$ to left of center of gravity.

6-59. $k_R = k_L/3$.

6-63. (a) $\omega_1 = 56.7$ rad/sec; $\omega_2 = 156$ rad/sec.   (b) ω_1, at 12.98 in. to left of center of gravity; ω_2, at 1.09 in. to right of center of gravity.

6-67. (a) 350 in. back of center of gravity and 4.3 in. ahead of center of gravity. (b) $k_{\text{front}}/k_{\text{rear}} = 1.264$.

6-71. $f_{ny} = 6.25$ cps; $f_{nzz} = 17.8$ cps; $f_{nxy_1} = 5.64$ cps; $f_{nxy_2} = 14.4$ cps; $f_{nzy_1} = 4.76$ cps; $f_{nzy_2} = 12.3$ cps.

6-74. $k_{y1} = k_{y4} = 25.5$ lb/in.; $k_{y2} = k_{y3} = 42.5$ lb/in.

6-76. (a) $f_{n_1} = 22.5$ cps; $f_{n_2} = 36.5$ cps.   (b) f_{n_1}, node at 75% of distance from rotor 1 to rotor 2; f_{n_2}, nodes at 28.5% of distance from rotor 1 to rotor 2 and at 23.8% of distance from rotor 2 to rotor 3.

6-77. $\Theta_1 = 0.212°$; $\Theta_2 = 0.107°$; $\Theta_3 = 0.1765°$.

6-86. $X_1 = 0.133$ in.; $X_2 = 0.138$ in.; $X_3 = 0.258$ in.

6-94. 59.4 cps.

6-95. $\Theta_1 = 264°$; $\Theta_2 = 115°$; $\Theta_3 = 0.090°$.

6-100. (a) $f_{n_1} = 47.2$ cps; $f_{n_2} = 69.0$ cps; $f_{n_3} = 199.5$ cps.   (b) f_{n_1}, node at rotor 2 (both 1 and 2 stand still); f_{n_2}, nodes at 53.5% of distance from rotor 5 to rotor 6 and at 53.5% of distance from rotor 3 to rotor 4; f_{n_3}, nodes at 12.4% of distance from rotor 1 to rotor 2, at 94.4% of distance from rotor 5 to rotor 6, and at 94.4% of distance from rotor 3 to rotor 4.

6-104. $f_{n_1} = 123$ cps; $f_{n_2} = 493$ cps.

6-110. 0.555 cps.

6-116. $\omega(\gamma/Gg)^{1/2}L \tan \omega(\gamma/Gg)^{1/2}L = \pi d^4\gamma L/32Ig$.

7-1. (a) $KG/(1 + HKG)$, (b) HKG, (c) $[(1 + HKG - KG)/(1 + HKG)]R$.

7-2. (a) $[G/(1 + HKG)]U$, (b) $-[G/(1 + HKG)]U$.

7-5. (a) $K_1K_2G_1G_2/[1 + K_2G_1 + K_1K_2G_1G_2(H_1 + H_2)]$.   (b) $K_1K_2G_1G_2(H_1 + H_2)/(1 + K_2G_1)$.   (c) $\{[1 + K_2G_1 + K_1K_2G_1G_2(H_1 + H_2 - 1)]/[1 + K_2G_1 + K_1K_2G_1G_2(H_1 + H_2)]\}R$.

7-10. (a) $\{G_2(1 + K_2G_1)/[1 + K_2G_1 + K_2G_1G_2(1 + HK_1)]\}L$.   (b) $-\{G_2(1 + K_2G_1)/[1 + K_2G_1 + K_2G_1G_2(1 + HK_1)]\}L$.

7-11. $4D^2c + Dc + 5c = 5r$.

7-12. $4D^2c + Dc + 5c = U$.

7-13. $4s^2 + s + 5 = 0$.

8-2. 7.67 lb/in.

8-5. 15.7 lb/in.

8-8. (a) 0.898, (b) 0.0337 cps.

8-11. 1.480.

8-15. (a) 0; 1.80 in.; ∞.   (b) -19.5 db.

8-17. (a) 2, (b) Ct^2, (c) -0.667 in.

9-2. <0.6.

9-5. >0.535.

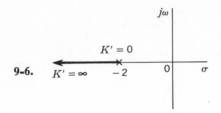

9-6.

Stable for all values of K.

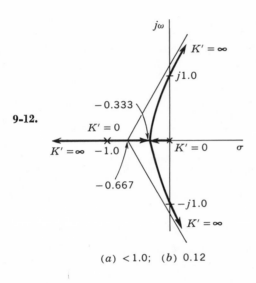

9-12.

(a) < 1.0;   (b) 0.12

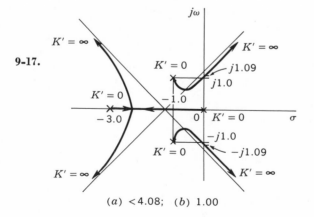

9-17.

(a) < 4.08;   (b) 1.00

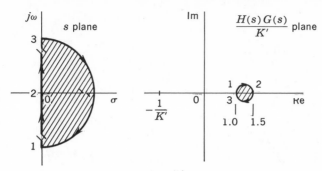

Stable for all values of K

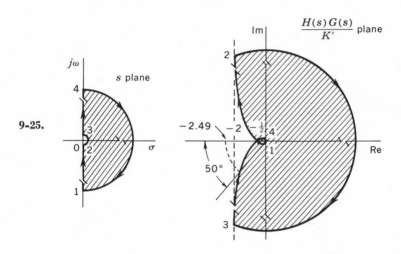

9-25.

(a) < 1.0; (b) 0.20

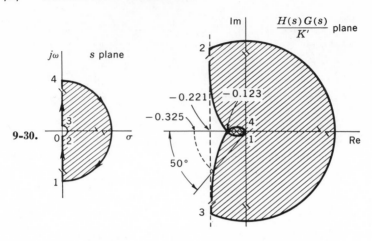

(a) < 4.08; (b) 1.54

9-33. (a) < 10.4, (b) 2.92.
9-36. (a) Reduce to 52% of original value (−5.7 db), (b) −24.7.
9-37. All.
9-42. (a) < 1.0, (b) 0.21.
9-47. (a) < 4.1, (b) 1.5.
9-50. (a) < 10.4, (b) 2.9.
9-53. (a) Reduce to 52% of original value (−5.7 db), (b) −24.7.

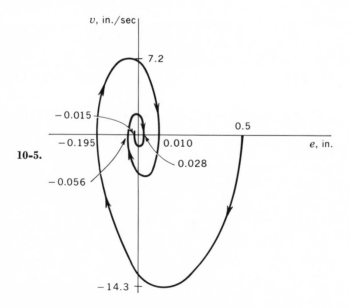

10-6. 0.055 sec.

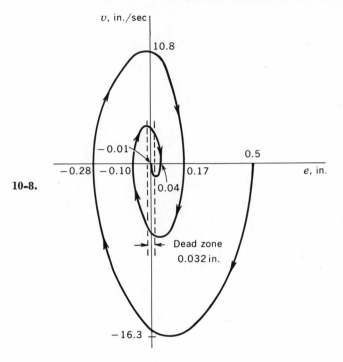

10-8.

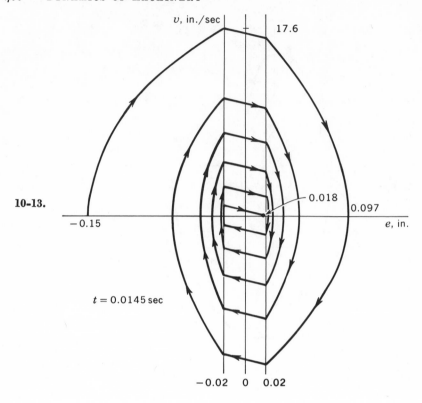

10-13.

$t = 0.0145 \, \text{sec}$

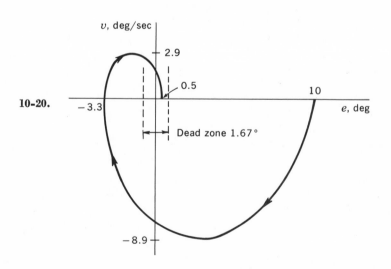

10-20.

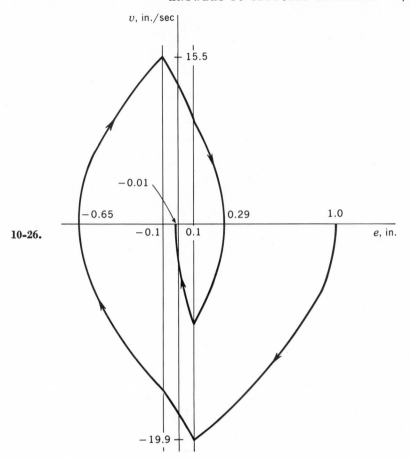

10-26.

INDEX

469